Beep to Boom

Drawing on decades of experience, *Beep to Boom: The Development of Advanced Runtime Sound Systems for Games and Extended Reality* is a rigorous, comprehensive guide to interactive audio runtime systems.

Packed with practical examples and insights, the book explains each component of these complex geometries of sound. Using practical, lowest-common-denominator techniques, Goodwin covers soundfield creation across a range of platforms from phones to VR gaming consoles.

Whether creating an audio system from scratch or building on existing frameworks, the book also explains costs, benefits and priorities. In the dynamic simulated world of games and extended reality, interactive audio can now consider every intricacy of real-world sound. This book explains how and why to tame it enjoyably.

Simon N Goodwin is an Interactive Audio Technology Consultant who lives and works in Warwick, UK. His game-development career started in the 8-bit 1970s, latterly including a productive decade as Principal Programmer in the Central Technology Group at Codemasters Software Company, where he designed and implemented advanced Ambisonic 3D audio technology in multi-million-selling games, including six number 1 hits in the UK and major EU territories and two BAFTA award winners, RaceDriver Grid and F1 2010. Simon has professionally developed games, tools, audio and VR technology for Amiga Inc., Atari Corp., Attention to Detail, Central ITV, Codemasters, Digital Precision, dk'tronics, Dolby, DTS, Electronic Arts, Quicksilva, Racal and Silicon Studio Ltd., and written regular technical columns for many magazines including *Amiga Format, Crash, Linux Format* and *Personal Computer World*. Simon is expert in console, mobile and PC audio and streaming system development, variously working as an inventor, sound designer, game audio programmer and audio systems engineer. He has been granted five US and UK patents, advises on AHRC and EPSRC research programmes and gives talks at GDC, AES and university conferences.

AUDIO ENGINEERING SOCIETY PRESENTS . . .

www.aes.org

Other titles in the series:

Audio Production and Critical Listening, Second Edition
Authored by Jason Corey

Recording Orchestra and Other Classical Music Ensembles
Authored by Richard King

Recording Studio Design, Fourth Edition
Authored by Philip Newell

Modern Recording Techniques, Ninth Edition
Authored by David Miles Huber

Immersive Sound
The Art and Science of Binaural and Multi-Channel Audio
Edited by Agnieszka Roginska and Paul Geluso

Hack Audio
An Introduction to Computer Programming and Digital Signal Processing in MATLAB
Authored by Eric Tarr

Loudspeakers
For Music Recording and Reproduction, Second Edition
Authored by Philip Newell and Keith Holland

Beep to Boom
The Development of Advanced Runtime Sound Systems for Games and Extended Reality
Authored by Simon N Goodwin

Beep to Boom

The Development of Advanced Runtime Sound Systems for Games and Extended Reality

Simon N Goodwin

NEW YORK AND LONDON

First published 2019
by Routledge
52 Vanderbilt Avenue, New York, NY 10017

and by Routledge
2 Park Square, Milton Park, Abingdon, Oxon, OX14 4RN

Routledge is an imprint of the Taylor & Francis Group, an informa business

About the cover images: The oscillogram shows output from a 1982 Sinclair ZX Spectrum's BEEP 8,48 command. The four-part 3D soundfield waveform below it depicts a Texan thunderstorm, recorded in 2009 with an experimental Ambisonic microphone hand-made by Dan Hemingson.

Library of Congress Cataloging-in-Publication Data
Names: Goodwin, Simon N., author.
Title: Beep to boom: the development of advanced runtime sound systems for games and extended
reality/Simon N Goodwin.
Other titles: Development of advanced runtime sound systems for games and extended reality
Description: New York, NY : Routledge, 2019. | Series: Audio engineering society presents . . . | Includes
bibliographical references and index. |
Identifiers: LCCN 2018044750 (print) | LCCN 2018046501 (ebook) | ISBN 9781351005531 (pdf)
| ISBN 9781351005524 (epub) | ISBN 9781351005517 (mobi) | ISBN 9781138543911 (hbk : alk. paper)
| ISBN 9781138543904 (pbk : alk. paper) | ISBN 9781351005548 (ebk)
Subjects: LCSH: Computer games—Programming—History. | Video games—Sound effects—History.
| Computer sound processing—History. | Sound—Recording and reproducing—Digital techniques—History.
Classification: LCC QA76.76.C672 (ebook) | LCC QA76.76.C672 G655 2019 (print) | DDC 794.8/1525—dc23
LC record available at https://lccn.loc.gov/2018044750

ISBN: 978-1-138-54391-1 (hbk)
ISBN: 978-1-138-54390-4 (pbk)
ISBN: 978-1-351-00554-8 (ebk)

Typeset in Minion Pro
by Apex CoVantage, LLC
Printed and bound by CPI Group (UK) Ltd, Croydon, CR0 4YY

Contents

Chapter 1 The Essence of Interactive Audio 1

Chapter 2 Early Digital Audio Hardware 11

Chapter 3 Sample Replay 27

Chapter 4 Interactive Audio Development Roles 35

Chapter 5 Audio Resource Management 43

Chapter 6 Loading and Streaming Concepts 51

Chapter 7 Streaming Case Studies 63

Chapter 8 The Architecture of Audio Runtimes 77

Chapter 9 Quick Curves—Transcendental Optimisations 85

Chapter 10 Objects, Voices, Sources and Handles 95

Chapter 11 Modelling Distance 111

Chapter 12 Implementing Voice Groups 119

Chapter 13 Implementing Listeners 125

Chapter 14 Split-Screen Multi-Player Audio 139

Chapter 15 Runtime System Physical Layers 145

Chapter 16 Mixing and Resampling Systems 153

Chapter 17 Interactive Audio Codecs 173

Chapter 18 Panning Sounds for Speakers and Headphones 183

Chapter 19 Ambisonic Surround-Sound Principles and Practice 197

Chapter 20 Design and Selection of Digital Filters 225

Chapter 21 Interactive Reverberation. 233

Chapter 22 Geometrical Interactions, Occlusion and Reflections 249

Chapter 23 Audio Outputs and Endpoints 255

Chapter 24 Glossary and Resources 267

Acknowledgements 277

Index 279

CHAPTER 1

The Essence of Interactive Audio

Let's pretend you are strapped in on the starting grid of a Formula 1 motor race. Twenty-four cars of a dozen designs roar as the lights turn green. Ninety-six tyres of various types and infinitely variable loads are poised to emit any or all of eight subtly inflected sounds, including peel, skid, scrub, bumps and roll transitions, not to mention brakes, suspension and other parts. Seconds later, they'll be jostling for position on the first corner.

In F1 and similar games, all this is orchestrated in absentia by the audio team, performed by the audio runtime system, conducted by the player. It might play on a single phone speaker, stereo headphones, surround-sound speakers or more than one of those at a time. Interactive audio might be rendered more than once, in a local split-screen view or remotely on the screens of competing players or in headphones for a virtual reality (VR) player and simultaneously on speakers for a watching audience.

Or imagine you are in a shell-hole—for practice, as penance or for kicks. Between the bomb-blasts, small-arms fire and cries of friends and foes, buffeted mud squelches round your calves as you squirm. Hundreds of explosions large and small echo above and around you. Sounds shift realistically in your speakers or headphones as you warily scan the parapet. The ground around you thumps with each impact. Will you know which way and when to jump? How soon and how surely can you know when an incoming shell has your number on it? Without accurate audio—not just cinematic immersion—you won't last long.

More prosaically, imagine you're trying to find a public toilet. Normal headphones, in conjunction with binaural sound synthesis, GPS, gyros and an augmented reality system in your phone, can tell you which way to go, whatever you're looking at, whatever you're trying to find.

A rally car splashes across a ditch. In Codemasters' *DiRT2* game, this "event" triggers 24 voices of fresh sound effects to play. Milliseconds later, as the positions, velocities, amplitudes and echoes of all those sounds are finely adjusted, another dozen voices chirp up. That's what happens when you give a sound designer a budget of hundreds of voices and need to sell millions of copies on all major console and PC platforms.

That's how modern games work. Whether the player is racing, exploring, shooting or playing a sport, convincing, immersive and informative sound is essential. The same techniques apply to virtual, augmented and extended realities, in education or training as well as entertainment. Each soundfield is tailored for a single listener who controls the camera, the view-point (first or third person, close or distant) and chooses the listening environment. All these parameters they can change on a whim.

Propelled by commercial and aesthetic competitive pressures, gamers represent a diverse and wealthy global community of early adopters, while extended reality introduces new markets in education and simulation. Interactive audio is a superset of prior sound technologies. This book explains what it needs and what it lacks.

Size of the Challenge

The author was Central Technology Group Lead Programmer on the *DiRT2* driving game, which has grossed around $200M since 2009, primarily on PlayStation and Xbox consoles; it's still selling on PC and Mac. More than 50 programmers worked on that game for more than a year, plus a similar number of full-time testers, dozens of designers and more than 100 graphic artists. This book directly addresses the roles of the eight sound designers and five audio programmers.

The concepts are relevant to anyone interested in creating interactive media and the many differences between that and the old passive media of TV, cinema and recorded music. This book draws on decades of game development experience, encompassing music, sport and shooting genres and even VR space exploration. It includes war stories, deep geek detail, analysis and predictions.

DiRT and *Formula 1* (F1) games are hardly the tip of the iceberg. Top-selling "triple-A" games, like Rockstar's *Grand Theft Auto 5*, cost hundreds of millions to develop but generate operating profits of billions of dollars.[1] In its first five years *GTA5* sold more than 90 million worldwide, at prices higher than any movie. Such success is only possible by designing it to suit all platforms, current and future, rather than picking one console or PC configuration.

GTA5 audio benefits from audio adaptations demonstrated by the author at the 2009 Audio Engineering Society (AES) Audio for Games conference in London. Such scalability is a prerequisite of sustained global sales, and it's been achieved by constant technical innovation, including the adoption and refinement of advanced techniques described here.

This is a book about managing complexity in a way that suits the customers, the designers, the platforms and current and future genres of entertainment and training. It's a book about curves, synergies and neat tricks. It's also about having fun making interesting noises and understanding and playing with psychoacoustic principles.

It's Different for Games

A misperception, slowly abating, concerns the superficial resemblance between games and passive media like TV and movies. Those are made for a mass audience, compromised to suit a generic consumer and set in stone before release. But each game is live, one time only.

The experience is never the same twice. The more it varies, the greater its lasting appeal and the more it involves and teaches the player. Whether it's limited to a few dozen sounds or the hundreds of concurrent samples modern computers can mix, it is a designed experience, dependent upon categorisations and decisions made long before but mediated by the player. There is no 'final cut.'

> The ear never blinks.

The runtime system necessarily embeds the experience of sound designers, engineers and live mixers so it can "finish" the product on the fly. This draws on the asset-selection and balancing skills of designers, just as movies might, but demands more variety, flexibility and dynamic configuration, because the players call the shots. Audio is more demanding than graphics because all directions are equally important in a game, the ear never blinks and there's no "persistence of hearing" akin to the persistence of vision which smooths out the flickering of video.

Perception is multi-modal. Georgia Tech research, published by MIT in 2001, established that high-quality video is more highly rated when coupled with high-quality sound.[2] Good graphics also make poor audio seem worse! Even if audio is not explicitly mentioned, it has a profound influence on perception.

Psychoacoustics

Sounds have multiple dimensions: pitch, intensity, spread, timbre, distance, direction, motion, environment. Our brains interpret these according to psychoacoustic curves learned over lifetimes in the real world. Each aspect, and the changes in each dimension, must be plausible, sympathetic to the whole and perceived to be smooth, progressive and repeatable, to maximise the informational content and forestall confusion.

Whether they can see it or not, audio tells the player the distance and direction of each sound, how it is moving, and what sort of environment surrounds it and the listener. More deeply, it identifies threats, opportunities, choices and risks which will influence future events, in whatever way the player chooses to interpret them. Learning by finding out is more personal and persuasive than following some other's story. It relies on coherence, consistency and flexibility, because any glitch might break the spell and destroy the suspension of disbelief which turns a simulation into a lived experience.

Like any artistic representation, the results can be symbolic, realistic or hyper-real. Games and VR often target a remembered dream state more than strict reality, in which auditory salience works as a filter rather like depth of field in a film. But whereas cinema auteurs pick a subset of the sounds to fit their pre-ordained story and spend arbitrary time compiling each scene, interactive media must continuously identify and include cues that help the listener create a new, unique story of their own, not once but many times over, without inducing monotony or revealing gaps that might burst the bubble of immersion and agency.

Strict realism is just a start. The most realistic *Richard Burns Rally* would only be playable by Richard Burns. In a real *F1* race you'd rarely hear anything but your own engine, but in a game that's not good enough. Interactive audio has the capability to be symbolic, extracting just the essence or key cues in a scene, or hyper-real—reproducing the remembered synaesthetic experience, augmented by imagination, not the prosaic reality. If you're ever unlucky enough to be in a car accident you may find the crash disappointingly dull by game standards—but games and VR are meant to be fun, thus all the more memorable.

What's New?

This book explains how modern interactive audio systems create and maintain consistent and informative soundfields around game players and consumers of extended reality products. It complements books for sound designers and game programmers by filling the gaps between their accustomed models of reality.

It's not a book about sound design, though it contains many tips for sound designers, especially related to interactivity. It's not a book for mathematicians, though it builds on and refers to their work. It's not even a book about psychoacoustics, though of all the related fields, that one informs the content most of all. There are many excellent books about all those subjects, as the references reveal.

This book pulls together concepts from those fields and decades of practical systems design and programming experience to explain how they fit together in modern games and extended realities. It presents a layered approach to implementing advanced audio runtime systems which has been successfully applied to platforms ranging from obsolescent telephones to the latest game consoles, arcade cabinets and VR rigs.

This is a practical book for designers, programmers and engineers, boffins, inventors and technophiles. The focus is on runtimes—the active parts that ship to the customer—rather than pre-production tools, though Chapter 24 surveys free and commercial tools for content-creation. It tells how to do things, with tested examples, but more significantly it explains why those things are useful and how they fit together.

Starting from the first computer-generated beeps and the basic challenges of volume and pitch control, it traces the development of audio output hardware from tone generators to sample replay, the layering power of multi-channel mixing and the spatial potential of HDMI output, building up to the use of Ambisonic soundfields to recreate the sensation of hundreds of independently positioned and moving sound sources around the listener.

The author has pioneered the interactive use of 3D soundfields but is well aware from research and direct experience that few listeners enjoy a perfect listening environment. One of the greatest changes taking place in media consumption in the 21st century is the realisation that there's no correct configuration. Figure 1.1 shows the preferred listening configurations of more than 700 console and PC gamers.[3] Every listener benefits from a custom mix, and interactive audio systems deliver that as a matter of course.

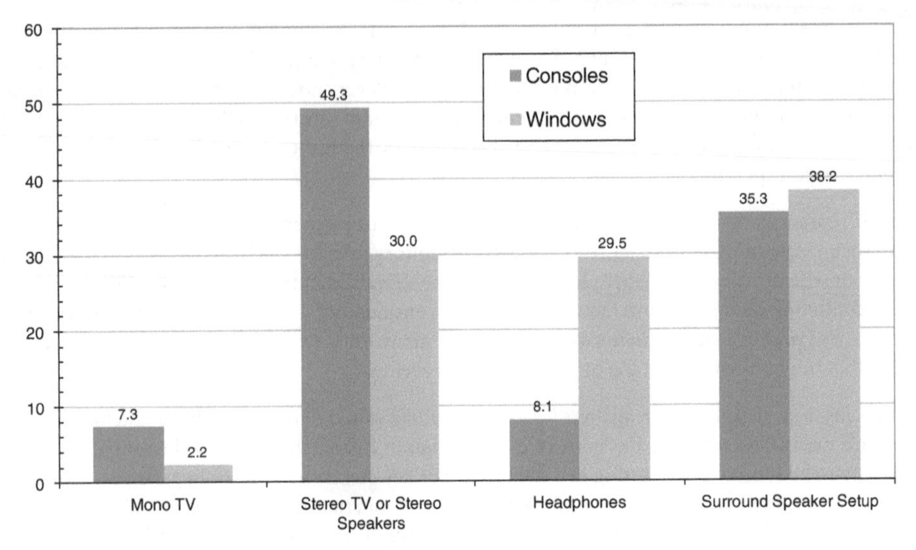

Figure 1.1: Listening preferences in percentage by output format

Since then phones and tablets have shifted the goalposts—although Apple TV and many Androids support HDMI and 7.1 output, most listeners use mono speakers. Live analytics provide specific usage information. Over 40 million sessions, barely 3% of mobile players of *F1 Race Stars* used headphones; 17% had the audio muted! The story was similar in *Colin McRae Rally* for iOS, with 8% on headphones and 12% unable to hear the co-driver. It's nice to know that 88% were listening, even if mostly in mono.

Single-Player Optimisation

Traditional movies deliver a generic mix for a mass audience in a cinema, aiming to create a sense of immersion so that those in the cheap seats still feel part of the party. Immersion is easy—it involves little more than spreading ambience around the listener. But interactive media is much more demanding. It focusses on each individual listener without requiring them to be locked in a halo-brace centred in an anechoic irregular pentagonal room with a wire-mesh floor—even if that would deliver the most technically perfect reconstruction available with commodity components to the golden ears of an ideal listener.

Some expensive aspects of VR—like head tracking, which adjusts the experience every few milliseconds to account for movement of the client's ears and eyes—are not needed for conventional gaming. Here the player's thumb directs the senses of their avatar; looking away from the screen is a recipe for disaster. However control is achieved, audio challenges remain. The "sweet spot" in which sounds are perfectly balanced may be small, especially if space is tight and speaker positions are compromised, but a lone player is strongly incentivised to move into it.

Playing and exploring together is still fun even if the shared environment compromises individualised perception. So this book reveals techniques to deliver multiple soundfields in a single listening space for split-screen multi-player games and ways to mix adaptively for multiple simultaneous listening environments, like a single-player VR experience shared by remote passive observers or others waiting their turn in the same room.

This concept of "multiple endpoints" marks out new media platforms and modern ways of listening. Custom mixing and spatialisation technique can create additional coherent soundfields at low marginal cost. We'll explore the pros and cons of headphone and multi-loudspeaker listening while showing how even a single mono loudspeaker can inform active listeners about the movement and relative position of 3D sounds in their vicinity.

Dynamic Adjustment

One of the greatest strengths of interactive audio—as opposed to TV, radio, cinema and other media designed to place a ready-made mix before the ears of many passive consumers—is the continuous conversation that takes place between active listeners and a dynamic audio rendering system. Interactivity is a cyclic process.[4] Unlike old media, interactive audio adjusts continuously to hints from the player as well as dynamic changes in the simulated world.

> All directions are equally important.

The player, not the director or camera operator, decides where to look and listen. All directions are equally important when all can be interchanged with a flick of the player's finger; our ears perceive sound from all directions, without blinking or blind spots, even though their response is coloured by direction and environment. Just as the brain integrates uneven responses to sounds in a complex environment by small head movements, filling in the gaps from short-term memory, sedentary players derive the same benefit by twitching their thumbs.

Learning Through Play

Even if they're incomplete or not quite correct, players learn to interpret multiple environmental cues to their advantage by experiment. The power of such learning stems from reinforcement, positive and negative, and the tight feedback loop between the input devices—gamepads, mice, track balls, accelerometers or keyboards—and output devices like screens, speakers, headphones and haptic emitters like rumble motors, butt-kickers and "force-feedback" steering wheels.

Optional sub-woofers are used in cinema for low-frequency effects and often to economise on the low-frequency response of more numerous speakers through "bass management." These deliver both sonic and haptic feedback. We still need to know what the client expects, drawing on their larger experience of real-life listening and their equipment's capabilities, to decide which aspects of the sound to model and prioritise.

Scalability

Rally is a motor sport that sets a single driver against the clock and a complex environment. It may be simulated on a handheld phone or a console rig that dominates a room. This book compares Rally games made in the same studio a few years apart, showing how they were customised for their platforms.

A current PC or console game might mix 250 voices—individual spatialised sounds—from an hour of short samples in RAM and hours more swappable or streaming from disc, all with filters and layers of reverberation driving a 2D or 3D 7.1-speaker rig. Contrast an obsolescent iPhone struggling to play a dozen sounds at once with no reverb or filters and in mono at half the sample rate and video stuttering along at a cartoon-like pace.

In the heyday of iOS 7 Apple would not list a title on their store unless it ran on their end-of-life iPhone 4, even though that model was about a seventh the speed of the next one up and 50 times slower than the latest iPad or phone. In the past decade game developers have switched from chasing tens of millions of high-end gaming PCs and consoles to supporting hundreds of millions of battery-powered handheld devices, where the best were 100 times more capable than the least. To reach such audiences you must innovate on the latest devices without forgetting the also-rans. Discoverability is all, in the biggest and most competitive fashion markets.

So *Colin McRae Rally* mobile was playable on iPhone 4, cheap Androids and even BlackBerry phones, yet the same game got a second Apple Store feature on the strength of its 60-Hertz video and 48-kHz stereo sound on high-end iOS 8 devices. That configuration also plays extra collision, environment and surface sounds, and allows the user's music to override the game's. The iPhone 4 was limited to 24 kHz mixing, 15-Hertz video, reduced effects and playing sounds from the two wheels nearest the listener.

Scalability matters. To reach the biggest market and to stay there you need to reach the minnows at the start without short-changing the nascent whales. You might find you're making the same game, in varying priority order, on iPhones and iPads, innumerable semi-compatible Android phones and set-top boxes like the NVIDIA Shield, PCs in myriad configurations, plus recent Sony, Microsoft and Nintendo consoles. Sometimes even macOS.

Memory capacities vary from half a gigabyte (GB), mostly allocated to other things, to many times that amount. Physical game media hang on for high-end consoles, while one Blu-ray disc can hold 200 GB of compressed data, but downloads will predominate in future and with no such limit. However much RAM you get, you can't load everything into it—not least because of the loading time—players are impatient people. So we discuss how to organise, compress, index and track game audio data.

This book explains how to transparently handle different formats—sample rates, codecs, multi-channel sets, resident and streaming—so game designers don't need to worry about platform details, yet the products make the best of each configuration. Even early in the 21st century it was not uncommon for a game to come with over 150,000 samples; all it takes now is a smart commentary system and internationalisation to drive this over a million.

Pervasive Interactivity

New forms of content delivery, 3D graphics and sound mean that even passive media is sometimes authored with game-centric tools like Unreal Engine and Unity. As customers come to expect custom experiences, dynamic systems find use in film and TV, and by architects, trainers and planners, not just those seeking to bottle up interactive thrills.

After years making do with hand-down technologies like MP3 and Dolby surround, designed for the average expectations of passive consumers, the tide has turned. Naturally interactive and scalable schemes like Ambisonics,[5] context-sensitive mixing and Virtual Voice management are being shared in the opposite direction, with the realisation that film and TV are subsets of interactive media. Even if you're doing nicely in the old media world, it pays to learn how you can empower your customers and ease your workflow, especially in pitching and prototyping, by borrowing ideas from games and XR.

Audio Runtime Systems

The audio runtime system is the part of a simulation which creates soundfields around its users, enabling them to find their way around and make sense of the simulated environment using the skills they've learned in real life. It sits under the "game code," called from the "update thread," and above the "audio pump" that spits out freshly mixed blocks of multichannel sampled sound every few milliseconds. It does the job of a live mix engineer as well as resource management.

How high above that pump the runtime sits depends upon the layering and the pre-existing audio capabilities of the host. This book shows how the layers fit together and how to evaluate existing components to see if they fit your needs. It explains how to test those and decide just what your product needs, rolling your own if desirable.

The main purpose of the layering is to provide a consistent set of advanced capabilities on every platform, large and small, yet tailored to take advantage of the unique features or limitations of each platform and take any help—in terms of standard audio components, like mixers, panners, filters, resamplers and reverbs—which is predictable enough to be of value. There are two deeper interfaces in this layer cake.

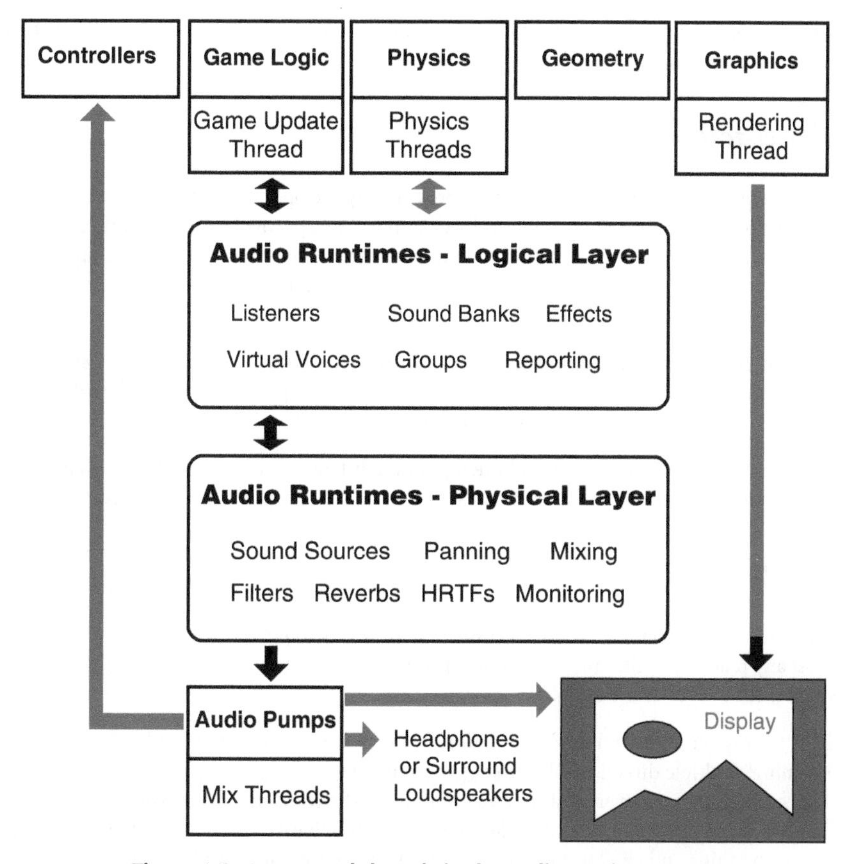

Figure 1.2: Layers and threads in the audio runtime system

Runtime Layers

The top layer is an object soup of Virtual Voices, which play samples generated on the fly, loaded piecemeal or in organised Banks. The voices behave rather like independent threads, all gathered together in Groups and mixed via Listeners—virtual microphones—to whatever audio outputs the player selects. All in 3D. This layer works out what to play when there's more than the hardware supports—hence the term "Virtual Voices." It handles distance effects, Doppler pitch shifts, frames of reference, memory management, profiling and reporting. It provides a common interface for game programmers so they can make one game for many platforms, which makes allowance for the strengths and weaknesses of each without dumbing down the strongest or over-taxing the weak.

This logical layer rides on top of a simpler interface which can be bolted in a few days onto any multichannel platform audio sub-system capable of mixing memory-resident samples and outputting them, like XAudio or OpenAL. At this point, if the platform is mature, you can take advantage of tightly coupled and optimised machine-specific implementations of basics like pitch conversion, filtering and reverb—providing you know how to translate the controls in a platform-independent way or can explain the differences in a way which will thrill rather than puzzle hard-pressed designers. You grab what works and fix what's left.

Typically, you get optimised hardware/DSP/codec and reverb access, quite often using hardware which would not otherwise be available to you. If all goes well, assuming good documentation or willingness to experiment and rely on the results, you get the benefit of several platform-experts working on your audio team, albeit distantly. And if not, there's always a fallback.

Platform Abstraction

If you lack all or some of this physical layer—generally provided by the operating system maker—you can augment or replace it with your own code. This book explains by example the steps involved in building a generic cross-platform mixer, as well as where to find them off the shelf, perhaps for free. The requirements are more relaxed than for a full application programming interface, since all we really need at this layer is multiple sampled voice replay with pitch and volume controls.

Software mixing won't be as fast as hardware, but it's easier to keep it working consistently across all the platforms. It is a stable core, decoupled from its clients by the logical layer, in which piecemeal or wholescale optimisations can be performed as your profile timings and production schedule dictate.

The customer-friendly stuff like Virtual Voices and microphones, distance and Doppler effects, event logs and priority systems is all done upstream, in the logical layer. It's most of the code on any platform and shared between them all.

What You Get

In other words, we give you, or your heirs and successors, a quick way to get a sophisticated, scalable audio system running on almost any platform—after the first—and a production-engineering framework to optimise the heck out of it afterwards without everything breaking.

This book explains the organisation and implementation of a system to manage hundreds of virtual 3D voices with Ambisonic positioning, multiple directional listeners, group updates and reporting, distance and Doppler effects, group and per-voice filters, delays, reverberation and optional asset streaming. The mixing stage supports as many voices as your hardware and other needs of the title can handle, with live decompression, parametric equalisation, advanced Ambisonic panning and click suppression, for mono, stereo, binaural and surround-sound outputs to multiple endpoints.

Key aspects are illustrated with short program snippets expressed in a slim subset of C++, the standard language of portable real-time systems, which can be converted to ANSI C, C# and related languages. The code has been tested with Visual Studio 2017 for Microsoft x86 and x64 architectures, Apple LLVM version 9.0.0 (clang-900.0.39.2) on macOS and GCC 5.4.0 on Ubuntu 16.0.4 compiling for ARM64 with -std=c++11.

The caret ^ symbol is used in this text to denote exponentiation, aping languages like Algol, TeX, Google's calculator and code representations of exponentials when a superscript is not available; 2 ^ 3 = 2^3 = 2 * 2 * 2. Unfortunately C has no exponential operator, relying on the library function powf(), e.g. 2^3 = powf(2.f, 3.f), which is why I opt for the commonly used ^ shorthand. The separator .. is consistently used, as in programming languages like Pascal and Ada, to denote a subrange. For example 1..4 denotes values between 1 and 4. The prefix m_ introduces variable values which are class members. Empty parentheses after an identifier indicate that it's a function.

Sections of This Book

The book is in four sections. First it traces the history of interactive audio generation, showing the fundamental audio issues of timing, mixing, volume and pitch controls and how they've been tamed in software and hardware. The following section considers what a modern audio runtime system needs to do and the team roles which make it possible.

The third major section, from Chapter 15 onwards, explains implementation options and their rationales in detail. Chapter 24 lists concepts and resources, free and commercial, which can help you build and maintain a flexible, portable, extensible audio runtime system without reinventing wheels. You don't have to use everything in this book to derive full benefit, but you should consider most of it.

The references cite many relevant books and research papers, but in such a new industry some insights are only available online. Inevitably links go stale in the lifetime of a book, so you may need to use Google to find some. The Wayback Machine[6] tracks old websites.

Topics are introduced in logical sequence, to minimise the need for cross-references, but once a term is introduced its meaning is taken as read. For best results, read the book first from start to finish. Later you can cherry-pick concepts and implementation details to suit your application and audience. There are plenty to choose from.

References

[1] *GTA5:* www.forbes.com/sites/insertcoin/2018/02/08/at-90-million-copies-sold-gta-5-further-cements-its-inexplicable-legendary-status/#429cd6de2e22

[2] *Interactions in Perceived Quality of Auditory-Visual Displays; Russell L. Storms and Michael J, Zyda, Presence, ISSN 1054-7460, Volume 9 Number 6, December 2000*

[3] *How Players Listen; Simon N Goodwin, AES 35th International Conference, Audio For Games, 2009:* www.aes.org/e-lib/browse.cfm?elib=15172

[4] *Understanding Interactivity; Chris Crawford, 2000, ASIN B0006RO8ZA, no ISBN*

[5] *Periphony: With-Height Sound Reproduction; Michael A. Gerzon, JAES, ISSN 1549-4950, Volume 21 Number 1, February 1973, pages 2–10:* www.aes.org/e-lib/browse.cfm?elib=2012

[6] *Wayback Machine*: https://archive.org

CHAPTER 2

Early Digital Audio Hardware

Audio output has been a feature of electronic computer hardware since the earliest days. The Manchester Mark 1, the world's first general-purpose commercial digital computer, had audio output and a built-in speaker, which could be made to "hoot," according to computer pioneer Alan Turing. His 1951 programming manual includes two short loops that can play a note an octave above middle C—the highest pitch the Mark 1 could reach—and the G a fifth lower.[1]

There are BBC recordings online of this hardware playing music in 1952. In fact the original music player was tacked on to a program interpreter and soon extended to play arbitrary tunes:

> *In 1948 the mathematician Alan Turing came to the University of Manchester and within a few months started using and working on the Mark 1. In 1951, Turing invited a friend of his called Christopher Strachey to write a program interpreter program for the machine. When it terminated Strachey had programmed it to use the "Hoot" instruction in such a way that the machine played the tune God Save the King.*
>
> *Strachey's algorithm was then used to create music programs that could load and play any tune. Later, it became traditional for Ferranti maintenance engineers to write a music program for each new type of machine.*

Chris Burton is a computer historian and has revised that quotation from a report on Manchester University's website,[2] explaining that the "hooter" applied 0.24 millisecond pulses to the loudspeaker. Simon Lavington's detailed explanation of the implementation dates it back to 1950.[3] Similarly early beepings are ascribed to the Pilot ACE, which Turing previously designed in Cambridge, and the antipodean CSIR Mark 1.[4]

So computer audio was in at the birth of digital computing, starting with a 1-bit output. Decades later the Apple][micro had similar audio hardware. Access to memory location 0xC030 caused the internal speaker to click, and repeated accesses generated a tone. The pitch depended on the short delay between pulses. Some of my first 8-bit games used that technique on Apple and ZX Spectrum micros.

Programmers of the slightly older TRS-80 home computer made noises by polling the hardware port connected to the tape drive motor. Hammering this caused the motor control relay to click repeatedly, till the relay contacts wore out at the unanticipatedly high duty cycle.

Early home computers, including IBM's original PC, used audio cassettes for program storage. This meant they were implicitly capable of both audio input and output if you connected them to an amplifier in place of the tape drive. But the output was digital clicks rather than notes or arbitrary waveforms.

Computer music research followed different lines at Bell Labs, where Max Mathews and John Pierce programmed IBM valve computers to re-express digitised scores as audio. The quality was good—the telephone company had cutting-edge digital-to-analogue converters—but the process was much slower than real time. This scripting of the synthesis of complex samples led to tech like MIDI (Musical Instrument Digital Interface) decades later, but it was hardly interactive in the 1950s or 1960s. Mathews gave his forename to the audio toolkit Max/MSP, still going strong.

Engineers were alert to the possibility of computer-generated sound more than a century before. The 1843 commentary on Charles Babbage's analytical engine by Byron's daughter, Ada, Countess of Lovelace, observed, "*the engine might compose elaborate and scientific pieces of music of any degree of complexity or extent.*"[5] But this was not predicting real-time sound generation, just one enumeration of the many possibilities of symbolic computation.

Polyphonic Stereo

Real-time multichannel audio arrived at Massachusetts Institute of Technology (MIT) a decade after the Manchester recordings. Members of the Tech Model Railroad Club gained access to a TX-0 prototype transistorised computer, precursor to DEC's PDP-1. Pete Sampson, one of the key figures in the development of the first interactive computer game, *SpaceWar!*, got it to play baroque music in four-part harmony.

The hardware was minimal—Pete got permission to hook four wires from existing front-panel indicator lights to passive analogue filters and a mixer which could combine them into mono or stereo for amplification. So even in 1964 multiple voices engendered multichannel output.

Each wire could be set to a high or low logic level according to the state of the most significant bit of an 18-bit number written to a flip-flop latch. The challenge was to find a way to update all four wires at different rates, corresponding to musical harmonies, using a simple, fast loop of code. The loop had to be short and fast to allow accurate replay of high notes; with a single processor capable of a maximum of 200,000 memory access cycles per second, there was only time for a handful of instructions between the updates of each output wire. Worse, to keep the whole lot in tune the replay loop had to be balanced so that conditional tests and output to one didn't affect the timing of other voices.

Another MIT student, Dan Smith, captured MP3 files of this software in action and explains the ingenious way Pete's "Harmony Compiler" managed to keep time. "*It was soon discovered that the best way to generate tones was to perform repeated additions, (letting the total repeatedly overflow) and copy the accumulator's high bit to the output.*"[6]

This approach yielded fast digital oscillators without any delays or hiccups caused by conditional execution. Such "branchless programming" remains a powerful technique in real-time system implementations; this phase-accumulator technique is now used to index samples as well as to directly generate periodic waveforms, as the higher-order bits of the accumulator describe a sawtooth wave.

The highest note attainable is limited by the total number of instructions needed to update all four voices and keep count of when one of them next needs to change pitch. The longer the loop, the longer the period between changes to the waveform. High notes demand fast changes.

The highest pitch is heard when the step added is half the word size, so alternate updates toggle the top bit. The deepest note is heard when the value added is just 1, and it deepens as either the machine word or update loop increases in length. Any voice can neatly be muted, at no cost and without disturbing other voices, by adding 0 each time.

This approach also guarantees a near-symmetrical square-wave output, as half the values in the accumulator will have the high bit set and half will have it clear, and modulo additions preserving the remainder will cycle between those sets of values at a rate proportional to the magnitude of the number that keeps being added. It's easier to understand this from a diagram, simplified here for an 8-bit word; values above 127 have the top bit set and are shaded so that the transitions between high and low output in each row, for a given step, are more apparent. Grey blocks are at high level, white ones low, and the 96-step wave is the odd one out—see how uneven its transitions are.

Table 2.1: Note pulse timings

Step	0	1	2	3	4	5	6	7	8	9	10	11	12
32	0	32	64	96	128	160	192	224	0	32	64	96	128
64	0	64	128	192	0	64	128	192	0	64	128	192	0
96	0	96	192	32	128	224	64	160	0	96	192	32	128
128	0	128	0	128	0	128	0	128	0	128	0	128	0

The 8-bit word can hold values from 0 to 255, so any overflow above this is lost. Hence adding 64 to 192 or 128 to 128 yields 0, modulo 256. Three octaves can be seen—the lowest uses the step of 32 and spends four cycles low then four high alternately. Doubling the step size to 64 gives alternate pairs of low and high outputs, and the maximum useful step of 128 plays an octave higher still, at the shortest possible pulse width and hence the top pitch.

These are the easy cases—the problematic ones are the in between step sizes, where the capacity of the word is not an exact multiple of the size of the step. Most musical notes and corresponding step sizes don't fit that neatly. The step 96 row shows the consequence. To get an average period of 96/256 (0.375 updates per pulse, or three clicks every eight iterations) the toggling must occur at an uneven rate, spending either one or two cycles high or low before swapping back to preserve the average.

As Smith heard, "*the result of all this is that the basic 'notes' produced by the PDP-1 were precisely on pitch, but had a rough, ugly buzzy background noise, lower in pitch than the note, and more obvious on the higher notes.*"

Such effects are subtler now because of higher sample rates but still happen whenever discrete steps are used to approximate arbitrary intervals. The single-bit output means that even if the step was expressed in floating-point format, such jitter would occur. It stems from the regularity of the output loop and small number of iterations between state changes, not the accuracy of the fractional representation.

The TX-0 and PDP-1 use 18-bit words, allowing 262,144 possible values (2^{18}) and hence the replay of more than 100,000 distinct pitches; a step of 1 would take 131,072 (2^{17}) cycles to toggle from low to high or back again. But the next available note is a whole octave higher in pitch—a step of two toggles once every 65,536 times round the update loop.

So the lowest step values are unusable for the replay of arbitrary notes, as the interval between them is so large, but greater values fall increasingly close to the logarithmically distributed intervals of the musical scale. Eighteen-bit precision, one part in a quarter of a million, gave the MIT hackers a reasonably accurate scale of notes over several octaves to play with.

The same add-and-output pattern can be repeated for the other three voices, using different output addresses and potentially different step sizes too. The "harmony compiler" works out step sizes appropriate for each of the four voices and a count of the number of cycles while all four continue without changing any of the step sizes. When a note needs to stop or change in pitch, this count is reached and the step gets updated, a count till the next change is loaded, and the tone-generation loop continues. There's a slight timing hiccup at such boundaries, but it's acoustically masked by the arrival of the new note.

Deliberate Detuning

The MIT real-time player had one other trick up its sleeve, which remains directly relevant to modern interactive audio. Smith explains, "*the music playing software actually had a deliberate detuning built in, a couple of cents for each of the four parts, because otherwise the notes tended to merge and sound like a single note with a funny timbre, instead of a chord.*"

Decades later, circuit-racing games ran into similar problems, solved the same way. When several cars in a race have the same engine and are controlled by the same AI system, it's likely that they'll play at the same pitch on the start line, down the straight and at other times when they're neck and neck.

This does not sound realistic—it's unnaturally precise, as the MIT hackers observed. Codemasters' Audio Manager Tim Bartlett used to deliberately offset the engine revs of non-player cars, and games like *TOCA RaceDriver* sounded all the better as a result. This half-century-old technique is still useful whenever there's a possibility of a flock of objects sharing audio assets in your game world.

Towards Game Audio

The first cartridge-programmable TV game, the Magnavox Odyssey, was launched in 1972 with no audio capability at all. Designer Ralph Baer had offered to implement sound, but this—and on-screen scoring—was turned down for cost reasons. Nolan Bushnell's *Computer Space*, inspired by MIT's *SpaceWar!*, arrived in amusement arcades contemporaneously, courtesy of Nutting Associates; it had four sounds, digitally switched but realised using simple analogue transistor circuits. Bushnell quit to found Atari and started work on the VCS 2600 games console, which brought in the microprocessor—an 8-bit 6507—and custom audio hardware.

A corner of Atari's custom-made TIA (television interface adapter) chip was dedicated to audio output; just 13 bits of hardware controlled each of two channels. The output was a stream of digital pulses, masked with ten types of optional regular or pseudorandom modulation to vary the timbre from a beep to a rasp. Pseudorandom means predictable only by mathematical analysis. Chapter 9 includes a section about "types of random."

A 5-bit divider circuit gave 32 different frequencies, erratically and unmusically distributed over a five-octave range, making them unusable for playing tunes. Four-bit linear volume control was a step forward, theoretically allowing sample output. But like the frequency control it used the minimum viable hardware, making it incapable of accurately tracking the psychoacoustic curves of human hearing.[7]

The neat thing about TIA was its ability to generate sound without processor intervention. It could carry on beeping the same pair of sounds automatically, responding at once to changes in the register values but otherwise leaving the processor to its own devices. Given the way the VCS display worked, directly controlling the TV output line by line, this was essential.

Processor-timed audio, as used on the MIT and Manchester machines, would have given players a choice of graphics or sound but not both at the same time. This and a dearth of memory also made VCS sample replay impractical.

Game audio hardware stepped up in the late 1970s when a company called General Instrument teamed up with Barbie doll firm Mattel to create the chipset for the Intellivision console, improving both pitch and volume controls by adding more bits, enabling smooth control. Mattel shipped over three million Intellivisions, and millions more AY chips ended up in later Atari, MSX, Sinclair and other home computers, as we'll soon hear. But there's more to be said about 1-bit output first.

Just a Bit

Single pulses might seem rather limited, but in principle any sound can be played at arbitrary quality with a single output bit. Philips Bitstream CD players proved this—you just have to toggle that bit very fast indeed to render a range of tones and levels.

Each doubling of the output rate is—after smoothing—equivalent to adding one extra bit of precision, but there is an obvious trade-off between pitch and (average) level control. To get fine control of both, hardware must sit in a tight loop doing nothing but toggling that bit millions of times a second. The first home computers had to make compromises in that respect, especially if they were to deliver animated graphics as well as sound. Dedicated audio hardware was needed to relieve the 8-bit processors of the tedious timing.

Even a 1-bit output can be made to play recognisable chords, as Matthew Smith's hit game *Manic Miner* proved on the ZX Spectrum in 1983.

```
OUT 254,32
```

moved the tiny speaker diaphragm one way and

```
OUT 254,0
```

moved it back again. By using different delay times alternately, the varying gap between the pulses is heard as two distinct, if somewhat distorted, tones. Later games like *Fairlight* and *Zombie Zombie* refined this technique by varying the mark/space ratio of the 1-bit output. Kenneth McAlpine analysed and explained the technique in *Computer Games Journal.*[8]

This method of direct audio output monopolises the processor, which must count cycles in a tight loop between pulses and calls for exact timing. Another early Spectrum game, the 1982 asteroids-clone *Meteor Attack* from Quicksilva, fell afoul of this consideration. The game attempted to speak its title in glorious 1-bit resolution, but the result was unintelligible even allowing for the dearth of bits.

This happened because the game code had been written on a "full" system with 48K of memory but adapted before publication to suit the cheapest Spectrum model, which had just 16K. Crucially, the first 16K of memory, on both models, was shared between the processor and the video output circuits. The video took priority, so that code running in its memory ran at uneven speed depending upon the position of the display beam, ruining the sample replay.

This priority juggling was itself a fix for an older problem—the TRS-80, Acorn Atom and the original IBM PC display circuits gave the processor priority, giving smooth timing at the expense of interrupting the display, causing spurious lines and dots to appear on screen when video memory was updated. As a result, Acorn Atom *Space Invaders* took place in a snowstorm.

PC Beeps

Such interactions encouraged the introduction of custom hardware for timing and sound generation. Sometimes this re-used components designed for non-audio purposes. The original IBM PC speaker was connected to an Intel 8253 interval timer chip, which contains three 16-bit counters. The pitch of the square-wave output depended upon the pre-programmed count—the larger the value the lower the note—and the processor was relieved of the task of counting out each delay.

This method of generating a time delay, and hence controlling the pitch of a sound, by counting down remains common and has one significant implication. The higher the pitch of the required sound, the smaller the count and therefore the coarser the frequency resolution. The steps between available frequencies get closer and closer together as the pitch descends.

Human hearing and conventional music follow a very different curve, a logarithmic one where each note in the conventional equal-tempered chromatic scale is 1.0595 times the frequency of the one below. This ratio is the twelfth root of 2 and means that twelve successive multiplications by 2 yield a doubling of pitch, an octave interval.

The inexact mapping between these two curves is the reason high notes in old computer games are often out of tune. There's plenty of resolution in the reciprocal curve at the bass end, so low notes can be pitched accurately, if not exactly, but the closest counts for higher notes are audibly sharp or flat.

Early arcade games copied electromechanical pinball machines, which used custom analogue circuitry to generate specific sounds. In 1978 Taito's *Space Invaders* bolted on Texas Instruments' SN76477 "complex sound generator" chip.[9,10]

This used analogue components to control the pitch of the sounds, making it unsuitable for playing arbitrary notes, but it was still a breakthrough in several respects, as Andrew Schartmann pointed out—it combined music, a simple four-note loop, with sound effects during gameplay and progressively increased the tempo as the player blasted away enemies, delivering both informational and emotional feedback to the player.[11] This was the epitome of interactive audio, vintage 1978.

Texas Instruments SN76489

Texas Instruments followed up with a digital version, the SN76489 Digital Complex Sound Generator, which was originally designed for their TI99/4 home computer. Variations of this appeared in a dozen arcade cabinets and many home computers, including the Memotech MTX and Acorn BBC Micro series, the IBM PC Junior, Coleco Adam, Sord M5, Tatung Einstein and Tandy 1000 PC clones.

Millions of compatible chips were licensed and built into home gaming systems, including the ColecoVision, Sega Master System and Megadrive (known as Genesis in North America) and SNK's Neo Geo Pocket handheld console, which was still in production 20 years after the TI chip made its debut.

The SN76489 family of chips is polyphonic, capable of playing three notes at a time without processor intervention, and includes a fourth voice of pseudorandom noise, sounding like radio static and a suitable base for chaotic sound effects like wind and fast-flowing water. This noise could be filtered at three pre-set pitches or—most flexibly—by linking its clock rate to the frequency control of the third note voice, allowing filter sweep effects.

The 10-bit digital pitch control allows over 1000 different frequencies, albeit with a non-musical countdown-based reciprocal scale. Chip revisions boosted the clock rate from 0.5 to 4 MHz, improving high-note resolution by three octaves. Four-bit digital volume controls allow the attenuation of each voice by 0 to 28 decibels (dB) in 2 dB steps, or muting of the output. Thus the tone from voice 3 could be silenced while it continued to control the pitch of the noise output.[12]

General Instrument AY-8912

The main rival to the SN76489 was the AY-8910 series from General Instrument of Pennsylvania, announced in 1978 and pervasive a decade later. Part of their appeal to console and home computer designers was their non-audio interfacing capability. These larger chips added one (AY-8912) or two (AY-8910) 8-bit ports, which were not part of the audio circuitry but took advantage of its processor interface and spare pins to add extra general-purpose digital inputs or outputs. These were often used to connect joysticks on gaming systems or home computer printers and sometimes even MIDI synthesisers, as on the ZX Spectrum 128 range.

The AY-8912 soon found its way into pinball and arcade cabinets, then the US Intellivision and Vectrex game consoles and multi-million–selling home computers, including Amstrad CPC and Oric micros and later models of the Sinclair ZX Spectrum, all designed in Britain in the 1980s. A Yamaha version of the chip, renumbered YM2149, provided the beeps for the Atari ST computer range and many computer models built to the Japanese MSX hardware standard.

The GI audio circuits generate one noise and three tone outputs, much like Texas Instruments' rival, but 12-bit pitch controls make the General Instrument chips more accurate in rendering high notes. A 5-bit counter controls the noise timbre, with no provision to link to a tone voice, and once again volume controls were 4-bit, allowing 16 logarithmic levels between maximum and muted.[13]

Figure 2.1: The AY-3-8912 provides Spectrum 128 sound and MIDI output

Digital Noise Generation

Like the Atari and Texas chips, General Instrument used a binary shift register with feedback to generate digital noise. This approach is economical to implement in hardware and sounds like high-frequency random noise when the register is shifted often—corresponding to low values in the 5-bit counter. With higher counts and hence slower updates, the sound becomes rough and grainy rather than resembling low-pass filtered random noise. It's still more suitable for drum and explosion sounds than square-wave tones. Its characteristic rasp remains a popular feature of the "chip tunes" nostalgic enthusiasts continue to compose.

The wiring of the shift register is ingenious. Whenever the 5-bit counter reaches 0, it tests the least significant bit of the shift register and flips the noise output level from 0 to 1 or vice versa only if that bit is 1. It then generates a new pseudorandom bit by combining that bit with one three places along, generating 1 if they differ and 0 if they match. Finally, it shifts the register one place to the right (discarding the least significant bit) and puts the new pseudorandom bit in the most significant place.

Over the next 16 updates that value will slide down to the output bit. The mashing up of bits 3 and 1 is enough to make the pattern pseudorandom, providing the initial value of the shift register is non-zero. The Texas and Atari implementations work and sound alike but differ in the internal feedback arrangement and the number of bits in the shift register. More bits allow a longer period.

The technique is rarely used for interactive audio now because of its harsh binary tone. Samples and software filters are more precise and easier to control. But it still finds a place in mobile telephony, randomising the frequency of transmissions to dodge interference. Qualcomm's Code Division Multiple Access (CDMA) uses a 42-bit linear feedback shift register, much longer than any sound chip. 32 and 64-bit pseudorandom number generators are presented in Chapter 9.

Opening the Envelope

Unlike the Texas chip, General Instrument made limited provision for automatically varying the volume of all the voice outputs in various sawtooth-like patterns at a programmable rate. This was somewhat generously described as an "envelope" control, though without the separate "ADSR" parameters labelled that way in keyboard-controlled synthesisers, and directly implemented in later audio replay hardware and software.

ADSR stands for "Attack, Decay, Sustain and Release." The four parameters set the dynamic change in the level as a note starts and stops, making it much easier to model the sound of analogue instruments like pianos. Two control signals, known as key-on and key-off, control the timing. A and D parameters set the initial attack and decay rates after key-on, S is the sustain level (rather than slope) while the note is held on and R determines the rate at which the sound decays after key-off.

While there's no way to directly implement such volume envelopes on these early chips, arbitrarily complicated dynamics can be simulated by periodically adjusting the 4-bit voice volume controls. This need not be done particularly fast; good results can be obtained by running a short routine 50 or 60 times per second, typically using the timing interrupt associated with display updates, to vary the volume of each voice on the fly.

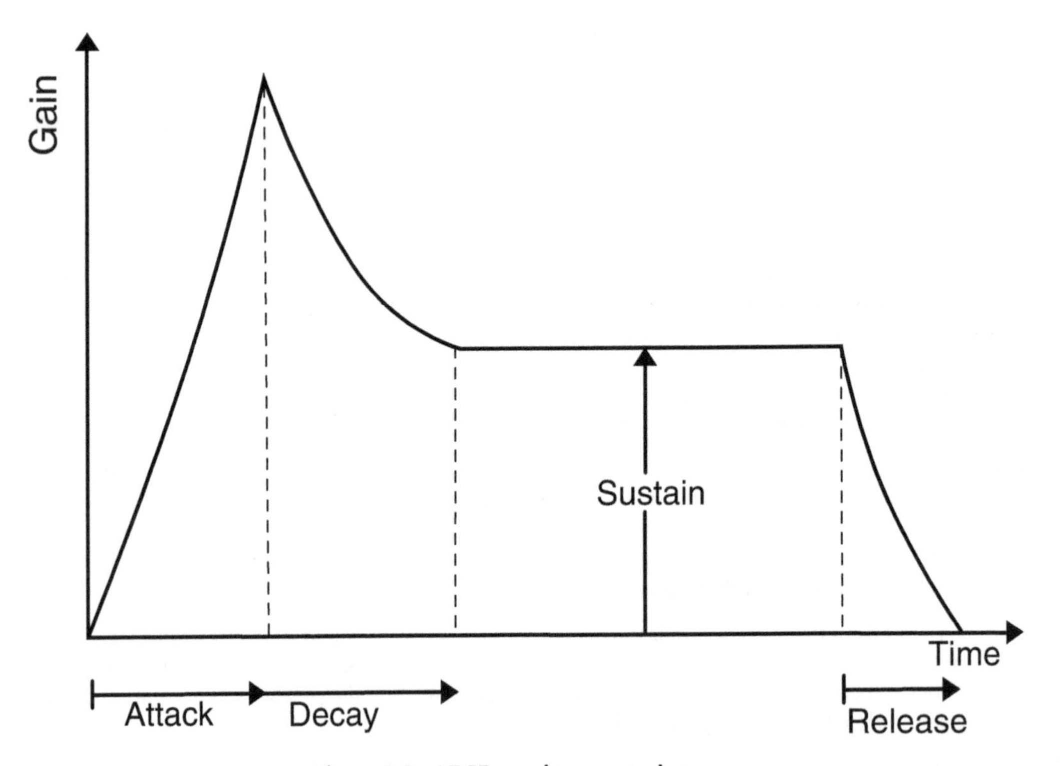

Figure 2.2: ADSR envelope control stages

Interrupt Updates

An interrupt is a ubiquitous hardware feature that briefly diverts a processor from its default program into a time-critical routine under control of an external signal. In this case, any signal that arrives at regular intervals of a few milliseconds will do the trick; there's a simple trade-off between the interrupt rate and the flexibility and processing overhead of the associated audio code.

This approach eats into your control over volume, as the same bits must be used to set the envelope as well as the overall volume of the voice, but otherwise it has several advantages over the limited pre-set envelope options:

- Different envelope shapes can be applied to each output voice
- The same mechanism can vary pitch as well as volume, allowing glissando and vibrato as well as tremolo and envelope effects
- Interrupt-driven fast pitch-modulation can compensate for a limited voice count by facilitating chord-arpeggios
- The player can interleave sound effects as well as music, reallocating voices to game triggers on the fly
- The hardware envelope system can be redeployed, at high rates, to vary the output timbre
- Deferred note and noise triggering can be built into the same player, making a stand-alone audio system

This thin layer of software over the soundchip created the role of game sound designer and explains why early practitioners like Rob Hubbard (composer for *Commando*, *International Karate* and many other celebrated game soundtracks) were necessarily programmers as well as musicians. Game music was supplied to the development team in the form of a "music player," a short tightly written machine-code routine, and a block of data, seldom more than a few kilobytes long, which would dish out updates to the sound chip each time the routine was called.

In its simplest form this approach mimics a player piano, where the position and length of slots in moving punched cards controlled which notes are played when. But software is a lot more flexible. Repeated sections and individual parts can be stored separately and combined on the fly, according to higher-level block replay controls, enabling multi-part tunes several minutes long. The tempo of the tune can be varied, independently of this data, as smoothly as the interrupt timing was controllable. Any control available on the chip, not just pitch and volume, is accessible.

This self-contained approach meant that—apart from haggling about memory and hooking up the regular interrupt signal, plus any sound effect "trigger bits" for the game to set and the player to read—the game designers and programmers could work independently of the sound people. As games grew in complexity faster than in budgets, this logical and functional specialisation facilitated remote multi-project, multi-platform working.

It's no coincidence that these players appeared in the early 1980s, as the MIDI standard was hammered out. Like a MIDI sequence, the Music Player data separates pitch and volume information from the associated sounds. Unlike MIDI, the player data is specific to each sound chip and the code is hand-written for each processor architecture.

Nonetheless this was the beginning of audio programming as a specialism and a multi-platform activity. The same player code runs on any Z80-based micro, so tunes were easily ported between Amstrad and Sinclair 128 machines. Only the code, not the data, changes for a different processor using the same family of sound chip.

Note and control data can be used on any AY-89xx-series chip and arithmetically converted (with some loss of precision) to TI models. Volume controls directly correspond. Timings, in the form of counts of the number of interrupts to wait between control changes, can be tweaked to suit a different interrupt rate, as when converting a video-synchronised title from European 50 Hertz to the 60-Hertz updates then standard in Japan and North America.

Two other early sound chips deserve special mention, as they introduced important features. They came from Atari and Commodore, seminal home computing hardware companies.

Atari POKEY

Atari Corporation's POKEY built on the experience of the VCS 2600 TIA chip. POKEY appeared in classic arcade cabinets, including *Centipede*, *Gauntlet* and *Missile Command*, and Atari's 8-bit home-computer range. Like the AY part it serves dual duty, managing digital peripherals as well as sound. The name stands for "POtentiometer Keyboard Integrated Circuit," not mentioning sound capabilities, but the core can generate four audio voices while handling input/output and timing duties for the three-piece 8-bit Atari home computer custom chipset.

Atari arcade titles like *Firefox*, *I Robot*, *Major Havoc* and *Return of the Jedi* pushed this capability up to 16 voices with a quad-core version of POKEY. The pitch control of POKEY was limited to 8-bit resolution, fewer than the TI or GI chips. It's the worst of the late 1970s chips for high-frequency accuracy unless you're willing to trade channels for resolution.

POKEY offers either four simple voices, two with filters or two with precise pitch control.[14] The hardware supports pairing consecutive (odd- and even-numbered) voices together into a single voice, giving 16-bit pitch resolution, ranging down to less than 1 Hertz and greatly improving the tuning but halving the voice count.

The first two voices have digital low-pass filters which can progressively remove bass content from their output under control of the other voice pitch controls. You can have two simple voices and one with either filter or 16-bit pitch control, but the hard-wired pairing means you can't have both those at the same time.

Rather than a single dedicated "noise" channel for hissing and similar sounds, each of the four voices can be optionally modulated by five timbres of noise, giving an exceptionally wide range of non-tone effects, especially in conjunction with the voice pitch control. POKEY uses binary delay lines, shift registers with feedback, to generate pseudorandom noise. The difference is that several sizes of shift registers can be selected, independently on each voice.

The 8-bit Atari compendium De Re Atari[15] lists more than a dozen combinations of pitch and noise settings which give a passable impersonation of many non-musical sounds, including wind, fire, steam, motors, guns, collapsing buildings and a Geiger counter! None of these are as realistic as sample replay, but they are simple to set up by just writing a couple of values per voice to the hardware and can all be varied as they play, avoiding monotony and giving the listener plenty of interactive feedback.

It's easy enough to impress a listener with computer-generated audio in the first few seconds and more rewarding but much harder to hold their attention for minutes or hours. The brain swiftly filters out continuous sounds, however beautifully recorded they might be. If you keep the cues varying over time they will remain interesting, especially if you integrate continuously changing information from the player's inputs or the simulated world—like the speed or orientation of the player—to modulate them.

Players learn to use this information to their advantage, homing in on the safe limits of what they can do, guided by their ears. Even if the mapping between the environment and the sound is simplified or the sound is more symbolic than realistic, the player's brain extracts the information it needs to interpret the scene.

This is the essence of interactive audio: user input and the game world interact, immersing and informing the player. Strict accuracy is not required. Consider the audio equivalent of cartoon-style simplification, extracting and emphasising key features most relevant to the player.

Log or Linear

Like the other chips of its era, POKEY's volume control is 4-bit, with a "volume only" mode for playing samples. But unlike the GI and TI chips, which have 4-bit logarithmic volume controls, in steps of 2 dB, POKEY uses linear weighting for the four bits. This means that the volume steps sound very large for low values and almost imperceptible at the top of the range. The signal level doubles between a volume of 1 and 2, a 6.03 dB step in voltage and quadrupling of power, whereas the same step of one unit at the top of the range from 14 to 15 equates to 0.6 dB, an increase so slight that untrained listeners will probably hear no difference at all.

A change of 1 dB is about the smallest volume fluctuation which can be detected reliably; any less and most people can't tell the difference. As 1 dB is a ratio rather than a fixed amount, this applies whether the difference is between 50 and 51 dB or 79 and 80. Since the ear needs to be able to sense small changes in very quiet sounds but similarly small changes in loud ones are insignificant, our ears and brains have evolved to have a logarithmic response to sound intensity.[16]

The graphs compare linear and logarithmic volume controls (0..15) with attenuation in decibels. The vertical axes show linear gain in the first case and logarithmic attenuation in the second. Logarithmic attenuation is generally preferable. This book uses dB attenuations throughout the Logical Layer and linear gains only in the Physical Layer while panning and mixing.

Both types of volume control have a similar range in dB between 1, the lowest audible level, and 15, the loudest. The step between 0 and 1 is infinite, since any signal is infinitely larger than none at all! But the only interval in which POKEY gives the same perceptual difference between two values as the other chips is between volume 4 and 5, which increase corresponds to about 2 dB on both scales. But the linear scale gives a 12 dB difference between volume 1 and 4, double that of the logarithmic control which adds a consistent 2 dB per step.

It follows that even though the range is the same, the curve is very different. Perceptually, the linear fade hardly changes at first, then speeds up, briefly tracking the logarithmic steps though at a much higher level, before it swoops down the plughole of reciprocal quantisation.

A key skill of specialist audio developers is a sensitive understanding and choice of curves. Convincing mixing, fading, direction and distance cues demand transfer functions which seem smooth to the ear, not a voltmeter. Broadcast studios use logarithmic "Peak Programme Meters" (PPMs) with 4 dB steps between marks on the scale, because this corresponds to the psychoacoustic sensitivity of the human ear in air, after adjustment for frequency, boosting the reading for sounds predominately in the sensitive speech band around 300..3000 Hertz. For specific curves, google "Fletcher-Munson.'"

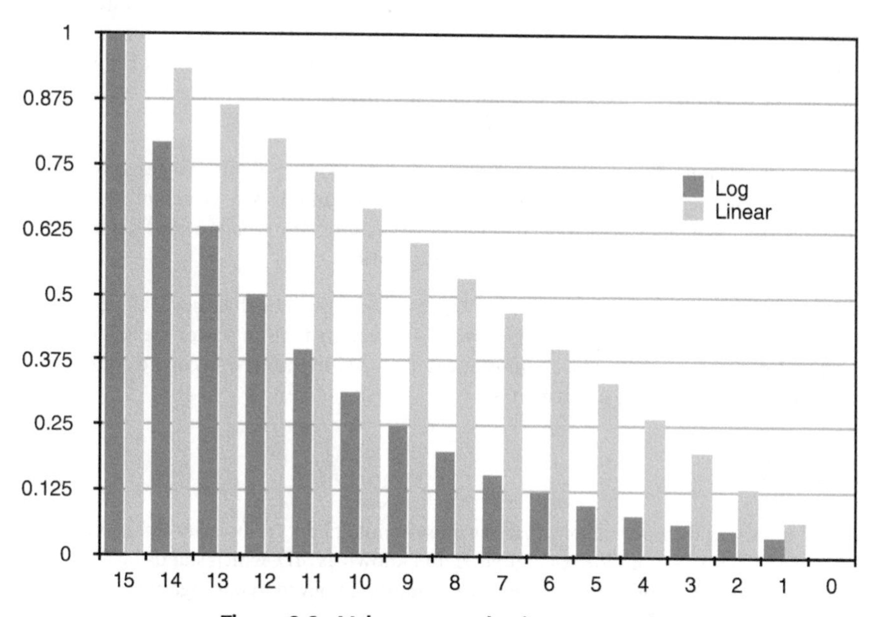

Figure 2.3: Volume control gains compared

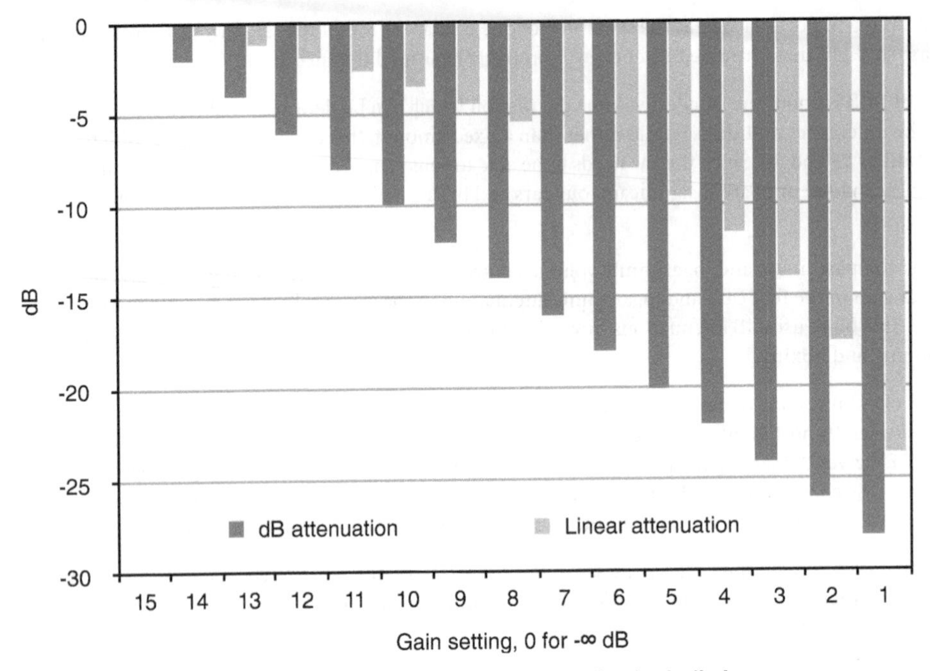

Figure 2.4: Volume control and attenuation in decibels

Overmodulation

POKEY has an interesting volume limitation which often applies to modern hardware and software mixers too. Although all four of the channels could be set to any volume in the range 0 to 15, with 15 meaning "as loud as possible," the highest volume that could be used on all channels **at once** was 8. If the total of all four channel volume settings reaches the range 33 to 60, the combined output becomes overmodulated; as De Re Atari puts it, "*The sound produced tends to actually lose volume and assume a buzzing quality.*"[15]

The problem is one of headroom. There's only so much capacity. If you allow all four voices to play at full volume, you limit the volume of any single one of them, so that monophonic audio is very noticeably quieter than multi-voice sound. This also applies when voices are paired for higher quality. Atari made a considered trade-off to balance the dynamic range between one and four voice configurations.

The more voices you have, the more this is an issue but the less likely it will be that they'll all peak at the same time. But unlikely things sometimes happen in interactive media—provoking them is one of the joys of gaming and VR. If the system breaks down under exceptional circumstances the break in immersion can be quite jarring. This issue will be examined analytically at the end of Chapter 3.

Commodore SID

Another influential 8-bit sound chip is the one that has most maintained its popularity since—Commodore's 6581 part, made for the best-selling Commodore 64 computer and known as SID, which stands for "Sound Interface Device." Like POKEY the volume controls are 4-bit linear, sounding uneven to our logarithmic hearing, but the decay and release envelope curves are properly exponential.

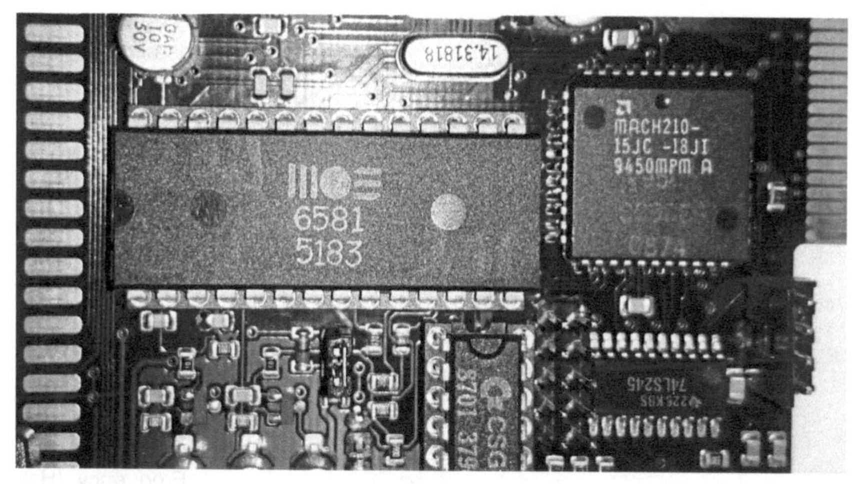

Figure 2.5: SID lives on, relocated from a Commodore 64 to a double-edged Amiga Zorro/Intel PCI card

SID was designed by Robert Yannes in 1981.[17] Two years later he left Commodore with other C64 designers to found Ensoniq, responsible for the Mirage, SQ and VFX synthesisers, among others, and ES137x high-quality, low-price PC soundcards. The ultimate Apple II model, the Apple IIGS, used the 5503 DOC Digital Oscillator Chip from Ensoniq synths and samplers.

SID is another three-voice part but much the most expressive of the 8-bit hardware tone generators. It is not limited to square waves and random noise output but can also generate triangle and sawtooth waveforms and vary the duty cycle of the pulse, much like *Manic Miner*'s 1-bit two-tone output. It combines these capabilities with a single powerful digital filter, which can smooth the waveforms, and synthesiser-style ADSR envelopes, allowing sounds to fade in and out over a few milliseconds or several seconds, with 16 linear "sustain" levels for the output to settle at between key-on and key-off.

The limited voice-count limit made it hard to play chords at the same time as rhythm, let alone sound effects too. This inspired a characteristic "chord arpeggio" sound which rapidly modulates the pitch of one or two voices to give the impression of more notes.

Any voice output can be routed through a fully parametric three-band filter. Three-band refers to the input sound being separated into low-, middle- and high-frequency bands. These can be selectively enabled, so a band-pass filter would use only the middle component. A notch filter, which suppresses midrange frequencies, can be set up by enabling the low- and high-band outputs.

Fully parametric means that the width and hence selective resonance of the middle band can be controlled, as well as the centre frequency. These "res" and "freq" controls are among the most expressive knobs on a classic analogue synthesiser; adjusting them while a sound plays changes the shape of the waveform and its timbre in exciting ways which earlier sound could not approach.

SID's filter ranges between 30 and 12,000 Hertz, covering almost all audible notes, with a similarly ear-friendly slope of 12 dB per octave, varying the signal level by a factor of four for each halving or doubling of frequency. Chapter 20 gives code to implement this type of filter and formulae for the associated control curves.

Other SID features let voices link and interact, generating even more complex harmonics, including bell- and gong-like sounds produced by ring-modulation between two voices. Sixteen-bit pitch and 12-bit pulse-width registers allow smooth control of each voice.[18]

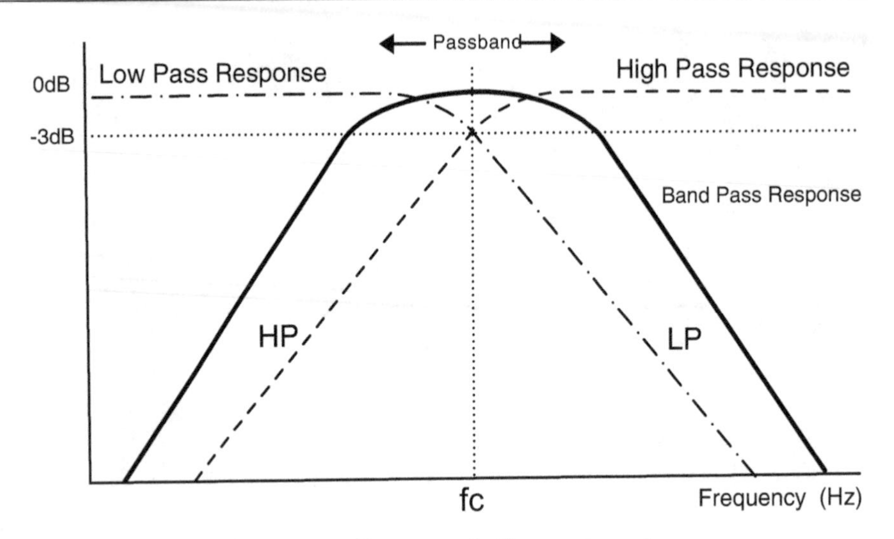

Figure 2.6: SID parametric filter configuration

Philips SAA1099

Dutch electronics group Philips BV made a late entry into the square-wave sound chip market with the SAA1099, which surfaced in MGT's SAM super-Spectrum home micro, some Creative Labs PC soundcards and Silicon Graphics Workstations. It's rather like a doubled-up, stereo version of the AY-8912, stripped of the peripheral ports.

You get six voices with 11-bit frequency control—three bits to set the octave, eight to pick a note within it—and 4-bit volume controls. Simple AY-style sawtooth envelopes could be combined with direct volume control, at the expense of one bit of resolution—effectively, splitting the dynamic range between program and envelope control. Two independent noise generators can be set to three pre-set pitches or smoothly controlled by tying into one of the six tone voices. Stereo panning is achieved by setting differential levels for the left and right outputs from each voice, and the envelope generator modulation phase can be flipped on one side to give positional shimmer effects.[19]

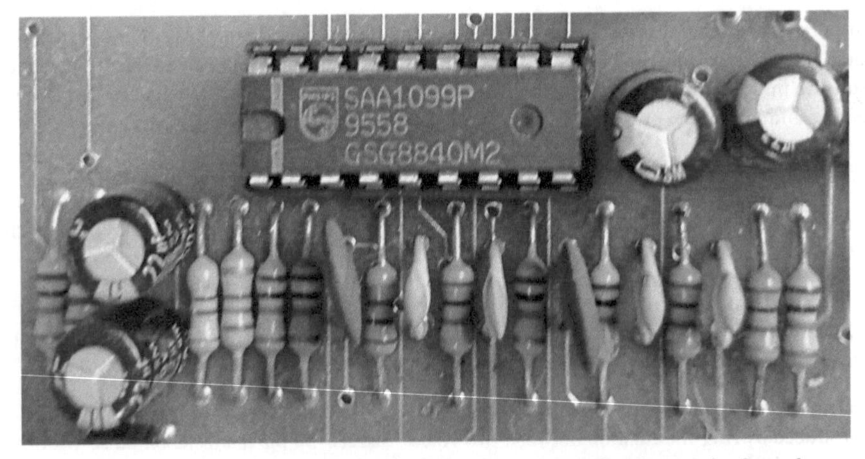

Figure 2.7: SAA1099, one of just six chips on the MGT SAM motherboard

The very audible weakness of early tone-generator chips was their fixed-shape square-wave output, giving a characteristic harsh, reedy tone, rich in odd harmonics. Strings, bells, flutes and other smoother timbres seemed out of reach—let alone speech synthesis.

Sample Output

Atari VCS programmers realised that 4-bit volume controls could be directly manipulated to fade a standing wave up and down, allowing arbitrary sample replay—but this was prodigiously expensive in memory and processor time. Just like the 1-bit sample replay in some Spectrum and TRS-80 games, everything else had to stop while the processor fed volume changes to the sound chip at sub-millisecond intervals.

About 3K of data was consumed every second, just to keep a wave of sub-telephone quality playing, on machines with a total capacity of 64K or less. It was a gimmick, good enough for snippets of speech at the start and end of a level and specialised apps like grainy drum-machine simulators but impractical to use interactively while a game was in full spate.

The volume controls of POKEY, TIA and SID supported 4-bit linear output in 15 even-sized steps. The AY, TI and SAA chips were superior in that respect because their volume controls used logarithmic 2 dB steps. Such non-linear volume controls are sensibly psychoacoustically weighted, so that each step up or down in volume sounded equivalent to the logarithmic human ear, but taking advantage of this requires that recordings made with a linear analogue-to-digital converter should be captured at much higher resolution and re-scaled as part of the conversion process.

This could be done offline, but few people realised the need at the time, further distorting and limiting the quality of 4-bit sample replay. Multiple voices, and even the AY chip's envelope generator, could be coupled together to increase the resolution, but not beyond that of a poor telephone line and only at the cost of pausing everything else while the 4-bit nybbles were spooned out to the digital-to-analogue converter (DAC).

The next step was to automate that process and in doing so eliminate pre-set waveforms. Thus the purely synthetic beeps of chip music were supplanted by recordings which could express any waveform, not just the few easily implemented with counters and timers. It was a shift as significant as that from cartoons to photography.

References

[1] *Turing's Manual:* http://curation.cs.manchester.ac.uk/computer50/www.computer50.org/kgill/mark1/progman.html

[2] *Mark 1 Music*: www.manchester.ac.uk/discover/news/first-digital-music-made-in-manchester

[3] *The Hoot Order*: www.computerconservationsociety.org/resurrection/res77.htm#d

[4] *Early Computer Music Experiments in Australia and England; Paul Doornbusch, Organised Sound, ISSN 1355-7718, Volume 22 Number 2, 2017, pages 297–307:* www.cambridge.org/core/journals/organised-sound/article/early-computer-music-experiments-in-australia-and-england/A62F586CE2A1096E4DDFE5FC6555D669

[5] *Notes by the Translator of L. F. Menebra's Sketch of the Analytical Engine Invented by Charles Babbage, Ada, Countess of Lovelace 1842; Volume III of Taylor's Scientific Memoirs, reproduced as Appendix I in Faster Than Thought, edited by B. V. Bowden, Pitman, 1953*: https://en.wikisource.org/wiki/Scientific_Memoirs/3/Sketch_of_the_Analytical_Engine_invented_by_Charles_Babbage,_Esq./Notes_by_the_Translator

[6] *PDP1 Polyphony*: www.dpbsmith.com/pdp1music

[7] *Racing the Beam; Nick Montfort, MIT Press 2009, ISBN 9780262012577, pages 131–132*

[8] *All Aboard the Impulse Train: A Retrospective Analysis of the Two-Channel Title Music Routine in Manic Miner; Kenneth B. McAlpine, Computer Games Journal, ISSN 2052-773X; Volume 4 Number 3, 2015, pages 155–168*: https://rke.abertay.ac.uk/files/8564089/McAlpine_AllAboardTheImpulseTrain_Author_2015.pdf

[9] *Space Invaders Soundchip Teardown*: www.righto.com/2017/04/reverse-engineering-76477-space.html

[10] *Space Invaders Analogue Sound Schematics*: www.robotron-2084.co.uk/manuals/invaders/taito_space_invader_l_shaped_board_schematics.pdf

[11] *Maestro Mario; Andrew Schartmann, Thought Catalog 2013, ISBN 1629218472*
[12] *SN76489 Datasheet*: ftp://ftp.whtech.com/datasheets%20and%20manuals/Datasheets%20-%20TI/SN76489.pdf
[13] *AY-8910 Chip Datasheet*: http://map.grauw.nl/resources/sound/generalinstrument_ay-3-8910.pdf
[14] *Atari POKEY Patent*: http://patft.uspto.gov/netacgi/nph-Parser?patentnumber=4314236
[15] *De Re Atari; Atari Inc. 1981, no ISBN*: www.atariarchives.org/dere/
[16] *Online dB/Voltage/Power Converter*: www.sengpielaudio.com/calculator-FactorRatioLevelDecibel.htm
[17] *Commodore 64 Design Case History; Tekwa S Perry & Pauw Wawwich, IEEE Spectrum Magazine, ISSN 0018-9235, March 1985, pages 48–58*: https://spectrum.ieee.org/ns/pdfs/commodore64_mar1985.pdf, www.commodore.ca/gallery/magazines/c64_design/c64_design.htm
[18] *Commodore 64 Programmers Reference Guide; Howard W Sams & Co Ltd, 1982. ISBN 0672220563, Chapter 4 and Appendix 0 (6581 SID Chip Specification)*
[19] *Philips SAA1099*: http://pdf.datasheetcatalog.com/datasheets/1150/493200_DS.pdf

CHAPTER 3

Sample Replay

It was time for a new approach to interactive audio output. Rather than create complex waveforms with timers, counters and random noise or sum them from sine waves by Fourier or FM synthesis, might it be possible to replay analogue waveforms directly, through a digital-to-analogue converter? The Australian Fairlight CMI (Computer Musical Instrument) showed the way, playing and mixing up to eight 8-bit samples in parallel.[1] Fairlight threw hardware at the problem, with at least 15 separate circuit boards, a custom CPU card and 16K of dedicated memory for each voice, but at £20,000 only a few hundred were sold.

Samples take a lot of memory, compared with state-machine register values, and they need to be output at very regular intervals. Even a 1% fluctuation in sample output rate reduces the effective resolution to less than seven bits, regardless of the potential precision of the DAC. Comparable audio quality did not reach gamers till July 1985, with the launch of Commodore's Amiga 1000 home computer.

Amiga Paula

Like the 8-bit Atari home computers, also masterminded by Jay Miner, the Amiga was built around three custom chips and an off-the-shelf microprocessor.[2] Audio was implemented in the 48-pin Paula chip, along with the usual rag-tag of miscellaneous peripheral controls, for game paddles, interrupt handling, serial interfacing and floppy-disc data. Amiga chips were given female names in the hope that Silicon Valley eavesdroppers would turn away, thinking that their neighbours were discussing romantic rather than technological opportunities.

Paula's audio implementation was simple to use and yet extraordinarily expressive compared with previous interactive sound chips. Four voices read 8-bit samples from specified memory addresses at specified rates and volumes. The first and last voices were routed to the right stereo output, the middle pair to the left channel.

Figure 3.1: Paula implements Amiga 500+ sample outputs, analogue joysticks, serial and disc ports

Just 44 bits control each voice, requiring only six machine instructions to set it up.[3] Two registers per voice held the address of the first byte of sample data and half the number of samples, as they were picked up in 16-bit pairs. Four extra bits in a shared "DMA control" register toggled the associated direct memory access channel for each voice.

Another 16-bit register sets the rate at which the voice advances through the sample data in steps of some multiple of about one third of a microsecond. Paula simply maintains the same sample level until the required period, typically several hundred steps, has elapsed, then moves on to the next. The abrupt transition between consecutive sample levels introduced audible noise, somewhat smoothed by an output filter, but there was no attempt to mask the sharp corners of the waveform by digital interpolation, fading between sample values, as found on more recent systems.

Each voice takes one of 64 linear volume levels, plus silent, giving a dynamic range of about 36 dB but very uneven steps, ranging from 0.1 dB at the loud end, from level 63 to 64, to 6 dB between levels 1 and 2. The maximum sample rate depends on the type of display, as two samples are picked up from memory for each channel during the "flyback time" reserved for old TVs to deflect their beam from one line of pixels to the next. This limited the maximum to 28,867 Hertz when a conventional TV was used but allowed much higher rates if a rare multi-sync monitor was connected.

Pulse-Width Modulation

Amiga volume controls use pulse-width modulation, slicing up the output from the DAC in 64-cycle blocks at 3.5 MHz and passing from 0 to 64 of those through to the filter. This explains why the range exceeds the obvious 6-bit range, 0..63. Above CD sample rate, with a multi-sync monitor, the number of effective volume settings falls as the period between output samples drops below 64 cycles.

Like most of the 8-bit chips, Paula also included facilities to link voices, so that one varied the volume or pitch of another for tremolo, vibrato or both. These were rarely used because they reduced the number of simultaneous sounds and could be faked, at a lower rate, with software updates controlled by a timer interrupt. A Music Player could use the same interrupt to trigger the selection of samples, as well as pitch and volume changes.

Hardware analogue filters smoothed out the digital noise and stepping artefacts of the 8-bit waves, limiting the bandwidth to 7 kHz and muffling the top octave. This was a good compromise for TV displays and low-rate samples but limiting for fast ones.

Later Amiga hardware revisions linked one of the filter stages to a spare digital control bit, making the filter less aggressive and letting higher tones—and more noise—through. This control bit was already linked to the brightness of the computer's power light, so that Music Players which toggle it blink that light in time with the cymbals or other high-pitched sounds as the filter is momentarily turned off to let their overtones through.

Another neat hack allowed stereo samples to be played at near-14-bit resolution, using two related samples, one played at full volume for the high-order bits, the other played at low volume on the same output, contributing the low-order detail. Circuit nonlinearity somewhat limits the precision, but the technique could be used on modern hardware if you're willing to accept the same trade-offs and can't directly output the full-precision sample.

Tracker Modules

A long sample could contain any combination of sounds mixed together but would use up over a megabyte of memory per minute, even for mono AM-radio quality. That's more than the total memory capacity available for sound, graphics and code on most Amigas. The original A1000 shipped with a barely usable 256K; years later, Commodore's custom-chip revisions pushed this up to 2 MB, but games graphics hogged the majority of that, typically leaving only tens of kilobytes for samples.

Complex instrumental tunes were squeezed into this space by repeating sections or individual samples at intervals controlled by a sequence for each voice. The whole tune or "module" was packaged into a small set of samples and timing data indicating which sample should play, at what volume and pitch, on which voice. Free programs known as "trackers" allowed non-programmers to create modules, and simple standard sample and data formats permitted users to copy sounds and sequences between modules, giving an easy entry into Amiga music making.

Tracker modules are still sometimes used in games when memory is tight. For instance, the rhythm-action game *Dance Factory* used 24-channel XM tracker music, composed by Allister Brimble decades after his Amiga debut but in similar style, to entertain the PlayStation user while CD audio tracks are being analysed to generate dance sequences. The tracker format allowed several minutes of decent-quality music to be played from co-processor memory while the optical disc drive and main processor were tied up detecting CD beats and sections.

Modules offer more control and variety than can be obtained by playing a pre-mixed track. Layers can be stripped and added, sounds can be swapped, and the entire tune can be sped up or slow down by just adjusting the trigger timings. As the name implies, one of the commonly used game audio runtimes, FMOD, started out as a module player.

MIDI protocols can achieve similar effects but deal only with timing and control, abstracting out the sample data. Nintendo game music often works this way, so it can speed up and slow down without changing key. A premixed stereo track could be streamed from disc instead and would probably sound more impressive to a passive listener, but that approach would be less interactive and would also limit the use of the drive for loading other game data.

Contrast iD Software's nine-voice *Doom* sequences, composed by Bobby Prince in the MIDI-based MUS format, with the later *Quake*'s CD-format compositions by Trent Reznor. The former is clearly sequenced, the latter more like a film soundtrack. The original nine-voice cap stemmed from limitations of the OPL2 synth chip on budget PC soundcards, which uses FM additive synthesis, mixing sine waves rather than samples.

Voice Management

The four-voice limit was another challenge for Amiga programmers and musicians, especially once a mix was broken down into lots of little samples to save memory. Bass drum, snare and hi-hat rhythms alone would in theory use up three of the four voices, leaving just one for melody or sound effects.

But bass and snare drum sounds typically alternate, so one voice could deliver both, as long as a third composite sample, mixing both the bass and snare sound, was available for moments when they coincided. Similarly, hi-hat sounds could be mixed in.

The combinations soon mount up; you might consider creating a one-bar loop with all the basic percussion sounds mixed and sequenced together. The trade-off involves comparing the length of each individual sample and necessary pair with that of the entire loop, which in turn depends upon the sample rate. When you combine several samples you need to pick a rate high enough to accommodate all the component frequencies.

Sample Rate Tuning

Bass beats can be stored and played accurately using a short recording at a low sample rate, but cymbal sounds need a much higher rate or they sound dull. The same applies to water, wind and breaking glass. Part of the skill of the sound designer is in setting asset rates as low as possible without obvious loss of quality.

As long as there's at least one sound with substantial high-frequency components playing—background weather sounds often play this role—psychoacoustic masking means that other sounds can be unobtrusively trimmed in size and rate. Such sounds also help to mask high-frequency noise introduced by literally grainy granular synthesis, weak anti-aliasing filters and fast but sloppy sample rate conversion. Such compromises are necessary when you're mixing dozens of voices on low-cost hardware preoccupied with non-audio work.

Audio systems programmers help by providing codecs that trade processing time for memory space. By such mixing of tricks *Colin McRae: DiRT* squeezed over an hour of bright-sounding samples into less than 20 MB of PS3 memory. Admittedly that's quite a lot of RAM, but the same data in CD format would've taken nearer 600 MB.

Earlier Rally games on PlayStation 2 (PS2) made do with little more than a minute of samples in about a tenth the space. The switch from the VAG ADPCM codec to ATRAC3, and some clever setup explained in Chapter 17, meant that 40 times as much sound could fit in 10 times the earlier space; yet Moore's Law[4] and concomitant memory expansion meant the total audio footprint was less, proportionately, on the PS3 than it had been on the PS2.

Mixing Samples

There's another way to mix samples which trades resolution for voices. The *SpecDrum* ZX Spectrum add-on was little more than an 8-bit DAC in a box but capable of playing three short percussion samples at a time from a 20K PCM sample buffer with remarkable quality for the time (1985) and price (under £30). It did this by mixing them on the fly in tightly written Z80 machine code and constraining the stored drum sample levels.

Seven bits were reserved for the bass drum and six for the other drums and hi-hat. The 8-bit DAC allowed levels in the range −128 to +127. The 7-bit bass drum used the range −64..+63, and the others were limited to the 6-bit range −32 to +31. This meant that one bass drum and two other samples could be mixed by simple addition and written without distortion to a single 8-bit output. Unsigned samples are harder to mix, as the offset varies with their count, but obsolete PC sound cards tried it.

The seventh bit lets the bass drum be 6 dB louder than the others; this was necessary because of the relative insensitivity of the ear to low frequencies and the importance of the bass beat as a reference for other parts of the rhythm. In theory four 6-bit samples—or even 16 4-bit ones—could be packed the same way if they could all make do with the same dynamic range.

In practice that would have overtaxed the Spectrum's 8-bit processor, capable of sustaining only about 0.5 8-bit MIPS with luck and a following wind. Mixing three voices in a tight hand-tuned loop took 50 microseconds. Any more and the 3.55-MHz processor would've been incapable of maintaining a 20-kHz output rate. Contrast this with the 25,600 MIPS throughput, in 32-bit floating-point format, of a single obsolescent PlayStation 3 (PS3) vector co-processor.

The tracker application *OctaMED* squeezed eight voices out of Amiga hardware intended for four by mixing pairs of 8-bit samples, discarding low-order bits and stuffing the result into the buffer being read by one of the four hardware voices. Timing was not critical, as the voice buffer queued mixed samples waiting to play. As long as the average mix time was less than the rate of sample consumption, the system could keep up.

The big snag, which the *SpecDrum* bypassed by concentrating on percussion, comes when the two sounds to be mixed are at different sample rates. There's no way to do this in real time on 16-bit Amiga hardware without introducing substantial distortion. The sound at the higher rate can be played correctly, but samples from the slower one will be misplaced in time, to an extent varying depending upon the ratio of the two rates.

These days, with hundreds of voices and megabytes of sample memory you might imagine that there's little need to pre-mix combinations, but like many old-school techniques this can still be useful. The alternative-reality squad-based shooter *Operation Flashpoint—Dragon Rising* combined speech samples in a similar way to get around a loading bottleneck. OFP kept tens of thousands of speech samples in five languages on disc, loading phrases on the fly using FMOD's rudimentary streaming system.

This was fine for sequential speech and conversations but tripped up when all six members of a squad replied to their commander's instructions. Loading the word "yup" in various accents from half a dozen places on an optical disc took too long, even before the game ran out of streaming buffers. My fix was to detect such combinations and play one of a handful of premixed samples instead of the separate parts. In game engineering only the worst case matters, so identifying and trimming such "spikes" is a key part of maintaining immersion.

> In game engineering, only the worst case matters.

Chapter 4 of the PS1 guide explains the original PlayStation audio hardware, the SPU (sound processing unit), not to be confused with the PlayStation 3's more general "synchronous processing elements" or SPEs. The earlier chip can mix and decode CD audio plus up to 24 ADPCM-compressed voices from 512K of dedicated sample memory, with 16-bit stereo output at the CD standard rate of 44.1 kHz.[5,6]

SPU has 14-bit volume controls and little-used keyboard-style envelope controllers for each voice. A facility to flip the phase on any voice facilitates analogue surround encoding, supporting schemes from Hafler to Pro Logic. A single basic delay/reverb stereo effects unit with ten pre-set room sizes can be selectively applied to any voice, at some cost in audio RAM for the delay buffer.

The PS2 includes a superset of the original CXD2922 SPU. It's effectively two copies of the earlier chip daisy chained, with four times the dedicated RAM. You get 48 ADPCM voices, two simple reverb-delay units, a couple of PCM inputs and everything mixed down in hardware to the slightly higher DVD standard rate of 48 kHz.

Blocks of 256 samples are mixed and output by the hardware 187.5 times a second (256 * 187.5 = 48,000, the fixed output sample rate), so voice triggers and parameter updates are applied every 5.33 milliseconds. This is a key factor determining the responsiveness of the system, along with the game update rate, usually rather slower, and further delays introduced by the output system, as discussed in Chapter 23.

Figure 3.2: Sony's SPU chip, on a Mac NuBus card made for PS1 sound designers

Headroom

The formula used for scaling up to 50 voices of decoded audio into a 16-bit stereo stream has not been disclosed, but it certainly involves limiting the output range of any one voice to fewer than 16 bits and aggressively clamping the output if it would otherwise overdrive the DAC. This means stereo music, or mono speech played on a single voice, can easily be drowned out of the mix as other voices overpower it.

There are two simple ways to address this. You could reduce the volume of all the other sounds or play the vital speech on two or more voices at once, a trick I learned from Lyndon Sharp at Attention To Detail Ltd (ATD). The first technique is normal practice for live-mix engineers: to preserve headroom, if you can't hear a voice it's wisest to work out what else to turn down. If the rest of the mix is already set up and your control systems lack group volume controls (oops!) that can be a fiddly and error-prone process.

The second approach needs some care, too. The headroom in the hardware mixer has evidently been chosen on the assumption that the waves playing on each voice will be decorrelated—they won't all peak at the same time, and when mixed there will typically be as many on the positive side of their respective waves as on the negative, keeping the average in a 16-bit range.

This assumption is not valid if you deliberately play the same sample several times simultaneously; each doubling up boosts the combination by 6 dB, so there's only so far you can take this before you hit the headroom limit. But it did the trick when the speech arrived late in development of ATD's *Salt Lake 2002* Winter Olympics game and is worth bearing in mind for similar circumstances. It's also an illustration of hidden assumptions any digital mixer must make and how to explore and exploit them.

You could also process the speech samples to make them perceptually louder, but that affects the timbre and intelligibility too. For commentary or remote colleague radio speech, that should have been done already, so it's rarely an option.

In theory you need to allow one bit of headroom for each doubling of the number of voices you mix. This means that output can never go out of range even if all the inputs use the full range but limits each sample to a 10-bit range when mixing 64 of them for a 16-bit output. Richard Furse calculates that **if the samples are uncorrelated** only the square root of the number of headroom bits is required: 2 bits for 16 voices, 4 bits for 256, and so on.

It's unlikely that all voices will play at full volume, and this seems to be how each SPU gets 24 voices of ADPCM decoded to 14-bit resolution into a 16-bit output. It's still important to apply a clamp, as detailed in Chapter 23, to minimise distortion in rare circumstances, when momentary correlations, filtering or sample rate conversion blips might otherwise burst the bounds.

Multichannel Phasing

The two-voice approach has one additional complication. If you just issue two commands to play samples one after another in the game update, they will generally be triggered as part of the same hardware update block so the samples are exactly superimposed. But if the unsynchronised output of buffers to the hardware happens to need service just after one play call and before the second one is fully processed, the samples are replayed staggered in time. For a typical 256-sample output buffer and 48-kHz sample rate, the second sample plays 5.3 ms later than the first. This causes a hollow, echoey sound, with less of the expected volume boost, and it's jarringly obvious from the timbral change that something has gone wrong.

A similar issue intermittently affects many games with stereo music and can still affect any system that requires or is configured to render stereo or multi-channel audio by playing several mono samples to specific speakers. The VAG ADPCM format used in the PS1 and PS2 toolchain and hardware only supported mono samples. Separate play calls to trigger the left and right channels, appropriately panned, usually sounded fine but periodically ended up out of sync, with the second play call a buffer behind. When played through stereo speakers this delay causes cancelation of low-midrange frequencies, audibly thinning the sound.

Sony later addressed this with codecs and voices which can encode and decode up to eight channels of interleaved audio in parallel. But it's still up to you to pick those rather than mono legacy formats and make sure that the implementation you have chosen is really processing and outputting all the channels you need to synchronise in parallel. ATRAC and extended VAG ADPCM offerings support up to eight channels; most Ogg Vorbis implementations support more than stereo, as do modern Opus codecs, but there are still implementations which skip or skimp on this.

Last-century codecs like XADPCM and MP3 are limited to mono or stereo. So similar issues apply if you try to reuse the stereo MP3 format for front and rear surround speaker pairs—mostly they'll sound OK, but unpredictably and rarely they'll be staggered in time, obviously affecting the sound. In this case you need to use a codec and mixer that can encode and decode multiple channels in parallel, with guaranteed synchronisation or a function like **alSourcePlayv()** which forces two or more voices to start in the same frame.[7] We compare codecs in more detail in Chapter 7 and Chapter 17.

References

[1] *Fairlight CMI Details: Norm Leete, Sound on Sound Magazine, ISSN 0951-6816, April 1999*: http://petervogelinstruments.com.au/public/Fairlight%20Computer.html
[2] *AMIGA Hardware Reference Manual, Commodore-Amiga Inc., Addison-Wesley 1991, ISBN 0201567768, Chapter 5*
[3] *Amiga Qdos Mouse Organ, Simon N Goodwin, Amiga Format Magazine, ISSN 0957-4867, July 1999*
[4] *Cramming More Components onto Integrated Circuits; Gordon E. Moore, Electronics Magazine, ISSN 0013-5070, April 1965, pages 114–117*
[5] *Original PlayStation Hardware Details*: www.elisanet.fi/6581/PSX/doc/Playstation_Hardware.pdf
[6] *Reverse-Engineered Sony SPU Details*: http://psx.rules.org/spu.txt
[7] *alSourcePlayv, Section 4.6.3*: www.openal.org/documentation/openal-1.1-specification.pdf

CHAPTER 4

Interactive Audio Development Roles

Three distinct job roles need to be filled—by staff, contractors or brought-in resources—to perform interactive audio development:

- Sound designer
- Application audio programmer
- Audio systems programmer

This chapter addresses interactions between those roles and the systematic understanding that team members need to share for best results.

Sound designers author assets, and application programmers trigger them in accordance with the player's input and events in the game. Audio systems programmers provide mechanisms to load, select, play and monitor whatever the others throw their way. They tailor implementations to the capabilities of each platform—memory, processing power, dedicated audio hardware—so the others don't have to worry unduly about the underlying hardware and system details. Well-written titles leverage those differences—such as extra outputs, better codecs, hardware assistance—and mitigate the weaknesses of platforms which fall below average capability. Those still need to be supported so the product can reach a wide market and thus deliver the essential return on investment.

Sound Designers

The roles are mutually dependent. Designers pass assets—organised sets or "banks" of sound samples—to programmers, who find ways to fit them quickly into the available RAM at appropriate quality. If there are several designers—I've worked on games with half a dozen—one will typically be "audio lead," but the others will have substantial autonomy on specific aspects or game modes. Sound designers typically come from an arts or music background, such as Surrey University's Tonmeister course or music technology education, which combines tech and art skills.

In a large team, morale, workflow and productivity benefit if well-defined chunks of the work are assigned individual owners. If a product has several sound designers some might specialise in interface or "front-end" sounds—those associated with the menus and screens before and after the core simulation. These are typically more iconic and less spatialised than the in-game sounds, usually mono or stereo and static in front of the listener. Other designers specialise in speech, weapons, local ambiences or vehicle sounds.

The Main Game

The front end of a modern title may be substantial, with event, character and equipment selection and levelling-up screens, tutorial, practice, competitive and collaborative modes, local and networked multi-player lobbies. But the core of a game—racing, fighting, exploring a 3D environment—usually consumes the most sound-design effort.

An effective subdivision of labour would assign one designer to the mobile sounds—vehicles, players, items they carry—and another to the static ambiences associated with specific locations in the simulated world. These might include crowds, water and fixed installations, sounds associated with surfaces like gravel, wood and mud, and patterns of reflections and reverberation associated with buildings and the terrain.

Games are typically designed in sections, known as levels, and the mobiles may appear in many levels but the statics are more likely to be level-specific. Asset management, memory allocation and loading times are easier to balance if budgets are set for each level and group. Even open-world games where the player explores a single vast map, such as *Grand Theft Auto*, swap assets associated with specific places and vehicles in and out of memory as the player moves between them.

To be sure a large game will be playable without having to complete all the design and implementation, publishers call for a "vertical slice" of the gameplay—a single map and limited subset of the mobiles intended for the final product. This is often done to tight deadlines and may determine whether the product is completed or canned.

Dividing the audio work into static and mobile components boosts productivity by allowing each designer to pick priorities and trade-offs within a coherent bank of sounds which they own. This minimises communication about details, which is costly and contentious, and frees the other team members to design around fewer more general constraints like the budget and schedule for the environment, per vehicle and interactions between audio and other game teams.

Game Audio Programmers

Sometimes the game (or interactive application) audio programmer is a jobbing coder who has chosen—or been ordered—to specialise in audio implementation. If there's more than one, juniors typically end up with responsibility for speech, music and front-end audio, leaving 3D and more dynamic systems to their seniors. All should be skilled at searching other parts of the game, especially menu and physics systems, to pick up trigger events and intensity controls that keep the sound mix in keeping with everything else. They work closely with the sound designers and other game programmers and intermittently with audio systems programmers.

> "Finding out why sounds aren't sounding is a large part of every game audio programmer's life."
>
> —Jon Mitchell[1]

Veteran audio programmer Jon Mitchell observes, "*finding out why sounds aren't sounding is a large part of every game audio programmer's life,*" listing 15 possible reasons.[1] Jon's insights inspired the **m_Display** toggles in Voice, Group and Listener classes described later.

It's wise to hang audio triggers from conditions that also affect the display, because it's more likely that others will notice the mistake if they get lost. A classic way to detect the need for bridge or tunnel reverb is to ask the game if it's suppressing rainfall. There will be a flag somewhere to inhibit rain visualisation while it's raining elsewhere but the car's under cover. This doesn't stop the rain or start reverb above the bridge, which a proximity test might.

Damage systems also provide useful triggers. When windows break or vehicles shed panels, wind noise and engine sounds increase for certain listening positions, and new sounds may be heard. In driving simulations it's possible to damage the exhaust so much that the back box falls off. This affects the sound of the engine as well as its efficiency. Delving into game physics we found an "exhaust efficiency" value which fell progressively from 1.0 as the exhaust took knocks; 0.81 was the lower limit, at which point you could drive over your own detached silencer. So the game used that value to adjust the exhaust volume and position and interpose ominous scraping noises.

Damage loops like these soon become monotonous, so I deliberately mute them when the engine load falls and retrigger them, from a random point, under acceleration. Loop pitches may vary with revs or road speed, depending on the location of the transmission fault. Wheel-bearing damage can trigger informative and wince-inducing audio at extremes of steering. Such intermittent feedback is more obvious and meaningful.

Sometimes triggers originate **from** the audio system. Backfiring detonations, when acceleration is cut, causing unburned fuel to ignite, are characteristic motor-sport sounds, associated with a visible flash from the exhaust. The mobile *Colin McRae Rally* deliberately suppresses this effect while the co-driver speaks to maintain intelligibility. Hence the audio system detects the condition, from speech, revs and accelerator inputs, picks a sound and then tells the graphics when to flash.

Automated volume adjustments make mixes more dynamic and emphasise transitions. It's commonplace to briefly fade down engine, crowd and surface sounds during speech. In *Burnout 3* Ben Minto's ducking system reduced the engine volume during cornering to release headroom for tyre noises. Sound effects are most salient when they change, which is why *Operation Flashpoint* designer Olly Johnson arranged for footstep sounds to be loudest when a new surface was encountered, progressively falling in level with familiarity.

Interactive music systems, beloved of classically trained composers, respond to changes in the game, which may be situational as well as location related. The proximity of an enemy may add a layer; defeating that selects another. The leaderboard lets us infer when a player has succeeded or fallen behind, triggering mix changes, as does being in first place or on the final lap or spinning out. Such adaptations sustain interest and involvement after the initial thrill passes.

Audio Systems Programmers

Often cross-platform game development will start on one "lead platform" chosen because it is the most stable or profitable. The audio systems programmer is beavering away adding new platforms while the others get on with adding and triggering content. In a small company this work might be outsourced to one of the game engine or audio middleware vendors, but it's still essential if your brand is to take full advantage of the platforms customers prefer—some of which may still be secret.

The sooner you know about alternative platforms, the better you can prepare for their differences. It is dangerous for an audio lead or systems programmer to accept assurances that platforms will not be added later, even if they're ruled out from the start, or that a title will never have a sequel and can therefore be hastily hard-wired for given content. Even if you reckon, for fine technical reasons, that a given iPhone is obsolete or a platform is not worth considering, once you have a hit you'll be under pressure to port it to everything. That's the audio systems programmer's job.

In small projects, such as the mobile version of *Colin McRae Rally*, one person may play all three roles. Game engines like Unreal Engine and Unity allow small teams to make competitive games and VR for multiple platforms. But the market is flooded with me-too products and cash-in clones. To stand out, new titles must do more. Add-ons and plug-in architectures allow the basic generic features of the off-the-shelf engine to be customised for specific products by anything from dropping an extension into a working title to creating new ones from scratch or pragmatically tweaking or extending components.

In Triple-A games, intended to sell millions worldwide at prices around $50 apiece, movie-scale budgets are normal. A single title may have specialist designers working on the recording of thousands of source assets, others editing and collating those for use and still others working to integrate them with game and system programmers and tools. This book deals most with what happens after that integration—how the assets are selected and used once the product is in the hands and ears of customers, beyond the reach of those who created it.

Speech and Localisation

A title with substantial pre-programmed speech will often have one designer working just on speech, especially the scripting, recording and "chopping" of thousands of lines of dialogue so that they can be played on demand and stitched together in a way that sounds natural. A convincing sports commentary may draw on over 100,000 phrases and names, pre-recorded in several languages and moods. This merits substantial programming support.

Adapting a title for multiple languages is known as localisation. It's often the latest and most expensive part of Triple-A product development, so it needs careful handling in terms of audio design, tech and organisation.

The *Club Football* series of sports titles shipped on DVDs, each with several GB of files. The rip-off versions touted on Glasgow's Barras market were trimmed to fit 700 MB CDs just by removing all the speech and a couple of promotional videos. Speech may not use many voices at runtime, but it has a big impact on disc and memory footprint.

The accepted minimum localisation for Occidental releases is five: EFIGS—English, French, Italian, German, Spanish—to cover Western Europe. Scandinavians are presumed to understand English. Spanish, English and French (hello, Quebecois) suffice for North America. Global products add Brazilian Portuguese, Korean, Japanese and Chinese options.

Russian and Polish are the most common localisations for Eastern Europe, and on PC titles those may be hacked in place of Western languages, often after the game is otherwise finished. It's commercially unwise to leave languages like English, French and German on those products, as they typically sell at a much lower price in poorer, piracy-prone territories. You're liable to be flooded with grey imports into wealthier territories—and perhaps short of disc space—if you leave the original translations alongside custom ones for niche markets.

Such translation work may be done by contractors, often hired by the overseas distributor, without access to the original source code, so it's vital that the system should be data driven. Typically, the game code will use symbolic names—enumerations, pre-processor symbols or similar constants—to refer to individual or small groups of samples.

It's unwise to use the literal text of the speech in the game source; that makes sharing, updating and translation tricky. Instead, symbolic keys are used to index into a data file, derived from a spreadsheet or custom speech-recording software, which relates them to lines of the script in each supported language and may also indicate how they're triggered, by position in world space or (again, more maintainably) a reference to a named zone, which can be updated when the art department move the start line to reduce the draw distance and maintain framerate. It's not unknown for triggers to need to be updated several times late in title development, so design your system to expect and ease such revisions.

This symbolic approach makes it straightforward to add languages in house or with the collaboration of specialist localisation firms like Binari Sonori SRL, who translate and record additional languages, using local knowledge to find appropriate foreign voice actors. This is often done right at the end of a project—after the scenario and script have settled down—and under intense time pressure. Specialist multi-lingual testers check the speech and the corresponding on-screen dialogue.

You should not assume that there will be audio and text translations for every language, selected together. To save time, data space and cost, some minority languages may have text-only localisation, using speech samples in another language.

Separate selection is also useful during testing, checking the consistency of lines by running the text in one language and the speech samples in another. Even if the player options or host environment only allow one language at a time, developers need *separate controls* for text and audio localisations. These can be implemented on an audio debug menu—stripped from the final release—so the public option configures the "graphics language," and audio folk can over-ride that when testing their own work.

Speech lines will vary substantially in duration, and hence sample size, when they're translated. English and Italian dialogue invariably needs less space than German, regardless of the speaker or translator. German text translations fill more screen space and are harder to hyphenate, to the point where long sentences might need to be split for display. This can have knock-on effects for audio in all languages.

Speech Splicing

Localisation also involves programming work, especially if lines are assembled from names and short phrases. The natural word order varies between languages—contrast "my aunt's pen" with "la plume de ma tante." The three English words end up in order (3) (1) (2) with an extra "la" on the front and the possessive suffix "s" changing into a "de" and moving to follow the subject.

Bits of this can be automated, such as the repositioning of adjectives, and the grammar of the speaker can be simplified—formal radio chat conventions help to excuse this, especially in a military context—but you risk compromised grammar, and a lot of special-case code to handle most languages, if you take this transliterating approach.

The longer the pre-recorded lines in your script the less of this you'll have to do, and the more natural-sounding the results. But since memory space and recording budgets are limited, and hundreds of proper nouns—names of people, places, teams, equipment, etc.—may need to occur in many contexts, some splicing is unavoidable. The usual way to handle this is to allow for a further level of indirection in the localised speech script file—an "adjectival phrase" marker might cite an adjective and a noun—"red" and "house," say—and invoke code to deliver "red house" or "maison rouge" as appropriate.

The splicing mechanism depends enormously on the platform and the genre. Weather forecasts, mission orders, air-traffic control, pit radio messages and rally co-driver instructions use standard phrases and ordering like *"four left over bridge, jump maybe,"* delivered in a staccato style for clarity; phrasing is less important than for scene-setting voice-overs. Even then, team- and squad-selection options may require slotting-in of custom words. This is the price of interactivity. The better you do it the more immersive the title will be, and the longer the player is likely to enjoy your efforts.

The original PlayStation had just 512K of dedicated audio memory for everything—ambience, physics sounds, streams, confirmations, engines and speech—so the co-driver calls for *Colin McRae Rally 1* and *2* were spliced together from a few dozen one- or two-word recordings, with baked-in gaps and a bit of code to even up the timing and sequence.

The extra audio RAM on Xbox and PS2 allowed longer, more varied and natural-sounding phrases on *Rally 3* and *4*, but the stages were also longer, requiring more RAM for all the lines, forcing the use of short lines and low sample rates. *Rally 5* stages included up to 90 calls to describe a track a few miles long, so I used a different technique to save memory, eliminate splicing and yet improve quality, exploiting the sequential design of rallies and circuit races and on-rails shooters for that matter.

My PS2 speech system used a sliding window to stream speech from DVD through two small consecutive half buffers of audio memory. All the samples for a given rally stage were assembled into a single audio file, with a small separate index in main memory, loaded at the start of the stage with the graphics and geometry, keeping track of the boundaries between phrases. As the player progresses the contents of the buffers are played, stopping automatically at the end of each phrase, typically in the middle of one of the half buffers, refreshing the half not currently playing with the next part of the speech file as soon as it had been played.

Say Again? Augmented Reality

This approach suits fast-action games but not AR (augmented reality), where competing demands for attention mean there must be a way to say "pardon?" and recall earlier utterances. You need two buffers, or one big enough for the two longest consecutive samples, loading to each end alternately.

The sliding window doesn't suit games where the player can go backwards round the track or take shortcuts, but it's ideal for point-to-point events like a rally. The appropriate tech depends as much on dynamic behaviour as it does on the assets and hardware.

Competing demands for attention mean there must be a way to say "pardon?"

Since the same drive was being used to load eight interleaved weather, crowd and ambience streams, pre-selected to suit the rally event, the speech and ambience buffer sizes were chosen to allow a half buffer of the other data to be reliably loaded while the current half buffers continued to play, allowing enough time to service all the streams with fresh half buffers before the currently playing ones ran out. For *Rally 5* this was around 700 milliseconds, determined mainly by the worst-case long-seek time for the standard Sony console DVD drive and the need to read co-driver calls from disc at any time without disturbing the ambience.

The co-driver speech system worked much like that of an old radio cartridge machine. Each trigger message caused the corresponding speech sample to play, regardless of length, till its end, leaving at least half a buffer of the next sample poised and ready to play immediately. Since a late message is useless—indeed, annoying—this worked better than loading the sample from disc and allowed any length of sample, including long introductory or final sentences before or after a stage, without requiring precious audio RAM to be pre-allocated for the worst-case sample. It also allowed more natural phrasing and less complicated triggering code, as there was no need for splicing.

The loading, replay and control for all the streams was handled by three eponymous threads on the IOP co-processor, leaving the game code to carry regardless. However, the Xbox version, with twice as much main RAM, could load the speech in one go and concentrate its buffer swapping on servicing the other streams, copied to the console's hard disc. This platform customisation is the domain of the audio systems programmer. The designer is busy enough getting the assets together and in the right grouping and order.

The game audio programmer just wants to say "play" at the right moment—determined by the car's location relative to an invisible trigger sphere or box in the game world, placed by the designer, or the nearest "cat's eye" marker on the track ribbon structure. Systems programmers decide when and where the samples are loaded, from the original optical media or copied to a hard drive if available. Original Xboxes had a hard drive, but Xbox 360's often did not. The reverse was true for the PS2 and PS3.

Platform Differences

Xbox One and PS4 both have hard drives as well as Blu-ray optical drives. PlayStation Portable (PSP) had an optical UMD (Universal Media Drive), with about twice the capacity of a CD but a propensity to halve the battery life of the handheld device unless the drive was spun down during gameplay. Later PSPs ditched this in favour of extra flash memory. We worked around this by drastically trimming the ambience samples in length and sample rate so they could all fit in memory and packing the speech for each rally stage into a memory-resident bank of samples preloaded at the start. This meant we didn't need to keep the UMD spinning.

System programmers make the best they can of the typical configuration of each platform. The game team should not need to worry about the platform differences, and the marketing team should be confident that the title makes full use of whatever capabilities it can rely upon in each case. Otherwise rival titles made for just one platform will out-perform any that only use resources they can be sure of finding everywhere, and you end up with "Triple-B" also-ran products that are unlikely to top any charts. Second best is no place to be in a fashion-led market. The number one game often out-sells the rest of the top five put together, and those five probably outsell the whole of the top 100, few of which make any money at all.

Platforms vary, so the approach to them must vary too. The next console generation brought better codecs—ATRAC3 and XMA—and enough memory to load an entire set of calls, typically around 1 MB, and a couple of minutes of clear speech—so this scheme was no longer needed. But within a few years I was porting the original games to new mobile devices, with more RAM but weaker codec support. So 2013's mobile remake of *Colin McRae*

Rally for Android, BlackBerry and iOS went back to the 20th-century splicing approach, but with all the samples preloaded and minimal compression.

It's the job of the whole team to make sure that the timing, phrasing, levels and timbre of stitched samples are well matched, so you don't end up with **the** *audio* ***equivalent*** **of** a *ransom* note, with ragged or robotic timing. The recording environment and dynamic compression must match for all the samples in a localisation set, even if some are recorded later in "pick-up" sessions to account for script changes or mistakes. These concerns matter less between languages, as you won't (please!) mix recordings made for one language with those for another, even if your vocal talent is multilingual—but it's still vital to make sure that the perceived loudness of the samples in each localisation set matches.

Triggers: What to Say, When

Once you've got thousands of lines of speech in the can, it's time to work out when to play them—and, perhaps even more importantly, when not to. This is mainly the responsibility of the game audio programmer, guided by the sound design and taking advantage of whatever systems the game and audio runtimes provide to discover and record context.

Sampled speech may be used to confirm selections in the game menus or to tailor or localise cut-scene videos. These samples can be triggered on demand easily enough, though you may need to implement a queue or skip some to prevent more than one playing at a time. Later, in the main part of the game, the logic to select samples gets a lot more complicated and game specific. Remote messages, like the pit radio from engineers to drivers in a racing game, may be triggered by changes to the state of the player's equipment, such as tyres or fuel, position on the track or events elsewhere, often out of sight.

The idea is to give the impression that the player is being watched, dishing out sympathy or encouragement at appropriate times. Timing such remarks is tricky, especially for long phrases. You may wish to congratulate a player on an amazing overtaking manoeuvre, detected from a sudden place change. But it's best to wait a few seconds in case the player just forgot to brake at a corner and your premature congratulations are voiced as they bounce off the tyre wall and out of contention.

Sometimes there may be several samples to choose from, triggered by various circumstances. Three priorities—Optional, Important, or Vital—help you sift those. An Important comment might drop in priority to Optional after a few seconds or if another is already playing or a Vital one takes precedence. An Optional remark might be suppressed by either or unless there's been a reasonable delay since the last utterance. Static priorities can be associated with each speech event—which may trigger one of several alternate and equivalent speech lines for variety—but you may find need to adjust these dynamically, too, in context-specific ways.

A realistic interactive speech system needs some kind of memory to prevent needless repetition within a session. This stuff needs a lot of testing, so it must be implemented and refined early on but might need tweaking later as game physics and track layouts are revised for other reasons.

For example, *RaceDriver GRID* keeps a record of when speech samples were played and two values per sample intended to keep things interesting. One sets the minimum time before a sample can be played again, defaulting to one hour. Another prevents samples playing unless a minimum percentage of the game had been completed so that some were held back for use later in the player's career progression. If either of those values prevented replay of a sample when otherwise it'd be triggered, the game picks an alternative one, if available, or simply says nothing.

> Do not annoy.

This reflects *Halo* sound designer Marty O'Donnell's mantra: "*Do not annoy.*" It's better for the game to stay schtum than for it to wrongly guess what's going on, or tediously repeat the same phrase. "*No runs there*" is a reasonable remark for a cricket commentator to make occasionally. Six times in a maiden over would be five too many.

Lip Sync

If a title includes characters speaking on screen, the audio runtimes should assist the renderer by providing information for "lip sync," so the characters' mouths move according to the sound. Clearly when the voice stops the mouth should stop moving, and the simplest model just uses an average level from the speech channel to control the animation. This does not work well, because we make nasal sounds like "mmm" with our mouths closed.[2]

For more realistic results, provide separate mid-, high- and low-frequency signal levels, using those to control mouth shape as well as jaw position. Low frequencies correspond to plosives like "b" and "p," where the lips are momentarily closed, pouted then opened, whereas high-frequency sibilant sounds "sss" and "ch" are made through open lips but clenched teeth. Timing and range controls for individual speakers help, but this still leaves a lot of work for the artist, which is why game designers sometimes introduce helmets or microphone gestures which obscure the speaker's mouth.

Lip sync data is best generated offline, using an automated process for a first pass, with the option to tweak it manually where necessary, prioritising key characters and localisations. Only around ten measurements per second are needed, so the filter dataset is small compared with the audio; use the time since start of sample replay for synchronisation, in case the display frame rate wobbles.

References

[1] *Game Audio Programming; Ed. Guy Somberg, CRC Press 2017, ISBN 9781498746731, Jon Mitchell, Chapter 14, Debugging*

[2] *Automated Lip-Sync: Background and Techniques; John Lewis, The Journal of Visualisation and Computer Animation, ISSN 1099-1778, Volume 2 Number 4, 1991, pages 118–122*

CHAPTER 5

Audio Resource Management

Interactive audio draws on pre-planned resources. Let's consider what resources we need to reserve, the dimensions (RAM, CPU and bandwidth) and the phases in which they're loaded, used and discarded. This analysis informs the configuration of the runtimes and which features are most needed on each platform.

Games typically allocate only 10% of their memory and processing resources to audio, despite *Star Wars* director George Lucas's declaration that "sound is half the picture."[1] This average varies by genre and platform. Console games made in the last decade, since Sony's PS3 introduced the high-definition multimedia interface (HDMI) to gaming, reasonably expect to be able to position and play a hundred or more sounds, but phone games may get by with barely a dozen.

The first PlayStation hardware could mix and play 24 voices in stereo. PS2 doubled that. Consoles keep improving, but meanwhile new and cheaper platforms come up—like mobile phones, toys, tablets and set-top-boxes—where the constraints of older gaming systems remain relevant. The audio-led *Dance Factory* game reserved more than 90% of the PS2's memory and processing power for audio analysis while still leaving enough resources to render a 3D game with 24-channel music while it converted CD tracks into custom dances. Some of the techniques used would be familiar to Amiga programmers decades earlier.

The most audio-hungry console racers, like *DiRT2*, may allocate almost 40% of total CPU across three hardware threads to audio processing in some configurations. To come close to the capability of a single SPE co-processor, used to implement audio systems on the rival PS3, it was necessary to dedicate one of the Xbox 360's three hardware cores to audio in that intense single-player game.

Each core has two hardware threads, but these share resources like memory bandwidth and caches, so it's counterproductive to use only one hardware thread for optimised audio code—other code on the associated pipeline is liable to block time-critical audio operations or wipe out the shared local cache. The potentially high cost and unpredictability of those events led Microsoft to advise that both hardware threads of a single core be allocated to XAudio processing.

With game updates running at 30 Hertz upwards and audio outputs requiring service 192 times a second—more than six times as often, a typical ratio across console generations and manufacturers—audio timing and worst-case load were more demanding than graphics updates. With just 5.33 milliseconds to mix and generate each block of 512 audio samples, versus 33.33 milliseconds for the update and rendering, a hiccup of a couple of milliseconds is far more likely to interrupt the audio than to disrupt the display, especially because if the graphics miss a frame they've still got the old buffer to display, whereas audio will cut out completely, much more obviously, if it's momentarily overloaded. VR games demand higher rates just to maintain immersion.

The main update and render loops run on the other two cores, with granular synthesis, audio-related physics (such as ray tracing to find reflectors) and the communication between game update and XAudio bringing the total audio-specific CPU allocation to 40% on Xbox 360, twice as much as on PS3 for similar results and two to four times as much as on PC. The main reason for this was the use of 128 filtered voices and four stereo reverbs on Xbox and the poor performance of the original XAudio system on the long-pipeline cache-constrained PowerPC-based Xboxes. 10 Mb of extra RAM was needed for reverb and XMA data.

Bottlenecks

Microsoft implemented Xbox 360 reverberation by porting the Eventide studio reverb, which was not an ideal choice, being limited to stereo (requiring a spatially compromised bodge for rear surround channels) and CPU-intensive with tightly coupled early and late reflection models. The original implementation consumed about 25% of one hardware thread on an otherwise-idle core. Months of exhaustive rewriting and hand-scheduling of machine-code instructions only brought this down to 14%. And the memory footprint of each unit was close to 1 MB, to allow for the longest permitted delay in each stage, with no way to trim it.

Much of the problem in using this potentially high-quality but costly implementation was that it was monolithic, a black box with no facility to bypass, separately control or slow down unwanted sections. *DiRT2* used reverb for left- and right-side reflections, as critical as a cat's whiskers to guide the player fast through tight gaps, the vehicle interior and the external environment. The memory-churning late reflection modelling was only really needed for the last case—the distant immersive reverb—but there was no way to turn that off when close reflections were a priority.

By the time we came to *DiRT3* and *DiRT Showdown* we'd spent a year working closely with Microsoft on XAudio2 and had persuaded them to split the early and late parts of the reverb, allow lower sample rates, and let us mix mono and stereo busses. Running sections at half speed means some loss of quality, but the full frequency range of the direct signal masked this, while 24-kHz mixing saves almost half the CPU cost.

Mono input—chiefly from the player's engine—and panned decorrelated stereo output saves memory as well as processing time, just like the early *Midiverb* and *Microverb* budget studio reverbs, which had mono input and processing, with decorrelation only for the stereo output. This gives half-price immersion at the expense of panning. They were limited by the total digital signal processing (DSP) power available; modern games have many times more total grunt, but they're trying to do a lot more at the same time, with audio demands in the minority.

Given the design requirement for more vehicles and split-screen multi-player gaming it was untenable for audio to consume so much, especially on the platform with least CPU power. Savings came through re-assigning one stereo reverb between internal and external simulation depending upon the player's choice of camera—in car or outside—with transitional volume changes and wind masking to hide the discontinuity.

The same principle applies to other processor architectures which share resources between hardware threads, such as x86 and ARM. Whenever possible, audio threads should be allocated to common cores. The "affinity" between threads and cores can be set in most operating systems when a thread is created, and the audio systems programmer has a responsibility to do this and negotiate with other programmers on the team to make sure all the threads play nicely together, observing real-time priorities and not getting in one another's way.

You need to be very careful when doing this and should not assume that output drivers will necessarily get it right. A bizarre and initially baffling bug turned up late in testing of the BAFTA-winning *RaceDriver GRID* game on PC.[2] On one specific AMD quad-core CPU, with one variant of the X-Fi sound card, the car rotating on the podium during vehicle selection would shake more and more until it started to break up into separate 3D pieces, which came apart completely over a few seconds and then shot off into the distance!

This doesn't sound like an audio bug, but it turned out to be. The fault was traced to the drivers for the Fortis chipset on the Creative card—a cost-reduced DSP replacement for the full X-Fi hardware—destructively interacting with the game physics code. The card uses a mixture of software and hardware and started by setting the processor rounding mode for best efficiency in audio processing. When a processor core is shared the operating system is meant to switch such low-level settings back and forth as it juggles software threads on each core.

We worked out independently that the error in the driver set the required rounding mode on one core at initialisation and then dangerously assumed that subsequent updates would take place on the same core. Our physics programmers were doing similar but different setup for their own threads, but the "black box" audio driver was messing with this, so that the audio sounded OK (but might take excessive CPU time) but the physics went mad. And this only happened on one particular CPU architecture with a specific sound card.

Creative Labs subsequently fixed the "X-Fi Xtreme Audio" driver, but with our gold-master deadline days away we could not rely on that correction reaching our customers, so we ended up reworking our physics code to be tolerant of the incorrect rounding mode. This is a good example of how audio systems programmers need to work especially closely with QA testers and the rest of the game team and avoid allocating blame to others—complex systems typically fail because of unanticipated interactions, and you may need a deep understanding of systems that might not seem obviously related to diagnose and fix such faults.

Diagnosis is the hardest part; *until you can reliably reproduce a bug, you've little chance of fixing it.* The skills needed to do this are platform specific—with Android and Windows hosts the most varied in detail and hence hardest to fully qualify—so you need great QA and compliance teams to narrow down the context in which a device-specific bug is triggered. This is the necessary price of releasing products for platforms with little or no hardware and system QA—you are freely admitted to the jungle, but so is everyone else.

Consoles are another matter. You know that millions of people will have identical hardware, you or the manufacturer can and will impose a particular version of the system software, and most of it will be documented in detail. But documentation is rarely complete or up to date, even mass-produced monolithic products have bugs, and you may end up with millions of dollars' worth of coasters on your hands if you and the console QA team miss one.

Bug Reporting

Console development is invitation only, with access to docs and devkits—debug-friendly versions of the console—controlled by the hardware manufacturer. There's much more support for developers than on platforms open to all, but it's still a scarce resource. If you can prove that a bug exists and has no easy workaround, I've found individuals at Microsoft and Sony willing to bust a gut to help, but only if they can reproduce the problem and you can get the message through to the right person.

Even in a multi-billion-dollar console development employing hundreds over years, there will be just a handful of people working on the audio, concentrating mostly on new features and internal work requests. Your first port of call must be the documentation. Only after you've scoured that and any white papers and overviews on the developer support portal is it wise to assume it may be their bug, not yours.

Most of the games I've worked on this century have involved at least a million lines of code written over several years by dozens of people. System libraries, graphics, physics, AI and audio middleware may be larger still and only available as binary or headers. Often disabling other sub-systems of the game—turning off physics or audio, in the above example—exposes an interaction. If the bug does not respond to this pruning, that still helps you narrow down the cause.

Make sure that the problem does not occur on other platforms. If it does, it's definitely your fault. Analysing the difference between configurations that work and those that don't will get you over the hump without bothering anyone else, and more quickly.

Test on the current version of the system software approved for release. This may involve updating or rolling back. Unless it's a beta feature that has bitten you, always test with trunk or release libraries. If the previous or next SDK release does not show the fault you may be given special dispensation to ship that or get a big hint as to what's caused the breakage. Version tracking is vital.

The next step is to raise a question on the support forum, carefully worded to describe the problematic situation in as general a way as you can, stripped of trivia specific to your game. Try to create a small stand-alone test bed that demonstrates the issue. You may find your own bug that way, and your bosses will not be keen for you to share larger parts of the source of "their" unfinished game code.

If the issue rings a bell with other developers, or even if it doesn't for them in otherwise similar circumstances, combine forces to identify the cause and potential workarounds. Only after that—and allowing a day or so for developers worldwide to chip in, regardless of time zone—is it worth considering raising the bug with the makers of the system.

Your professional reputation is on the line. If you make a fuss and get through to the expert, only to find that it's your problem all along, you'll dent your credibility in a highly visible way inside your organisation as well as theirs.

Goading the Dragon

Each major console project has a support manager allocated by the platform vendor who will liaise with your producers rather than random members of the dev team; this manager is unlikely to be an audio specialist but will probably have had enough technical experience and hopefully good enough contacts to escalate your report to people who can really help. Such relationships may be vital at the time and valuable in future—but don't waste their time, even once you've got a clear channel to the expert. Be specific, be wary, check everything you say and try to provide a demonstration, trimmed to the essentials, so they can quickly reproduce the bug and use their internal tools and knowledge to help you over it.

Past solutions to system bugs have involved custom libraries, relaxation of submission rules or access to private configuration structures. But the safest option at short notice is usually a workaround, based on a detailed understanding of when and why the problem occurs. Such bugs often crop up late in development, with duplication booked and the end of a six-month marketing window impeding. Getting permission to share the whole title in a debuggable state, getting custom systems software (which still needs lots of testing by you and the supplier) or agreeing to special waivers all takes time and involves the approval of non-technical people who will need convincing of the issue in their own terms and might blame you, however solid your technical evidence to the contrary.

Here's a specific example. One high-performance title for a new console showed an erratic and worrying tendency to leak voices. This meant that some playing sounds would fail to update properly or, worst of all, seem still be playing after they had been told to stop. Such leaks are the bane of Virtual Voice audio systems, because the dynamic resource allocation system will keep trying to make use of the voices still available to it, culling more and more of the optional sounds to keep the vital ones playing, so it may be several frames or even a few seconds before the leaks become obvious. By then intermediate updates have masked the original problem.

Finding Leaks

It's useful to have code to validate and, if necessary, clean up physical voice assignments in a game. This traverses all the voices, looking for zombies which the low-level system thinks are playing but the logical level is no longer bound to. If the physical voice keeps track of the asset it was originally asked to play it may be possible to identify a common factor, such as use of a specific sample or bank. Each voice plays asynchronously, effectively by direct memory access, and non-trivial codecs fail or freeze if the data fed into them is not as expected. Dynamic resource allocation makes it hard to work out which died, when and why.

Unloading a bank of sounds that may still be playing—or could yet be un-paused—often causes voice leaks if the data loaded into the space they once occupied is not suitable for the codec. With luck, you'll hear a horrible noise.

If not, and the codec is imbedded in hardware or system firmware, you may lose a voice or a decode context. Complications like looping and multi-channel processing mean that there's not necessarily a 1:1 correspondence between voices and sub-systems to decode them, so switching assets to a simpler codec may make leaks seem to go away—for a while.

Voice leaks can have timing-related causes, making them hard to predict or reproduce. Those include unsafe threading code, such as non-re-entrant routines or data being accessed by a second thread before a previous one has finished, lost or ignored callbacks, and early exits from status polling routines which prevent later voices being updated.

A subtle understanding of the allocation and sequencing of downstream resources can be very helpful. The console problem only occurred when the system was heavily loaded, ruling out a minimal demonstration. Most low-level voices are controlled via a "handle"—an index or opaque pointer returned when each voice is allocated. These are allocated from a pool rather than at random, since the unsynchronised starting and stopping of samples at erratic times would otherwise cause memory fragmentation, especially when large but varying-sized samples are involved.

Voice re-use means that indexes and pointers are not dished out sequentially for long. But patterns in the voice handle values may still reveal useful information—grouping of leaked voices, for example. After trying many failed workarounds, including a continuous voice collector to check for leaks, similar code triggered at key points (e.g. bank loading) or periodically stopping everything and (hastily) re-establishing the mix, I noticed that all the indices of zombie voices ended with the binary pattern "111." The brutal fix was to request 12.5% more voices than we anticipated needing at the start, then ignore any which came back with an index which was 7 modulo 8.

The problem went away! The game shipped on time. There was a cost in memory for the unused voices, but this could be handled from the existing audio budget without disturbing the rest of the team by tightening voice caps and downsampling some sounds.

Armed with this very specific characteristic of the bug, the console system programmers worked out that there was a subtle direct memory access timing problem which they'd not spotted before. To improve performance, voice updates were being packed together in groups of eight for DMA between main memory and the audio co-processor. Immediate but otherwise unrelated memory transfers could clobber the end of the voice update packet, making the last voice in each group of eight unreliable on a busy system. The bug was fixed for later products, but the game shipped with the workaround and went to number 2 on the charts; *Grand Theft Auto* is hard to beat.

Audio Memory Allocations

Audio memory budgets have grown like everything else in games. 16-bit consoles and soundcards offered just 64K for uncompressed samples; the original PlayStation pushed this to 512K and added compression to fit almost two million individual samples—90 seconds at 22 kHz—into that space. PlayStation 2 pushed this to 2 MB. Later DirectSound3D cards and the original Xbox could cope with 8 MB of uncompressed sample data before running into memory-management problems. The following generation of consoles had 512 MB of main memory; audio-heavy games devoted some 5% of that, 20..30 MB, to sound, some of which was used for reverb and mixing buffers rather than pre-authored assets.

Then new mobile platforms arrived, with similar total RAM but less space for applications and far less dedicated audio hardware. Phones and tablets typically have a single stereo music decoder, with everything else done in software on a low-power 32-bit ARM7e processor. Table 5.1 shows the audio memory breakdown in 2015's *Micro Machines Mobile* for Android and iOS systems.

Table 5.1: ***Micro Machines Mobile*****, resident soundbank sizes**

Soundbank/Platform	Android	iOS
Power ups	6,085	4,131
Horns	1,709	1,709
Resident music	1,151	1,332
Surfaces and collisions	1,552	1,552
User interface	3,329	3,298
Generic ambience	155	155
Track or arena	518	447
Generic vehicles	454	454
Vehicles (total, 4 at once)	1155	1,050
Total (kilobytes)	**16,108**	**14,128**

Each track and vehicle had a custom soundbank and access to a shared set. The table shows figures for the largest track and four largest vehicles—as usual, in game capacity planning, only the worst-case matters. In game, tracks and vehicles re-use the space occupied by the Front End "User interface" bank. Banked samples are typically 24 or 32 kHz ADPCM, with the mixer running at 32 kHz for performance reasons.

A further 13 full-rate music files are streamed, in AAC format on iOS and Ogg Vorbis on Android. The "resident music" bank contains short uncompressed "stings" played from RAM between the streams to avoid leaving a gap, or the overhead of momentarily decoding two at once.

AAA Game Memory Footprint

When asked about the memory footprint of a typical Triple-A PC game, in November 2017 Matthew Florianz, Frontier Developments lead sound designer, told the specialist VGM community: "*I feel that targeting around 256 MB runtime is sufficient but if your engine guys let you have more, all the better.*"[3] By the time you read this, if Moore's Law holds fast, a lead designer with an audio budget under a gigabyte might feel short-changed. But it's been the case for a few years now that the limit on sound sample footprint is constrained by loading times more than it is by the assets or available RAM. Mobile products can trade off CPU time and RAM, especially regarding codecs.

For this reason, and because consoles and particularly phones tend to be much less profligate with memory than desktop PCs, it's still important to handle audio memory efficiently. This involves choosing appropriate codecs to speed loading and save space on disc, in RAM, or in the download package, and bundling samples together so that it doesn't take more time to find them all than to load them. We must decide when it's cost effective to load an entire sample or set into memory and when it's more efficient to play it piecemeal, streaming sections of a large file between disc (or other slower devices) and RAM.

Even in a single game, these decisions may vary between platforms. If the game data is on an optical disc already in the drive, it generally makes sense to fully use it, unless this is a mobile device like PSP (Sony's PlayStation Portable), where stopping the drive may double battery life. Nintendo's Switch imposes tight limits on memory-card access to prolong the life of the flash media, and this affects the choice of codecs and location of audio data for that platform.

By default recent consoles like PS4 and Xbox One install everything to hard disc, which is a mixed blessing, as it doesn't take many Blu-rays to fill up a writeable drive, and much of the content of many game discs is localised speech. Only a fraction of this—one language out of many—really needs installation.

Table 5.2 shows the sizes of the main resident soundbanks in the 2017 release *Micro Machines World Series* in *megabytes*. Another 36 large audio files are streamed from disc. The Xbox One sound design was complicated by an undocumented limit in Wwise for Unity of around 30 MB for all banks containing XMA-compressed samples, with the rest PCM.

Table 5.2: *Micro Machines World Series*, resident soundbank sizes

Soundbank/Platform	Windows etc.	PlayStation 4	Xbox One
Persistent bank	11	11	2
Surfaces and collisions	115	115	115
User interface	60	61	79
Vehicles (total, 12 types at once)	72	72	72
Weapons	60	61	81
World ambiences	50	51	13
Total (megabytes)	**368**	**371**	**362**

Windows and PS4 use Ogg Vorbis and ATRAC9 codecs, respectively, for speech and music, with all the rest uncompressed PCM. Linux and macOS releases share the Windows layout.

2006's *Micro Machines 4* banked just 12 MB of PCM in total on PC and barely 3 MB of VAGs on PSP. We've come a long way in a decade.

References

[1] *George Lucas*: http://cinearchive.org/post/65158356705
[2] *GRID Bug*: https://m.youtube.com/watch?v=Y4ZdIPtw_9A
[3] *Matthew Florianz, private communication.*

CHAPTER 6

Loading and Streaming Concepts

Streaming is the process of playing large audio files from relatively slow devices without first loading the entire sample into memory. This chapter and the next explain how to implement streaming and the implications for memory and file usage. It also shows how to be confident that files will play without interruption from all sorts of secondary storage devices, including optical and magnetic discs, flash memory and remote network files.

We'll consider the complications associated with looping samples, which must play continuously, returning to the start after reaching the end, and techniques which reduce the time spent seeking data and increase the proportion actually spent reading it. We'll quantify the practical limits on the number of streams which can be played from a given "spindle," with examples from games loading from CD, DVD, Blu-ray and magnetic media, and explore techniques that allow non-audio data to be efficiently read at the same time as samples are streamed from the same or other drives.

The alignment and block structure of data files on typical devices are also explored and the implications of streaming audio which has been compressed using fixed or variable-rate compression. Most streams read pre-mixed audio, but its sometimes necessary to *write* streams on the fly as a game runs, for instance for replays or YouTube uploads, so the complications of this extra work are also noted.

Why might you want or need streaming?

- To play arbitrarily long samples from limited memory
- To play things which are happening "live"
- To play a sample without having to load it all first
- To fairly share loading capacity between sub-systems

Do You Need Streaming?

Streaming is risky and fiddly. Game development is considerably simpler and less stressful if streaming is avoided unless the game design makes it essential and the cost of alternative approaches—mainly RAM at runtime—outweighs the saving of having a simpler, more robust game, especially during development while periodic update stalls and transfer bottlenecks are to be expected.

So whenever streaming is considered the saving in RAM should be weighed against the increase in total loading time—since seek delays rather than loading ones will predominate when files are loaded in relatively small pieces rather than one big one, from optical media—and the risk and complexity of a streaming solution. The pros and cons of streaming are summed up in this phrase: *All at once, or little and often?*

> All at once, or little and often?

Streaming was almost essential on the original PlayStation, with 3.5 MB total RAM and over 600 MB disc space, reliant on an audio codec capable of only 3.5:1 compression. Less than a minute of high-quality samples, in total, could be held in audio RAM. Sega's Saturn was similarly CD based, with only a little more RAM and less compression. On PS2 the ratio and rationale were similar, with the same codec, 40 MB total RAM and up to 4.7 GB of disc space, and maybe two minutes of samples crunched in audio RAM.

Times change. Default codecs are now two to six times more effective. Total RAM increased by a factor of about ten in the next generation of consoles—Xbox 360 and PS3—and mobile devices, and as much again since. For a decade this combination of changes has enabled us to pack an hour or more of short samples, typically a few thousand, in RAM at any time, swapping banks as contexts change and streaming larger samples from disc. But the optical drives have not kept pace—transfer rates and seek times are little changed, while the game size limit (on single-DVD PC or Xbox 360 titles) has not even doubled.

Arguably, streaming is 20 times less justifiable than it was on earlier optical-disc consoles. We do it anyway if it benefits the user. I'm assured that Rare's audio team put a lot of effort into streaming on 2018's Xbox One hit *Sea of Thieves*, and it's the mainstay of the open-world *Grand Theft Auto* series.

That said, streaming seldom smooths the development experience. Some of the most serious product-threatening bugs I've fixed relate to streaming problems, which often show up only late in development once the full set of data is being tested on secondary media rather than local networks, which inevitably glitch—internet protocol was never intended for hard real-time deadlines.

Problems are often introduced or exacerbated by hacky fixes introduced late in development by game programmers who may pile in initialisation tweaks, replay and profile updates without concern for what else the system is doing . . . after all, it all works fine, till it doesn't, and then failure is caused by resource contention, which is multi-factorial but most obvious to listeners, whose ears can't blink and process "history" in a less forgiving way than other senses.

Streaming is prone to glitch—halting or repeating sections—if the slower medium cannot keep up with the rate of replay. This is obvious and annoying as soon as it happens to audio with a beat or a message, such as music or speech, but often goes unnoticed, at least the first time it happens, for more stochastic background sounds such as rain, wind, cheering or applause. Even in those cases a second glitch within a second interval will be obvious to most listeners—it's as if they need to hear it go wrong twice to be sure it's the system not their attention at fault. In any case glitches are best avoided because they detract from the sense of immersion such background audio is intended to create, but it's worth noting that the obviousness of a glitch depends upon content as well as timing.

Disc Layout

Optical discs are recorded as a continuous spiral, unlike the concentric tracks of a hard disc, from the hub outwards. Compact disc audio discs originally spun at varying speed, always supplying data at a constant 150K per second whether reading the inner or outer circumference of the disc. As processing power increased, adjusting the mechanism proved more expensive and complicated than adapting to a variable bit rate. Modern optical drives spin at a constant speed, in rotations per minute, while still allowing a spiral track to be read. The inside of the spiral, nearest the hub, is shorter than the outside, on the edge of the disc, so files located there are read relatively slowly—around one third as fast as those on the outer edge.

The same "constant angular velocity" principle applies to most DVD and Blu-ray drives in PCs and consoles, and Sony and Nintendo's proprietary optical media. Console product authoring tools allow the position of every file on each layer to be specified. Sometimes files are deliberately duplicated, either between layers to eliminate the need for the optics to slowly refocus on a different layer or across the surface of a disc to minimise seeking. Since discs have a fixed capacity, and customers don't like to wait, you might as well use any spare to speed things up.

For instance, several different-looking vehicles might use the same engine audio data. To avoid loading each in two steps, one for the shared data and one for the unique material, you might duplicate the engine assets in each relevant block of vehicle data. Providing this doesn't affect the worst-case footprint, and after allowing for dedicated audio memory on some systems, it's reasonable to persist this duplication in memory too. After all, the game designer or player might select any combination of vehicles, and the largest set might not share anything. It's still worth duplicating the data to keep things quick and simple, especially when there are lots of different vehicles in a level.

Alternatively, you might have a dozen areas in your game world, each of which uses a mixture of shared and unique environmental weather assets. Batching those together to minimise loading time makes sense if the space is available and it takes more time to seek or select another file than it would do to pick up a clone of its data from the current read position.

The faster the drive, the smaller the lumps can be, but there still comes a time when it's worth duplicating them, unless solid-state rather than cheap rotating media is employed. This copying should be done as part of the last stage of mastering the disc image or installation package—it'd be both wasteful and error prone to carry duplicate data around earlier in the build process.

Just Fast Enough

It might seem logical to put the large files you wish to stream on the outside of a DVD so they can be loaded at the top speed, but that's better reserved for the data which is needed before the game can start so players are not held up when there's nothing else to entertain them.

The streamed files should be on a part of the disc that can be read at least as fast as they will be consumed but not necessarily any faster. Putting them on a faster part would be a waste, and typically tutorial and intro videos go on the slowest part of an optical disc, compressed enough so there's plenty of time to read them, even if the disc is dirty and the drive's old and tired, like most QA equipment.

When and Why to Stream

Since the earliest days of computing there has been a tension between speed and cost of storage. Big stores tend to be slower but cheaper per bit compared with random access memory (RAM). Secondary stores have variously been implemented with arrays of magnets—tapes, rings, discs, drums and bubbles—optical media like CD and Blu-ray, punched cards and paper tapes, even electrostatic displays and acoustic delay lines made of nickel or mercury. Primary storage, RAM, is faster but more expensive, so there's commercial advantage to be had by keeping bulky data on secondary media as long as possible, only transferring it to RAM when it's urgently needed, even though this introduces substantial complications as well as benefits.

The maximum duration of the sound and the number of channels and layers within it depends only upon the capacity of the secondary storage devices, invariably many times larger than total RAM capacity. Table 6.1 shows the increase in RAM and optical disc capacity of successive PlayStation models over two decades. A similar progression affected consoles from Microsoft and Nintendo.

Table 6.1: PlayStation primary and secondary capacities, in megabytes

PlayStation model	PS1	PS2	PS3	PS4
Optical disc capacity	650	650/4,330	4,330/47,732	122,072
Usable RAM capacity	3.5	40	480	5,120
Ratio, max media/RAM	186 *	108 *	99 *	24 *

The first PlayStations used CD-ROM media capable of holding almost 200 times more data than could fit into all the available RAM, which was divided into three areas optimised for CPU, graphics and audio access, in descending order of capacity. Music, 3D geometry and graphics textures stream into different areas of memory as games are played, often almost continuously. With few exceptions, such as NanaOn-Sha's 1999 monochrome line-art game *Vib-Ribbon*, which derived game data from the user's audio CD collection, the game disc had to be available for loading of fresh data at any time.

In 2000 the PlayStation 2 arrived with ten times more RAM, in similar specialised silos, and supported proportionately larger DVD as well as CD media. It was only practical to use single-layer DVDs, as testing of multi-layer discs involved manual disc swaps with one prototype disc per eventual layer, a QA nightmare. The optical media could still hold 100 times more than the total RAM capacity, yet games with lots of graphical levels and localised speech in many languages soon soaked up the extra capacity of DVD. The PS2 hardware supported an internal hard drive, but as this was an optional extra, games could not rely upon it, and very few took advantage. It was however very useful during development to save time transferring finished files from a host PC network before the full game was compiled onto read-only optical media.

Seven years later, PlayStation 3 titles came on either DVD or Blu-ray discs, with a further tenfold increase in capacity even though only one or two of the eight potential layers of the Blu-ray were available for game data. Microsoft's rival Xbox 360 persisted with single- and dual-layer DVD media for another generation, sacrificing substantial space to the disc copy-protection system but clawing some back with non-standard disc layouts.

This time Sony hardware had magnetic as well as optical drives, but Microsoft reduced costs by making the Xbox 360 hard drive optional, and again cross-platform software developers learned not to rely on it. An attempt to avoid Sony royalties by backing HD-DVD rather than Blu-ray backfired as few optional HD-DVD drives shipped; those were only usable for video rather than interactive media, and the format failed in the marketplace. Many game developers continued to aim for the common-denominator DVD capacity, so the PS3's extra space and hard drive was of limited benefit to cross-platform titles. They had to be designed to fit and load at acceptable speed for either platform.

Nintendo were late to move from ROM cartridge to larger and cheaper—but slower—optical disc software delivery, skipping the CD-ROM stage followed by Commodore, Sony and Sega. The GameCube introduced small versions of DVD media, akin to audio CD singles but with over a gigabyte capacity; the follow-up Wii console used full-sized DVD media, while the Wii-U made the jump to Blu-ray, like Microsoft's Xbox One.

The transfer speed and the time needed to seek any particular block of data on the disc have not increased in line with the extra capacity. This created a new bottleneck, along with the steady increase in total memory to be loaded and the graphical demands of HDTV, so Sony and Microsoft arranged for games to install large chunks of their data to the internal hard drive, where it could be found and loaded more quickly—after a new and unwelcome hiatus waiting for the installer to find and copy the key bits of the title to the hard drive. Thus the time for a typical game to start was kept to the irritating but bearable minute or so typical of floppy-based home computers decades older.

Optical Disc Performance

Like many other "benchmarks," raw figures for peak performance are a poor guide to the actual rate at which we can load data from optical discs. When compact disc was introduced the data rate was constant at about 150 kilobytes per second, just enough for 16-bit stereo audio at 44.1-kHz sample rate, and greatly simplifying the read electronics.[1] Nowadays almost all drives are constant linear velocity (CLV) rather than CAV. This means that they read data progressively faster as they advance from the start to the end of the disc.

Optical ROM data is recorded in an outward spiral, like a vinyl record running backwards, rather than the concentric tracks you'd find on a magnetic disc. Each circuit of the optical disc is thus a little longer than the one before, and the outer edge of a nearly full disc holds more than twice as much data as the inner one, as it's more than twice as long and the pits representing data are evenly printed. The same spiral principle applies to DVD and Blu-ray, but later drives traded software for hardware and simplified the mechanism, using electronics capable of adjusting to a variable bit rate, so that there's no need to slow down the drive as the amount of data per revolution increases.[2,3]

As part of its automatic dance-generation process, *Dance Factory* used the PS2's CLV drive to read and analyse standard audio CDs. This was nominally a 24* drive, capable of reading 24*150K per second—but only from the outer edge of a full disc, where the data track is longest. The peak transfer rate, in that exceptional case, was about 3.5 MB/s. But the same motor speed delivered only 1.1 MB/s at the start of a disc, and few music CDs were so full as to reach the peak rate at the end of a 70-minute disc.

The actual rate reading albums remastered from vinyl was about half that and even worse for short CD singles. It would have been a waste of effort to optimise the complex audio analysis code to keep up with the maximum speed when the data typically came in at about 40% of that rate.

The Xbox 360 dual-layer DVD drive reads almost 16 MB/s with luck and a following wind, but only from the outer edge of either of its two layers. It makes sense to put large files which will be read as a single block, such as the game executable, out there, even if you must pad the disc with dummy data to get there. But there's no point using the fast parts of the disc for data which will only be consumed at a lower rate—such as cut-scene and introduction videos.

Multiple layers complicate things. DVDs support one or two layers, and Blu-rays can have up to eight, though current Xbox One and PS4 consoles can only read the first four. This still gives them about 16 times the capacity of a packed DVD. To exploit CLV on a multi-layer disc you need to know whether it indexes disc blocks from the outside on the second layer, so that data is allocated from the inside outwards on the first layer, then back the other way for the second (at decreasing speed), or from inside outwards again, which is consistent but means the speed follows a sawtooth rather than triangle-shaped graph as you traverse the entire capacity. DVDs support both arrangements, as OTP and PTP, respectively. If this is not documented—or you don't trust the documentation—check this by experiment.

Either way, there's a cost to switching layers; this is drive dependent but typically of the order of 50..100 ms, accounting for the momentary hiccup partway through the replay of long movies. Video discs typically have just a few very large files, whereas games often comprise tens or hundreds of thousands during development, and a major part of the final "gold disc" mastering process involves merging, compressing and placing those optimally on the disc so the player doesn't have to wait several minutes for all the disjoint data files to be located and loaded. Manufacturers often impose a limit of a minute for initial loading to the interactive menu or for the transition from menu to gameplay, but even if they don't make such conditions part of the pre-release test submission rules (variously known as TRCs, TCRs or lot-checks), a long wait before the action starts will frustrate listeners.

Data compression increases the effective rate at which you can get data off the disc but rarely helps for audio samples, as they've much less entropy—slack to be squeezed out—than compiled code, 3D geometry, or XML configuration data. Combining or juxtaposing files which you know will be read sequentially can massively improve loading times but not until the data and associated "flow" are locked down. Loading optimisations must be done and rigorously tested right at the end of the development process, at the same time as final bug-fixes and up against a hard release deadline. It's no surprise that this is sometimes left as an afterthought or compromised in the rush to ship product.

The time to seek from one point on a disc to another depends upon the drive mechanism and the distance on-disc between the end of the last file and the start of the one you want to read next. If this distance is substantial, the cost in terms of delayed reading may be much greater than the time spent actually reading the data once it's been located. A modern console can read at least a megabyte of contiguous data in the time it takes to seek from the end of one file to the start of another arbitrarily placed on the disc.

Laser-disc seeking is physically implemented in two stages, which you can exploit by carefully arranging your data. The laser travels on a "sled" which is mechanically wound back and forth between the edges of the disc. This mechanism alone would not be enough for the drive to find and lock onto the spiral track of data, since its exact radial position varies continuously, because of the spiral, inexact centring of the disc hub and axial wobbling of the disc as it spins, which is particularly influential on the notionally fastest outer edge, where the hub is most distant.

The sled motor positions the laser only approximately. A second mechanism typically uses electromagnets to focus and fine-tune the laser upon the sled, compensating for all the other variability. Total seek time depends upon at least four things—the time to roughly position the sled (long seek), usually measured in high tens or hundreds of milliseconds, plus any layer-swap delay, plus the time to lock in on the data, perhaps 20..30 ms, and then the time to locate the exact start of the required data.

As it's recorded in a spiral, this disc turns continuously, and positioning is inexact, the drive firmware must aim to start reading a safe distance before the location of the data predicted by dead reckoning. If it locks on even a byte late, the drive must seek back and start again. So there's another delay of a few milliseconds for this and perhaps tens of milliseconds for a "short seek" back if the first guess cut the timing too fine.

Long and short seek operations account for the noises the drive makes as it loads data. The obvious noise is the long seek moving the sled, while short seeks produce much quieter "ticks."

Interactive applications place much greater demands upon a drive mechanism than replay of a single music or video stream. Open-world games like *Grand Theft Auto* hammer drives as they funnel context-sensitive graphics, geometry, music, speech and sound effects through RAM. They often seek several times a second, whereas film replay will seek once from the menu to the start then typically read sequentially for minutes at a time unless the viewer intervenes for a pause or replay.

Console manufacturers set strict specifications for the drives they buy in for their consoles, spec'ing these to sustain five seeks per second for hours at a time. This is one of the less-cynical reasons that pricier consoles require hard disc installation, even if a game is supplied on disc rather than by download. It still helps to understand optical disc layout to install games at a good speed, and similar seek distance and calibration considerations apply to magnetic discs, even if they seek an order of magnitude faster. You still want to cluster related files together, pack small ones into groups and use appropriate decompression on the fly, if you don't want customers twiddling their thumbs.

Pathological Seeking

Once I was called in to fix horrendous console loading times on an unfinished game, *Worms 4 Mayhem*, which Codemasters had obtained the rights to publish imminently. Most of the development work had been done on PC, with fast drives and loads of memory; the Sony one was glacially, unbearably slow once untethered from a host PC and asked to read everything from DVD, and the Xbox version was struggling despite taking advantage of extra RAM and hard disc.

Upon investigation, I discovered that the game data was organised in chunks, much like an IFF or WAV file, with two four-byte values at the start of each chunk—a "4cc" text marker like 'hats' or 'musc' followed by another four bytes giving the offset to the next chunk. The game located specific blocks of data for the chosen map, teams and weapons by scanning through the hunks, skipping over most of them to find just the data it needed—necessarily as RAM was very scarce.

This approach is standard practice for IFF files in memory but madness if the data is scattered across a DVD. Each check to skip an un-needed block involved seeking and loading a 32K disc block and discarding all but eight bytes, then moving on to the next. It would have generally been more efficient to read all the intervening data and skip the seeking. At an average of 125 ms per seek and load (allowing for a mix of short and long seeks) the drive was delivering an effective transfer rate of 64 bytes per second. That's about the speed of my first 1.7 MHz tape-based micro and a third the speed of loading from audio-cassette on a 1982-vintage ZX Spectrum.

The solution is obvious. De-interleave the table of chunk names and offsets from the data it described, and store that separately where it can be scanned without seeking.

> Audio is often the first thing to go obviously wrong.

Drives are faster now, and we've got more memory. But even if no seeking is needed, reading a single contiguous file, it would take *several minutes* to load all the memory of a PS4 or Xbox One from Blu-ray. Staged loading (e.g. front end, scene selection and equipment load-up stages), decompression, duplication and algorithmic data generation are all needed to keep the time the players spend waiting down to a (barely) acceptable half minute or so. The low-level understanding and hard real-time requirements of sound streaming mean that the audio systems programmer plays a key role in determining the loading strategy of triple-A games, and audio is often the first thing to go obviously wrong if this is not well enough planned and implemented throughout the game.

Practical Disc Layout Tips

When you control the layout, as for a game, it makes sense not to split files across layers and to group files that are likely to be accessed sequentially on the same part of the same layer of a disc. For instance, in a speech-heavy sports game, with thousands of commentary samples in many languages, data for each locale should be stored contiguously and on the same layer as other data loaded on the fly by the game, as opposed to data loaded at initialisation or from the front end as a map is set up.

It's worth exploiting short seeks, especially when you know you'll need to load several short samples in rapid succession but not concurrently. The squaddie speech system in DVD versions of *Operation Flashpoint* clustered together the samples and emotes for each language, so they could be arbitrarily sequenced, stitching in character or location names, without requiring a slow long seek that would break the flow. The amount of data that can be encompassed in a short seek depends upon the drive and disc location, but it's typically 30..60 MB on a DVD and substantially more on Blu-ray.

If in doubt, perform experiments. Manufacturers do change drive specs through a production run of tens of millions of consoles, but they take pains to ensure that updates are also upgrades, with performance at least as good as the specification released to early developers. Occasionally the loading of other game data or streamed music may disrupt such optimisations, but it's in the nature of speech that rare hiccups are unlikely to break immersion—as with momentary re-buffering of stochastic background sounds.

Continuous looped ambiences and seamless music replay may require inaudible tweaks to the asset to make sure they don't stutter at the end, when an incomplete block is loaded and must cover the time for a long seek back to the start. These techniques are explained in Chapter 7.

To improve performance on consumer systems which read files piecemeal, drives often read several blocks beyond the one requested into a local buffer. A typical five-block buffer on a DVD drive holds the next 160K of contiguous data. This helps an intermittent reader using small buffers to keep up, but in general, especially for audio, it's more efficient to load the entire file directly to the memory from which it will play.

As a title is finalised testers may identify pinch points where the loading of short samples disturbs the flow. There are tricks to hide such edge-case problems without requiring script changes or re-recording sessions. A short period of silence or one from a small set of generic emote sounds like a breath or gulp can be added at either end to pad an exceptionally short sample, or a short memory-resident sample can be used as a "shim" to keep the stream from starving.

Providing the number of combinations is manageable, pairs of short samples can be concatenated so you select and play one longer pair rather than two tiny snippets. The implied duplication will eat up some disc space, but no more RAM will be needed if the combined pairs are no longer than the largest sample that might otherwise play.

Conventional disc systems separate the map of file locations from the file data to simplify lookups while allowing dynamic changes. This is inappropriate for game data, especially on slow media, because of the need to seek back and forth between the directory and file contents. Games typically pack data together in one huge file, with an optimised index which can be loaded entirely into memory, eliminating directory lookup seeks thereafter.

Error Correction

Further considerations stem from the way optical disc error protection works. To prevent microscopic flaws or dirt invalidating the bitstream, the bits for each block of data are distributed around the disc surface, with additional redundant bits to allow the automatic correction of small errors and detection of gross ones.

A gross error causes the drive to reposition the laser and try again, usually several times, first with short seeks and then, if that doesn't do the trick, by winding all the way to the edge, recalibrating the sled position and a long seek back to the required data. Anyone with a dirty disc or drive will have heard the associated scrabbling noises. If you want to be extra sure that a stream will work even under such circumstances, allow enough time for a couple of long seeks—perhaps 300 ms—when you compute your worst-case buffer sizes. If the data is simply unreadable this won't help, and you should consider muting the stream rather than annoying the player with endless cheap-CD retry attempts, but allowing such "slack" gives you extra margin without wasting much RAM.

The error-correcting code (ECC) cannot work a bit at a time—it needs to collate and check an entire "block" drawn from various parts of the disc surface before it can return valid data. The error-correcting block size for CD was just 2K, which means that low-level combined seek-and-load operations usually address a disc in sections of 2048 bytes. If you want to read from an offset which is not a multiple of 2K—say, 1620 bytes in—the drive must read the partial block containing that offset, discard the first four-fifths and use just the last 404 bytes from that block before moving on to the next.

On CD this wastage is hardly significant, but the increased capacity of DVD and Blu-ray discs brought with it an increase in the error-correcting block size—to 32K for DVD or 64K for Blu-ray. This means that the minimum amount of data that can actually be read from a Blu-ray is 65,536 bytes, and then only if the data you request fits entirely into a 64K block, without straddling a block boundary.

If you ask for just eight bytes from offset 65,530, as a hard-pressed *Worms* programmer might, the drive mechanism will be forced to load the entire first 64K block to get its last six bytes, discarding the rest and the entire following 64K block just for the first two bytes at its start. We've asked for eight bytes and had to read and decode 131,072 to get them. Without even allowing for seek time, we've reduced the speed of the drive to 0.006% of nominal.

This is a contrived and pathological example—the degradation factor would be "only" 0.012% if the required data didn't straddle a block boundary. But it does illustrate that you can make more efficient use of any drive which uses ECC blocks if you position the start of files and seek locations within them on boundaries which are an aligned multiple of the ECC block size. It's a small improvement, but every little bit helps.

Note the word "aligned." On our first Blu-ray product, my colleagues and I went to great lengths to pad files to multiples of 64K and to make sure that they were packed tightly together. Unfortunately, we failed to account for an odd-sized system file which Sony's disc layout tool added at the start. Consequently, *every single* file or offset thereafter was misaligned. It was worse, slightly, than it'd have been if we'd not bothered with ECC block alignment at all.

The loading speed suffered, but luckily the other optimisations, particularly the merging of small files, minimised the damage. The mistake added between 7 and 25 ms to the effective seek time for each load transaction. Over all the customers, that's about three hours of time wasted for each block of data loaded, or ten days per complete level. I won't say which game this was, but if you've got a stopwatch and a PS3 you could work it out. Of course we fixed that in subsequent releases.

Beyond Optical Media

Manufacturers continue to juggle the balance between cost, capacity, loading and seeking times. Broadband means that some titles bypass the optical disc platform, but discs still have the edge for retailers and the millions with slow internet connections or small internal hard drives. Nintendo continued to shave costs by leaving out the hard drive which first Sony and then Microsoft had deemed optional. The same disc layout considerations applicable to games loading direct from optical media enable us to trim initial installation times, and even though magnetic discs find data faster, file organisation still makes a great deal of difference to the amount of time the user spends waiting for things to load.

Amusement arcade games like *Sega Showdown* run entirely from RAM loaded from flash memory, almost eliminating seek delays but placing a relatively high cost when writing data, as current flash memory works by erasing and rewriting substantial blocks—a process much slower than reading and limited to a small number of re-write cycles before the media becomes worn out. In the arcade extra RAM rather than flash memory is used for temporary results, and flash-based consoles like the Nintendo Switch developers strictly limit the rate at which titles can write data. This introduces further trade-offs between memory and CPU usage. Similar considerations affect disc-less phones, set-top boxes and tablets.

It's certain that manufacturers will continue to pile more storage into devices while cutting costs by introducing other trade-offs. So however their audio and game data reaches the player, developers of cross-platform interactive media will need to understand and juggle memories of varying speed to make efficient use of each. Localised speech, music, surround and environmental ambiences will remain a major part of the memory footprint, and audio systems programmers will need to tailor their systems to fit the strengths and weaknesses of each platform.

Cross-platform system programmers need to bear in mind these considerations:

- The capacity of each type of memory and minimum size of blocks which can be read and written
- Bandwidth in each part of the system, including contention between audio and other sub-systems
- Response time to find data or switch between data sets on each medium
- The total amount of time the user can be kept waiting for things to load or store
- How to minimise or recover from intermittent or persistent read errors (e.g. dirty discs)

During development the concept of a "translucent file-system" can save lots of time. This involves duplicating part of a directory tree in lower-latency media, such as RAM disc, network or flash memory, as a "patch" for the bulk of the data on a slower drive. Attempts to load a file look first on the lowest-latency medium, where recent changes reside, falling back to old data on slower drives if no patch is found.

Double Buffering

The key technique used to implement streaming is called "double buffering." A buffer is a single area of contiguous memory. While streaming, data is alternately played from one area of memory as it is loaded into another, swapping the roles of the buffers between "read" (or reload) and "write" (replay) as each is consumed.

Some notification mechanism is needed to tell us when a buffer has fully played so we know to queue another one. This can be implemented several ways, as we'll soon see. For now, just assume that we can find this out quickly enough to keep up with the demands of the player and the latency of the loading system.

Assume we have two buffers of the same size, labelled A and B. It's unsafe, in the sense that audio artefacts are likely, to use the same area of memory for both loading and replay: incoming data might clobber samples that are still waiting to be played. So we make a rule that each buffer must be in one of three distinct states: waiting to be loaded, or "fallow"; loaded and waiting to play; or playing:

```
enum BUFFER_STATE
{
  BUFFER_FALLOW=0,     // Waiting for data to load
  BUFFER_LOADED,       // Data loaded waiting to play
  BUFFER_PLAYING,      // Data playing
  BUFFER_STATE_COUNT   // Number of states
};
```

Before we can play anything, we need to load a full buffer of data. As soon as we've got that we can start playing it. While that plays, we have time to load another buffer. Providing that's ready before the first is consumed, replay is continuous. In simple terms, we load A, then play A as we load B, then play B as we re-load A and so on. In practice things are a bit more complicated, as we need to take account of queuing and consumption notifications. Queuing occurs when a buffer is fully loaded and we can tell the underlying system it's ready to play once the previously queued one is consumed. Consumption notifications occur when a buffer is fully played and therefore can be reloaded. So the sequence goes:

Table 6.2: Double buffered streaming sequence

Step	Action	Buffer A	Buffer B
0	Initialisation	FALLOW	FALLOW
1	Load buffer A	LOADED	FALLOW
2	Queue buffer A	LOADED	FALLOW
3	Play buffer A	PLAYING	FALLOW
4	Load buffer B	PLAYING	LOADED
5	Queue buffer B	PLAYING	LOADED
6	Buffer A consumed	FALLOW	PLAYING
7	Reload buffer A	LOADED	PLAYING
8	Queue buffer A	LOADED	PLAYING
9	Buffer B consumed	PLAYING	FALLOW
10	Continue from step 4	...	...

and so on. The second column in Table 6.2 Double buffered streaming sequence shows the action underway, and the third and fourth columns show the state of the buffers *after* that action is completed.

The size of the buffers is pre-set to ensure that each reload request can reliably be satisfied before the alternate buffer is consumed, so the output is continuous. This involves a careful analysis of factors that could delay loading:

- The speed of the drive (which may vary)
- The physical location of data on the disc
- Concurrent demands on the drive to load other data
- Time needed for error recovery and recalibration
- Time needed for notification of buffer consumption

Let's first assume, for old times' sake, that we're streaming uncompressed 16-bit PCM samples recorded and re-played at 48 kHz. This is the standard output rate on consoles and many PCs, phones and tablets. It's rare to apply Doppler shift to speed up or slow down streams, since they tend to be non-positional, but if you must do this, use the highest sample rate which Doppler might demand so that this worst case doesn't invalidate your arithmetic. The principle and modelling of Doppler shift, an important interactive audio motion cue, is explained in Chapter 13.

Each stereo channel requires two bytes (16 bits) per sample, so each stereo (Left and Right) frame is four bytes. Data is consumed at four bytes times 48,000 samples: 192,000 bytes or 187.5 kilobytes per second. If the worst-case time to find and load a buffer of data is 350 ms—a reasonable assumption for a CD drive doing nothing else—we need at least 67,200 bytes (65.625K) of sample data in each buffer to be sure we won't run out prematurely:

```
67,200 = 192,000 * 350 / 1000 (ms)
```

Rounding up for safety and allowing for the fact that data on a CD-ROM is recorded in blocks of 2K, we need to allocate two buffers of at least 66K to stream this data reliably.

How Streaming Sucks

Jon Holmes, a former colleague now at Microsoft's Rare studio, recently confirmed that they're still using audio streaming heavily and reminded me of a tip I gave him: "*When streaming, it's always better to suck than to blow.*"

> It's better to suck than to blow.

Streaming systems synchronise best when traced from the output back to the input—the fixed-rate output demands (or sucks) a steady stream of samples. Working from the source, such as a granular synthesiser, "blowing" forward to the output might seem more logical, but systems designed that way tend to be slower and more complicated. As often in interactive media, the key to understanding and implementation is finding the most suitable frame of reference.

Interleaving Streams

Streaming is an ideal way to introduce extra layers of music, background ambiences for weather and specific locations, applause and other crowd reactions, without overflowing the memory or working the drive too hard. The dominant problem, especially when running from optical media, is the amount of time wasted between the reading of blocks of sample data while seeking from one block to the next. Whenever possible streams should be "interleaved" on disc so that one seek and one load replenishes the data for several voices, even if they may not all be playing at once.

This interleaving simultaneously decreases the number of seek operations required to service all the voices and increases the size of the block of data to be read, to maintain audio output for a given amount of time till the next buffer update. The stream interleaves data for several voices, boosting the efficiency of loading, expressed as time spent actually loading sample data divided by total time spent both seeking and loading, in which seeking often predominates.

Chapter 7 compares the implementation of interleaved streaming on PCs and four console platforms, using a mixture of simple Mu-law and ADPCM encoding and more modern XMA, WMA, Ogg Vorbis and ATRAC codecs, tailored to suit the hardware of each system. The choice of codec depends upon fashion, licence fees and the capabilities of each platform. With ample processing power I'd use the modern AAC or Opus decoder in place of those, unless "free" multi-channel decoding firmware or hardware was available. Later Microsoft Xboxes, Sony's PSP, phones and some other portable devices have dedicated hardware decoders, locked down for just that purpose—use those if available.

References

[1] *Digital Audio and Compact Disc Technology (2nd Edition); Baert, Theunissen & Vergult, Newnes 1992, ISBN 0750606142*

[2] *DVD Demystified; Jim Taylor; McGraw-Hill 1998, ISBN 0070648417*

[3] *Blu-Ray Specifications*: http://blu-raydisc.com

CHAPTER 7

Streaming Case Studies

This chapter builds upon the concepts introduced so far to illustrate stream optimisation by platform. It draws examples from large and small games, on consoles, PCs, web and mobile devices.

We've already seen interleaving in action in our simple stereo example—if the same music was held on disc as two mono samples, one for the left channel and one for the right, it'd be harder to play. Since the pair must always play synchronously, the only reason to do this would be because the available codec was limited to compressing or unpacking a single channel, as the basic Sony and Nintendo ADPCM software and hardware were.

Even then, if you want to play more than a couple of streams at once, it's worth complicating the offline data-preparation and runtime loading and replay systems to pack channels together on disc and de-interleave them on the fly before submitting mono channels to the hardware decoder.

Colin McRae Rally 5 (hereafter *Rally 5*) played up to nine streams at a time, making almost eight times more audio sample data audible in-game compared with *Rally 4*. It organised the data to make best use of the console drives available. The approach varied between Xbox and PlayStation 2, as the original Xbox had a hard drive, whereas the PS2 version had only DVD to read from. These are vintage platforms now, but their hardware is no longer secret, and most of the techniques used remain relevant to modern consoles, computers and appliances.

One of the key skills of an audio systems programmer is to identify and use suitable resources that generalist team members will ignore. Sony's embedded system approach and need for backward compatibility with the first PlayStation gave the PS2 separate loading, audio and graphics output hardware. Few developers were more than vaguely aware of the two 2 MB areas of dedicated audio and I/O memory, but audio programmers made full use of them.

Getting your lead programmer to give up main memory for audio or DSP time on one of the main processor cores or graphics time for GPGPU sound processing is a much harder sell. Later we'll see the different approach required on Xbox and Windows architectures; the PS2 approach is particularly relevant as an example of how efficiently and predictably—hence reliably—an embedded system can be programmed when you're confident that millions of potential customers will have identical hardware.

The PlayStation optical drive system loads blocks of at least 2K (preferably 32K, from DVD) directly to the 2 MB of I/O memory dedicated to the MIPS R3000A co-processor or IOP. Programs there, invisible to programmers of the main system, forward data from disc, memory cards, game controllers and other peripherals. Sony provide libraries to do the basics, and developers add more to meet the needs of their game.

This is the only chip with direct access to the optical drive, and it runs at about a tenth the clock rate of the PS2's main 128-bit R5900 Emotion Engine but compensates by having excellent DMA capabilities for passing data from disc to any other part of the system and a fully pre-emptive multitasking operating system which supports lightweight threads.

Streaming Threads

A well-partitioned streaming system uses three simple threads in parallel. One just handles reading, one is for de-interleaving and copying data from the load buffer to the required destination and a third "control thread" of highest priority manages those, accepting commands and reporting results. All these threads spend most of their time asleep, but I've learned it's easier and more reliable to implement streaming by separating it into those three parts, with well-defined interfaces, than to try to write one complex polling loop which interleaves code to cope with loading, copying, queuing and cancelling of pending operations, let alone error-recovery and reporting.

> It's easier and more reliable to implement streaming by separating it into three threads.

Rally 5 streaming used two layers of buffering to reconcile the slow drive with the fast transfers possible between I/O memory and scarce, dedicated audio RAM. Two 4K buffers per channel were enough for this last stage, with the loader parcelling out eight 4K chunks from each disc block before asking the drive for more. All the streamed data on the PlayStation 2 version was in Sony's "VAG" ADPCM format.

The speech streamed first through a 64K loading buffer, split into two 32K sections to perfectly match the drive's ECC block size. This could have been smaller—each half buffer holds enough for several seconds of mono samples—but the space was uncontended, so we used it. The buffer duration is calculated as follows:

- 2048 frames of 16-byte VAG data fit in 32768 bytes
- Each ADPCM frame decodes to 28 16-bit samples
- Each buffer decodes to 2048 * 28 = 57,344 samples
- At 22,050 samples per second, it lasts 2.6 s (57,344/22,050)

The 4K blocks of audio RAM were updated eight times as fast, every 320 ms, while the speech was playing. This audio RAM was allocated as a single looping area of 8K, the hardware wrapping from end to start just as if it was an unchanging memory-resident sample. The low dialogue sample rate meant I might get away with 4K, in two

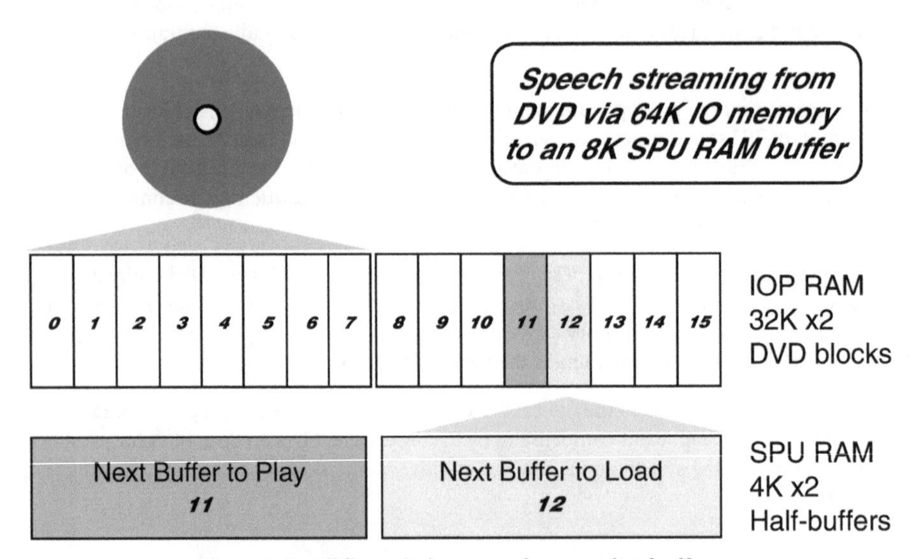

Figure 7.1: Sliding window speech streaming buffers

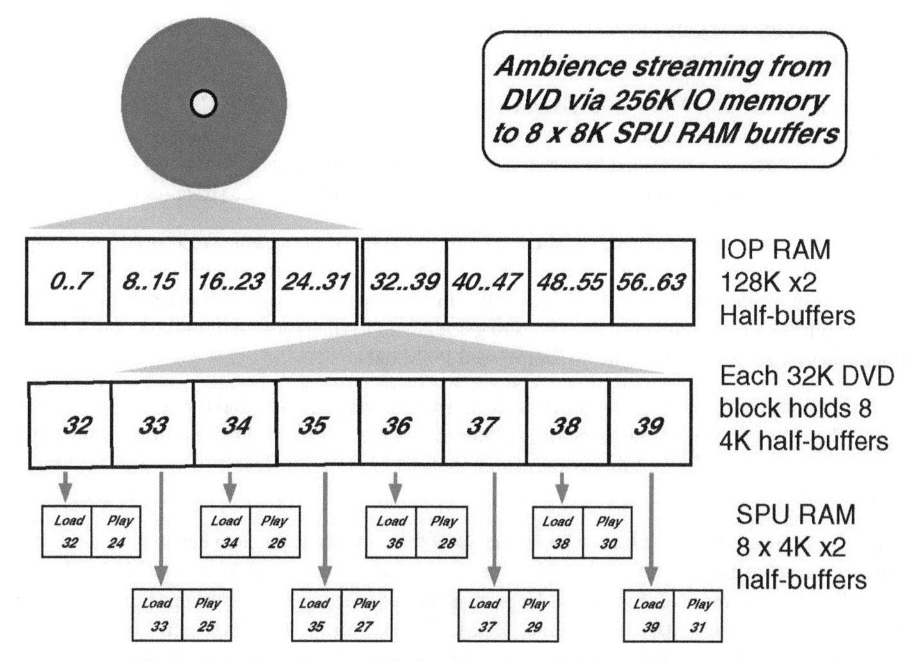

Figure 7.2: Loading and de-interleaving of eight ambient streams

2K half buffers, for the speech, but this was not necessary in practice. I could alternatively have upped the speech sample rate to 44.1 kHz, improving quality at a slight cost only in disc space, but it was already much clearer than lower-rate memory-resident speech on earlier games in the series. The dialogue was already bandwidth limited by the headset simulation used to compress and realistically distort the co-driver's instructions.

The playing thread monitored the hardware to work out when the speech channel crossed a 4K boundary and reloaded that memory slot from the larger load buffer as soon as it'd been played. Every eight blocks it asked the loading thread to replenish the next 32K half buffer (replacing 4K speech blocks 0..7 with 16..23) and carried on chewing through the other 32K (8..15) in the meantime. This re-loading is less time critical because the system starts by pre-loading the first 64K (blocks 0..15) of the speech file, so there's always 32K of data still to play at the point any half buffer needs to be reloaded.

The other eight of the nine streams got packed together, interleaved so that they could all be updated at once. A 256K double buffer was used to load 128K chunks of four DVD ECC blocks, 32K each. Each contained 4K for each of the eight streams, then eight more sets of 4K (bytes 4096..8191 of each ambience) and so on. These 4K chunks feed into each half of the eight 8K replay buffers, so the nine-channel streaming system used a total of 72K of sample memory, in nine pairs of 4K half buffers, plus 256 + 64 = 320K of IOP buffer memory, used for other purposes when the streaming system was not needed, in the front-end menus of the game and while loading graphics and related data for a new stage.

Magic 28,672

Each of those eight streams was a 70-second loop. This duration was chosen after experimenting to see how long a loop needed to be before the repetition would be non-obvious. To make sure that the streams looped smoothly, ambience designer Dan Gardner stuck a Post-it note reading 28,672 on his wall. Each ambient stream was authored to be an exact multiple of this number of samples to exactly fill disc sectors as well as the 28-sample VAG ADPCM frame. Designers are used to making loops a multiple of a certain number of samples, so there's no half-full ADPCM block at the end, which would cause a click to be heard on replay.

The usual multiple is 28 for Sony, 9 for Nintendo and 64 (among others) for XADPCM. For all but the shortest samples, a little resampling is enough to ensure this without audibly affecting the pitch. If the raw data is slightly too long to fill the last block, it's resampled to a very slightly higher rate to squeeze those odd samples into the previous block. If it's not quite long enough, resampling to a very slightly lower rate stretches it to fit.

The difference is so tiny that there's no need to adjust the nominal rate of the sample. You could use modern time-stretch rather than resampling techniques, especially as this is an offline process, so CPU cost is insignificant, but you'll probably find the simpler technique sounds as good if not better. If the sample is exceptionally short it may be necessary to duplicate it consecutively a few times to get it closer to the required boundary and reduce the need for stretching or shrinkage—or just play the uncompressed PCM data.

I gave designer Dan a much bigger magic number, 28,672, because we wanted to make sure the interleaved blocks of disc data, as well as the compressed samples for each channel, were just the right size to match the larger loading buffer and the disc ECC blocks. The number comes from analysis of the prime factors of all the numbers involved: 28,672 is 1024 * 28, or 2^12 * 7.

It helps to sift out the powers of two, as binary computers prefer these by design. Drive manufacturers have long settled on block sizes which are one of these powers. The original micro operating system CP/M used 128-byte blocks, while early Unix systems favoured 512, as large block sizes wasted space on systems built upon many tiny text files. Memory management units, used to remap RAM addresses and swap data to and from magnetic discs, almost always work in chunks of 4096 bytes (4K), and we've seen that 2K, 32K and 64K blocks are normal on optical media.

The fly in the ointment is the factor of seven in the Sony ADPCM output block size. It's balanced to some extent by the 16-byte input packets—XADPCM has the reverse asymmetry, with 64 samples in a 36- (not 32-) byte compressed packet. You may need to take account of the factors for the input as well as the output data. Sometimes this requires compromises—the factors of 44,100, the CD audio sample rate, are 2^2 * 3^2 * 5^2 * 7^2, whereas the modern standard 48,000 Hertz divides up a little more evenly, 3 * 5^3 * 2^7. Ears are conveniently less sensitive to slight variations in sample rate than computers are to incomplete or overflowing data blocks.

In this case everything fits beautifully, and the worst-case demand for bandwidth is identical to the average, if the eight interleaved samples are an exact multiple of 28,672 bytes. The eight-channel file will therefore be a multiple of 28K * 8 bytes long; this value, 229,376, is an exact multiple of the DVD ECC block size; seven 32K blocks hold exactly eight channels worth of blocks of 28K. Another advantage of this approach is that you never need to split a block of compressed data across disc block boundaries. Coping with such splitting is fiddly and error prone, so it's best avoided by design—streaming's already complicated enough, so switching some of the intricacy from the code to the data is a wise policy.

You don't always need to be this precise, especially on faster hardware—you might, for instance, settle for an 80% full final block, with appropriate changes to the buffer size or allowable latency. But if you want to be sure of the worst-case performance of your product or tailor a system to use minimum resources under well-known constraints, this analysis will stand you in good stead.

This factor of 28,672 would not always be a reasonable demand of the sound designer—some samples, especially music loops with a regular beat, can't easily be made an exact multiple of 0.65 (and a bit) seconds long. But in this case the loops were being made from field recordings by overlaying and crossfading sections of applause, weather or similar sound, so it was straight-forward to aim for a very specific duration, as the requirement was clearly understood from the start.

It's interesting and deliberate that the 70-second loop length is substantially more than that of a typical verse/chorus pair in a pop song; presumably those, and the 12-bar blues structure, have evolved to that length to make sure that the chorus is still fresh in the memory of the listener when it recurs. In this case we use a longer period so that repeating patterns of applause, cheers, coughs, air horns and similar rally crowd ambience are not obvious.

This worked well, partly because it's rare for a competing car to stay in earshot of any crowd or local ambience for that long and also because the track design spatially interleaved mono streams from the six available, so the combination of audible streams and their directions varied continuously while still roughly corresponding to the visible count and distribution of spectators.

Several combinations of these eight loops were stored on disc, as separate "stream bank" files. The combination chosen depended upon the weather, location and stage of the rally, local ambiences for forest, town, industrial, water-side and similar settings, and the player's ranking. One pair provided a continuous background bed of weather effects, playing much the same role as music in less realistic games.

Analogue Surround

This pair was not strictly stereo, though it played as such on two-channel outputs. It was actually encoded as matrixed analogue surround using an offline Pro Logic plug-in. This takes a mix with discrete front and rear channels and stores the front channels directly as stereo then mixes into them copies of the rear channels shifted out of phase so they can be approximately extracted by comparing the two channels and pulling out the anti-phase component.

This is an old technique, not exclusive to Dolby, aimed at extending stereo to provide greater envelopment without complicating the media player hardware. It was commonly used on the first two PlayStation and Xbox models—in fact, it was impossible to turn it off on the Xbox 360, forcing me to flip the coordinates of rearward sources to the front for stereo listeners only. Without this fix, vital sounds behind the player were almost inaudible on a 2.1 speaker system, as the anti-phase rear signals in the shared large speaker cancelled out.

The rise of discrete surround, with six or eight separate channels, has made analogue surround obsolescent. Only Nintendo currently offers a matrixed analogue surround output option, on the Wii-U console. It's still supported by almost all AV amplifiers for "stereo-compatible surround" film soundtracks. Proprietary encoders use +90° and −90° phase shifts to reduce artefacts for stereo listeners, but basic Dolby-compatible encoding can be achieved by simply flipping the sign of input waves intended for the rear speakers.

All you need to do to position a sound directly behind the listener, for a Pro Logic decoder or a compatible system like the wire-only Hafler surround arrangement, is to add the samples you wish to play from the rear equally to both left and right channels but with a 180° phase shift on one side or other. This is achieved by inverting the polarity (negating the sample values) on one side or the other. For an in-vehicle listener this is a trivially simple but effective way to separate the rear exhaust sound from that of the engine at the front. This will cause phase cancellation of low frequencies for listeners with stereo speakers, so it should only be done when they've indicated they have a two-channel analogue surround decoder, through the product's setup menu.

Rally 5 was also designed to pick bigger and louder crowds for a player high on the leaderboard, giving a sense of feedback and progression between stages. The sound designer authored around a dozen streams corresponding to different intensities of crowd. One or two of each was positioned in the game world wherever a crowd was visible to the player, so that it moved in relation to the listener. The multichannel weather effects were stationary.

Each stream bank would contain four or six of those crowds. The stadium-set "super special" stages used more crowd layers, as two were otherwise reserved for local ambiences on the longer tracks. Thus several stages with similar weather and a mixture of urban and country locations might share a stream bank, while a stage set entirely out of town would, for example, swap the urban loop for one more appropriate to its setting.

The stages of the game were marked up with crowd and ambience zones, areas in which these streams could be heard once the bank combinations were worked out, and simple scripts controlled which mono or stereo streams were built into each one. A combination of a data file and information about the player's progress allowed the right bank for each stage and level of crowd arousal to be selected at runtime.

Each bank was different, but many samples appeared in several banks. The data duplication was not a problem, as there was plenty of space on DVD; each bank occupied about 30 MB on disc but only 64K of audio sample memory. The benefit in loading just what was needed for every contingency more than outweighed the cost in data preparation and duplication and imposed no extra work upon the sound designer.

Coarse Interleaving

The internal file structure was chosen to match the loading and replay buffers, with each mono channel represented in the file by a 4K byte block of 28 * 4096 / 16 = 7168 samples. So the file was coarsely interleaved: rather than interleave samples for each channel, as might happen in a stereo WAV file with left and right channel samples immediately alternating, the data was collected into groups of samples for one channel, then a same-sized group for the next channel, and so on.

This did not mean that the individual ADPCM streams needed to be encoded in many short sections. In fact, each channel was encoded separately (including the two-channel pair) into a set of eight mono files, and the combined stream bank was created by appending 4K sections from each mono file in turn, creating a succession of eight-channel interleaved 32K blocks.

The relevance of this technique is obvious when the eventual hardware decoder works with mono packets for each channel. It's essential to keep the 16-byte VAG packets for each channel distinct, as they only make sense when decoded as a unit.

But the same principle holds for many other parts of an audio runtime system, even when working entirely in memory and with uncompressed data. Consider how you'd mix or filter an interleaved stereo sample like that cited in our first streaming example. Alternate left and right sample values are held in pairs, so processing them independently involves two passes over the data, each one skipping half of it. This wastes bandwidth and requires extra instructions or code variants for each multi-channel setup.

Three terms are useful when discussing and implementing this sort of interleaving: group length, frame and stride. Group length is the number of contiguous samples for a given channel which are packed together. For the VAG streams discussed above, this would be 7168; for the stereo WAV, it'd be 1. Note that this value is in samples, not bytes. Each set of samples, one per channel, is called a frame, so a 512-byte buffer of 16-bit stereo contains 128 frames, however they are ordered.

The stride is the interval in bytes between the end of data for one channel and the start of the next data for that channel later in the stream. This is most useful when writing code to pull channel-specific data from the interleaved stream. The stride is one less than the channel count, multiplied by the group length, multiplied by the number of bytes per sample. Thus the stride for tightly interleaved 16-bit stereo is 2. For blocks of 256 left then 256 right samples it becomes 512. Today's Web Audio API uses this "planar" arrangement to speed-up processing of individual channels.[1]

Memory Map

Table 7.1 overleaf contrasts the amount of memory needed to service the streams, compared with the banks of memory-resident samples, in *Rally 5*. Comparisons with newer games follow. The 2 MB of dedicated memory directly accessible to the audio co-processor is divided into three chunks of roughly equal size—for per-car sounds, per-stage surfaces and everything else. Streaming uses only 72K even though it adds more than 30 MB of samples to each stage. The game uses about 600K for car sounds and a little more for surface sounds; how much of each is used depends on the car model, country, stage, game mode (e.g. single or multi-player) and the weather.

Table 7.1: PS2 audio memory allocation in *Colin McRae Rally 5*

Bank contents	Bank size
Vehicle engine	400
Vehicle mechanism	100
Collision sounds	400
Damage effects	100
Kick-up	120
Surfaces	660
Weather effects	90
Other resident samples	42
Hardware reverb	64
Interleaved stream buffers	64
Speech stream buffers	8

Sizes in Table 7.1 are in kilobytes and would be three times higher without ADPCM compression. Modern games might allocate ten or a hundred times more space for similar sub-systems, but that's mainly to allow more variety and higher sample rates. The ratios reflect the importance of each set of sounds to the player, which is relatively constant.

"Other resident samples" includes the beeps used to confirm selections in the pause menu, notifications played when the player passes a checkpoint and similar non-diegetic sounds. The beeps used in the front-end game menus, when cars and rallies are selected, are the same sounds as those used in game, but at a higher sample rate. Most of the memory at this point is used for front-end music, which consists of interlocking loops or grooves which start and stop around a basic framework to vary the music depending on the menu page in view.

Alastair MacGregor's 2014 talk about the sound of *Grand Theft Auto 4* and *5* includes charts showing the subdivision of memory in those games on newer consoles.[2] While the proportions stay consistent—in fact, a switch to granular synthesis for player vehicles slightly reduced the car engine audio slice—the number of assets rises inexorably.

In 2008, *GTA4* included samples and metadata for 25,000 "audio objects" plus 50 hours of streamed audio for music and cut-scene animations and 80,000 lines of dialogue. 2013's *GTA5* tripled the number of objects, almost doubled the number of speech lines to 158,000 and added another 40 hours of audio streams. That still leaves about three-quarters of Blu-ray disc capacity free for even bigger games in future.

Channel Synchronisation

Every so often you may notice a phasing effect thins the sound. This is a bug caused by the left and right channels being one mix update out of sync. It's a common problem on PS2 games and other keyboard-derived mixing and decompression systems which only work in mono at the lowest level; the hardware has no concept of a "stereo channel," so each side of the stereo pair must be triggered independently. It's possible that one starts just before an output buffer swap, leaving the other delayed by 5.33 ms, the duration of a 256-sample output block at 48 kHz. Similar issues occur on newer systems if you try to use two stereo pairs for front and rear samples. It's helpful if the mixer sets and clears a flag during the critical region while it's mixing a fresh block to help systems upstream synchronise.

alSourcePlayv() works that way. Another OpenAL extension, not always implemented, allows mix updates to be suspended and resumed. Calling alcProcessContext() once, just after updating all voices, allows OpenAL to catch up on all the updates that frame in one go, which is more efficient and predictable than doing it piecemeal. Between times, call alcSuspendContext(). These optimisation hooks are worth emulating even if you write your own mixer.

Custom Soundbanks

Engine and mechanical bank contents depend upon the player's choice of car, and each combination is tested—and tweaked or specialised—to fit within the budget of 500K for both sets. This is part of the general process of shoehorning as much variety as possible into a limited space.

A similar approach applied to specific countries, so certain features were trimmed to make room for more types of road surface in stages set in Japan. Sound designers and programmers work together to identify these groups and trade-offs as the game design is completed, concentrating on the lead platform (normally the one with tightest constraints and highest sales potential—the two often coincide for obvious commercial reasons) but not forgetting the others.

64K of the PlayStation's audio-specific memory gets reserved for the hardware reverb, used to tweak the mix realistically near bridges and tunnels. We could have used a second hardware reverb for interior sounds—and did on later games—but had better uses for that 64K in *Rally 5*.

As well as the streams, the bank of resident kick-up sounds could also be updated from disc on the fly. This was a feature from earlier games in the series. It happened when the player switched from the in-car to external view; the alternate banks had the same number of samples, but they were differently treated acoustically to sound right for each camera position, more muffled for the interior view. There are actually four of these banks—rallies set in Sweden get a dedicated pair, as befits those snowy stages.

The stream buffer sizes were chosen to allow this transition without interrupting the streams so that in practice, up to three load requests could be outstanding at any time. The kick-up bank swap was the lowest priority, since it's more important to keep the speech flowing, then the ambiences, than to accept a fraction of a second's delay, worst case, in the change of the kick-up tone. This is one case where cinema conventions do us a favour—vision and sound are so rarely cut in sync that it's not obvious if occasionally we do the same, especially when user action triggers the transition—the camera swap, in this case.

Few modern games would try to cram all their audio into 2 MB, even if the decompression hardware made this the equivalent of 7 MB of 16-bit samples, but the ratios are still valid for games of this type. There's just a lot more variety in modern titles; a generation later we aimed for about an hour of memory-resident samples, in between 20 and 30 MB of RAM, but still relied on dynamic loading for interactive speech and music layers which could otherwise easily overflow that space and delay the start of the game.

Microsoft Variations

The previous explanation deals with the PS2 implementation, which was both the lead platform and the most technically demanding—as is often the case. The constraints were lack of memory, the need to take advantage of powerful platform-specific hardware (dedicated audio RAM, dedicated I/O RAM, hardware mixing, decompression and reverb) and the requirement to work well without a hard drive.

The Xbox version took advantage of twice as much main memory and a built-in hard drive so that the co-driver speech, in XADPCM-compressed banks of around 2 MB per stage, could be loaded and played directly from memory, while the ambient streams were installed to hard drive. Those were supplied as a mixture of two-channel weather and mono crowd and ambient streams, compressed with Microsoft's preferred WMA codec, which was not capable of handling an eight-channel interleave in any case. Small pre-allocated buffers allowed up to seven separate streams to load and play at once; the fast seek time of the hard drive made this feasible. The PC version was ported out of house, using the same basic approach and DirectSound3D interface as the Xbox version but with even more memory.

WMA was not an ideal choice of codec for this purpose, but it was well supported and came free with the devkit. The main snag, from an interactive point of view, was that it was designed to play a single mono or stereo stream without glitches on a slow and busy PC. We found that the decoder ran infrequently, only about seven times per second, delivering thousands of samples each time. It also used the main processor, contending with the rest of the game code, rather than the Xbox's embedded Motorola 56300 DSP, which was mainly used to implement 64 voices of Sensaura 3D audio—though my colleague Jon Mitchell also used that for platform-specific engine distortion.

Investigation in PIX, Microsoft's excellent real-time monitor, showed the WMA decoder demanding spikes of CPU time several times per second for each playing stream. This meant that the amount of main processing time needed for audio fluctuated, depending upon how many streams were playing and when they were started. In most game update frames they'd play from the decoded buffer, but periodically one or more would run out of decoded samples and need to unpack a new block of WMA. Extra buffer memory reduces the demand for disc access but doesn't eliminate the risk that all the streams might get hungry at once and cause the game update to fall behind the graphics.

The way the ambiences were distributed across the map reduced this risk, and it was further addressed by limiting the number of streams that could be triggered to start in a single game update frame—a fraction of a second delay in the onset of an additional crowd roar was not noticeable to the player but reduced the risk that the spikes lined up. It helped that they all ran at the same sample rate, without Doppler.

Another subtle difference came from the musical heritage of WMA—it was not capable of encoding a looped sample that played continuously, introducing a short silence when the sample wrapped back from the end to the start. To hide this and reduce disc activity, the samples did not play continuously, as on PlayStation, but only when they needed to be heard. This was a trade-off—the gap was audible only if the player stayed in earshot of a stream for more than a minute, but the first part of each sample was heard far more often than the final few seconds, reducing the variation. On balance, we got away with it!

Pre-Roll Stubs

Several games have used a mixture of memory-resident and streamed audio to deliver an instant, pre-crafted response to player activity more intricate than they could mix on the fly. The *Club Football* series reproduced the sounds of soccer matches, with tens of thousands of lines of detailed commentary and hefty crowd responses, to simulate a live TV broadcast.

To avoid a tell-tale lag after a goal or a near miss and deliver an instant yet appropriate roar from the crowd, scores of stadium audience reactions were recorded. A suitable subset of these, matching the teams and their status, was selected at the start of each match, but only the first second or so of each was pre-loaded into audio memory. This arrangement meant the system could start playing a crowd stream instantly, from the "stub" in memory, then seamlessly stream the remainder of that reaction sample, using the stub to cover the initial stream loading time.

This requires an audio system which can switch a playing voice between a sample held in memory and one held on disc without a glitch. Thus, in Chapter 10, the logical layer of our audio system uses the same type of voice handle, and voice, for both streams and resident samples. Once a voice is playing, the game code should not need to know where the data is coming from to update its position, volume or other properties. Only when new sample data is submitted, through a Play, Chain or Steal Voice call, does the sample location need to be stated.

This also helps with platform tuning. It might be that one console platform is using stubs and DVD streams while another streams directly from hard disc and a third keeps all the samples in RAM. This was the case for the platform-transition game *Brian Lara Cricket 2007*, which came out simultaneously on PS2 and the new Xbox 360, which had more than ten times as much RAM. That game used two-channel analogue Pro Logic–encoded streams on PS2 and discrete four-channel XMA-compressed surround-sound samples entirely in RAM on the generously endowed Xbox 360. Either way, once the sample was triggered generic cross-platform game code could update it.

Such techniques need not be confined to music, crowds and ambience. Adam Sawkins, my game-audio counterpart on Codemasters' *DiRT* and *RaceDriver GRID* games, explained the streaming system of his previous Criterion Games title, *Burnout 3*, in a January 2017 Audio Engineering Society UK talk.[3]

After allowing audio memory for music and "crash" stream buffers, the designers of *Burnout 3* found they had only 74 seconds of instant-access audio RAM available for everything else. Most of the samples were encoded at low rates, like 12 or 16 kHz, with an exception for glass smashes, which went in at 22.05 kHz to preserve some of their sparkle.

Streamed music, at 44.1 kHz, padded out the top octave most of the time, but during *Burnout*'s spectacular set-piece crashes the music was instantly swapped for one of 20 intricate pre-authored streams, ten fast and ten slow. For an instant response, the start of each stream was preloaded, like the *Club Football* roars, and the remainder of the 30-second crash mix was streamed, using the initial stub in memory to cover the stream initialisation time.

To avoid those crash streams becoming monotonous over the average ten hours of gameplay Criterion expected, there were far more crash streams on disc. The game design included a three-second "respawn" delay after a major crash, and *Burnout* used that time to select and load a fresh batch of crash stream stubs. It also mixed shorter samples over the backing stream. About half the sample memory was used to load a single 800K (compressed) bank of about 200 individual mono one-shot and loop crash sounds, which played as context-dependent sweeteners overlaying the stereo crash stream.

Interactive Music Streams

Moving on a generation and introducing more advanced codecs, another example streaming system implemented the eight-channel music used in *RaceDriver GRID*.[4] In a newspaper interview I explained the difference to Ben Firshman:

> *We don't just take the pre-recorded stereo music track. If you play a game like RaceDriver GRID, we get it in the form of what's called "stems," which is to say individual components. So, there will be a basic rhythm track, then you'll have various levels that can be mixed in on top of that, other instruments and so forth.*
>
> *In GRID all the music was broken down into eight separate layers, which we could play separately in surround as the game went on. For instance, when you overtook another car an extra level would come in regardless of where you were in the tune at that point, so as to celebrate that.*

Other musical transitions took place after a heavy collision, when the mix was trimmed down to just bass and rhythm, or when the player was in the lead and in the final lap. The mix varied dynamically in response to the player's success and failure, much like a pre-scripted film score but without forcing the viewer down a single linear path.

This system loops eight channels of 48-kHz music, with custom codecs for each platform, described in Chapter 17. On PC, with plenty of RAM and no guarantee of any audio DSP to unload the CPU, I used 2:1 Mu-law compression. On PlayStation 3 I used the Sony-supplied ATRAC3 codec, in its high-quality 132 kbps mode, which packs blocks of 1024 samples into 192 bytes. Since the music comes in eight layers, each block submitted for decoding comprises 8 * 192 = 1536 bytes.

As for *Rally 5*, I wanted to keep things simple and quick by aligning the data on disc to its natural boundary—64K, twice the DVD value, for the Blu-ray disc error-correcting code. The factor of three in the underlying block size

(192 = 3 * 2^6) pushed the half buffer size up to 192K, so the double-buffer size was 384K—rather more than *Rally* needed on PS2, though the sample rate was slightly higher and the quality, from the psychoacoustic codec, quite audibly improved.

The PS3 version of GRID squeezed about an hour of resident samples into 20 MB, so this allocation was significant but worthwhile to ensure robust streaming. With 1024 samples per block and 1024 blocks per channel in each eight-channel half buffer, the time between buffer updates from disc was over 5 seconds:

```
1024 * 1024 / 48000 * 8 = 5.46 seconds
```

This is not a particularly convenient size for authoring, but if your stems are specially made for the game and the composer is co-operative, it may be practical to build them to a multiple of that duration. Given the long latency between buffer swaps, in practice it's fine to leave the last block of the looped music only partly full—even if it's only 20% occupied, that still allows more than a second to wind back to the start.

The extent to which the final disc block needs to be fully occupied depends upon what else the drive is doing and how long it might take to finish that and get back to audio streaming. An analysis and general workaround for this degenerate-final-block problem is explained later, when we explore the implementation of *GRID*'s music on Microsoft's console. But first we should consider what else might be keeping the drive busy.

Sharing the Spindle

Streaming is not just for sound—it's also one of the techniques which gave *RaceDriver GRID* its exceptionally detailed graphics. To understand this properly we need to step away from audio, briefly, into the 2D world of textures and the 3D domain of vector graphics.

Each doubling in spatial resolution increases the size of associated graphics textures by a factor of four. To avoid considering many more pixels than anyone could see when rendering a far-away object, each texture is provided in several different resolutions, and the biggest, highest resolution source textures are only used for large objects close to the player. This prevents the blockiness apparent in early textured 3D games like iD Software's *Doom* and *Hexen* but at a great cost in memory and loading time, which increases with graphics resolution.

GRID's innovation was to keep all but the most detailed textures in memory, like rival games, only loading the highest resolution on demand as the player drove round the conveniently linear racetrack, for just the large objects visible at that point. If this works, the player gets graphics as detailed as they'd be on a console with substantially more RAM and without the long delay associated with loading all those needed for a long race circuit.

By scaling the tracks and cars around this expectation, we arranged for all this data shuffling to be almost transparent to the player. Under rare circumstances, such as right at the start of a race, when the game's still loading resident assets, you might not see the highest-resolution textures, but only for a moment. Next time you play *GRID* you may notice a subtle momentary flicker as the resolution is increased within a second of the scene unfolding, but this is only apparent at the start. The system fails safe—even if part of the data is unreadable due to a worn or dirty disc, gameplay is not interrupted—everything in the scene is still drawn, albeit at the slightly lower resolution of memory-resident games rather than texture-streaming ones.

It was critically important to the game that this system worked well, so we had a short run of test DVDs manufactured, using the full glass-master production process to get the same performance as the final product. Most testing is done with one-off locally recorded DVD-R media, but those are not identical—the capacity is lower, read errors are more likely, and the distribution of data varies—which is why we made a point of commissioning test pressings. These enabled us to identify the actual performance we could expect, including variability introduced by seeking and file placement, and set our audio buffer sizes conservatively to suit.

Variable Bit Rate

The Xbox 360 version of *GRID* stepped up from WMA to a new codec better tuned for interactive uses. Microsoft rolled out the new XMA codec for Xbox 360, with fewer channel-count restrictions and looping. The current version, XMA2, supports up to 64 interleaved channels, with seamless looping across the set, and is decoded by custom hardware on both Xbox 360 and Xbox One.

This dedicated audio hardware is not just a convenience—it's crucial to the performance of the Xboxes. In 2017 Microsoft Xbox core platform group program manager Kevin Gammill told IGN that software XMA decoding would have consumed two or three of the six Xbox One cores. The XMA hardware potentially doubles the system's performance even on non-audio tasks.[5]

XMA encodes samples into 1K or 2K chunks; these power-of-two sizes are easily matched to the 32K DVD block used on Xbox 360, which can conveniently hold two 2K sets of eight channels. But unlike ATRAC3 and ADPCM, which encode a constant number of samples into a fixed amount of memory, the number of input and output samples in each chunk depends upon the input sample value and, in particular, their frequency distribution—the more complicated the spectrum, the lower the compression ratio. This "variable bit rate" (VBR) approach saves disc space but is inconvenient when working out the compressed data rate required to sustain a given rate of output, yet cautious experimentation and measurements, adjusting the buffer sizes for real data, will yield predictable results. Err on the side of caution.

Another challenge associated with VBR streaming concerns the last block of a stream which is expected to play as a loop, like the *GRID* music layers. The last block must be full enough of data to allow time for the first block to be reloaded before it runs out, despite any other concurrent loading, for instance to refresh textures, for speech or for a sample-bank swap.

The solution is to perform speculative encoding, varying the parameters passed to the encoder till the total length of the output indicates that the last block is nearly full. As for ATRAC3, there's no need to be precise, providing you come close enough not to compromise your margin to wind back to the start.

Most VBR codecs support a "quality" parameter. In XMA this is a value from 0 (highest compression, lowest quality) to 100 (top quality, least compressed). Sound designers set a minimum value for this for the data they want to encode and the target sample rate. Small increases to this will change the amount of data in the last block without giving the designer or your customers bleeding ears. Depending upon the codec and the data, it takes a few tries to find a setting which packs the last block of the file sufficiently fully to ensure reliable operation. This speculation is easily automated, as the encoding parameters and output file size are readily available to the script or utility used to bake samples for use in the game.

If quality tweaks don't do the trick—and sometimes changes to the "magic" quality number make no difference or take you on to the next barely used block boundary—there's another way to finely adjust the encoded size of a sample stream, though it does involve a slight increase in runtime complexity. This is to resample the input to a slightly lower rate before encoding it, reducing the size of the input and output data with corresponding, though not exactly proportional, changes to the size of the encoded file. Again this is easy to automate, though it does slow the asset-conversion process. This can be minimised by caching the encoded data once suitable parameters are found and only re-running the search when the input sample length or contents have changed.

The runtime complication is that the replay system can't assume the same replay rate for all streams used in a given context. It needs a table of sample rates, one for each encoded stream, so that it can still be played back at the correct pitch. The reduction in bandwidth and cost of runtime resampling back to the output rate—generally 48 kHz—will slightly compromise quality but not enough to be obvious, as only small sample rate changes, of the order of a few hundred Hertz, are needed; the longer the sample the less the rate will need to be altered.

Notice that we always reduce the sample rate to get a good fit for the buffers—increasing it would run the risk of introducing aliasing artefacts into the sound, though the "guard band" associated with the original capture or subsequent resampling means you'd probably get away with a slight increase. Whichever way you bend the sample rate, make sure that you replay the stream at the exact rate to which it was resampled by the offline block-fitting exercise. Your customers won't notice a slight difference in bandwidth or sound quality, but they're entitled to object if the music ends up playing out of tune.

Bending the Rules

Quite a few codecs, including Sony's ATRAC3, do not officially support arbitrary sample rates. The encoder contains tables which set out the sensitivity of the ear to specific frequencies, and these tables are optimised for certain rates: usually 44.1 and 48 kHz, sometimes simple fractions of those, like 32,000 or 11,025. Millions of ears have proven that you can still use those codecs effectively for other rates, which is just as well, as Doppler effects would otherwise cause problems for moving sources.

You must bypass the encoder's check for a limited set of "valid" sample rates, or the input wave is likely to be rejected for not having one of the exact rate values expected. The offline sample packer just picks the closest permissible rate to the one it prefers, passes the samples through the encoder as if they were at that rate by patching the input wave header and then re-instates the exact rate required in the data which will be used to play the sample later. Even a discrepancy of an octave, such as supplying a 24,000-Hertz sample to an encoder tuned for 48-kHz output, sounds fine in practice; there's no way that I'd have been able to pack hundreds of samples, with total duration about an hour, into the limited audio allocation of *GRID*'s main memory otherwise.

Average of What?

You may also come across "constrained VBR" or "average bit rate" (ABR) codecs, which appear like they'd be the best of both worlds—dynamic allocation of bits to save space when possible yet with predictable worst-case input and output block sizes. These are worth investigating if your encoders and decoders support them and give you fine enough control over the number of input samples and encoded bytes. It's not much good to us if the "average" is computed over the entire length of the sample.

Some implementations of AAC, MP3, Ogg Vorbis, Opus and WMA support variants of ABR. At best, this can enable you to treat them like constant-bit rate (CBR) for buffer sizing and worst-case latency calculations. At worst, you'll need to use the same techniques outlined above for XMA and other pure VBR codecs.

Constant bit rate is well suited to streaming, but higher compression ratios are possible—for any given audio quality—if the codec adjusts to the complexity of the input wave, packing silence and simple sounds into less space than would be required—and in these cases, pre-allocated—for the most complex mix. The usual alternative is a VBR codec, such as Microsoft's XMA.

Deeper Buffering

In theory, any number of buffers greater than one could be used, but in practice, going beyond two adds latency, undesirable in interactive systems, and generally more complication than it's worth.

Deep buffering, with three or more blocks of data enqueued en route to the listener, is typically only used now in non–real-time systems like Microsoft Windows, where the destination output buffer size is pre-set and small and dynamic adjustments may be needed to cope with uneven or hard-to-predict variations in output latency. Blue Ripple's *Rapture3D* driver for Windows does this cleverly, monitoring how close to a buffer-under-run it gets and adding buffers—and, inevitably, delays—when things run close to the wire or removing them if there's plenty of slack.

The lower-level access, better documentation and greater control available on game consoles, set-top boxes and embedded systems means that the performance of double-buffering systems is relatively predictable there, so it can be tuned before delivery. It's still wise to leave it easy to tweak in case QA testing or late changes to other systems reveal a problem.

Mobile Music

Mobile phone games have tighter constraints—they may have far more RAM than early consoles, but they lack most of the dedicated audio hardware, and low-power ARM processors are ill-suited to make this up in software. What they and most tablets built on similar chipsets do have is a hardware music player, intended to decode and play heavily compressed stereo music with minimal overhead. If you're making a mobile game it's sensible to use this for backing music—if the user does not already have their own playing—or background ambiences like weather or audience sound.

The snag is that this hardware is designed to support a single track at a time. Even crossfading between two songs to avoid a gap between them can cause a big processing spike at the point when one song's fading down and the other is starting. Tolerate the gap, edit the fade or consider using a short sample—a radio station ident works well—to glue the two songs together without overlap.

Users may prefer their own music to yours—indeed, the Xbox XMP feature entitles them to replace any game music, however precious it may be to the composer. PlayStation developers are encouraged, though not required, to follow suit. It's not a good idea to play two unrelated tunes at once—their tempo and pitch will clash—and expensive too. Apple's AVAudioSession "secondaryAudioShouldBeSilencedHint" identifies when an iOS music player is already busy, so you can avoid both the clash and the processing spike of playing two beds at once.

References

[1] *Web Audio*: https://developer.mozilla.org/en-US/docs/Web/API/Web_Audio_API/Basic_concepts_behind_Web_Audio_API
[2] *The Sound of Grand Theft Auto; Alistair MacDonald, Game Developers Conference, San Francisco 2014*: www.gdcvault.com/play/1020587/The-Sound-of-Grand-Theft
[3] *Criterion Game Audio, AES Talk*: www.youtube.com/watch?v=e9FIvG8kQ78
[4] *GRID Layered Music*: https://theboar.org/2010/03/interview-simon-goodwin-codemasters
[5] *Xbox Compatibility*: www.ign.com/articles/2017/10/23/the-untold-story-of-xbox-one-backwards-compatibility

CHAPTER 8

The Architecture of Audio Runtimes

This section explains how the components of an interactive audio system fit together and what additional sub-systems can be implemented cheaply to make the foundation more useful in practice.

The recommended design splits the implementation into two layers—one which is generic to all platforms, written in a portable subset of C++, providing a rich and consistent interface for the game or application audio programmer to work with, and an inner "physical" layer which tailors the output of the platform-independent "logical" layer to the most suitable hardware or software interface provided by the platform vendor.

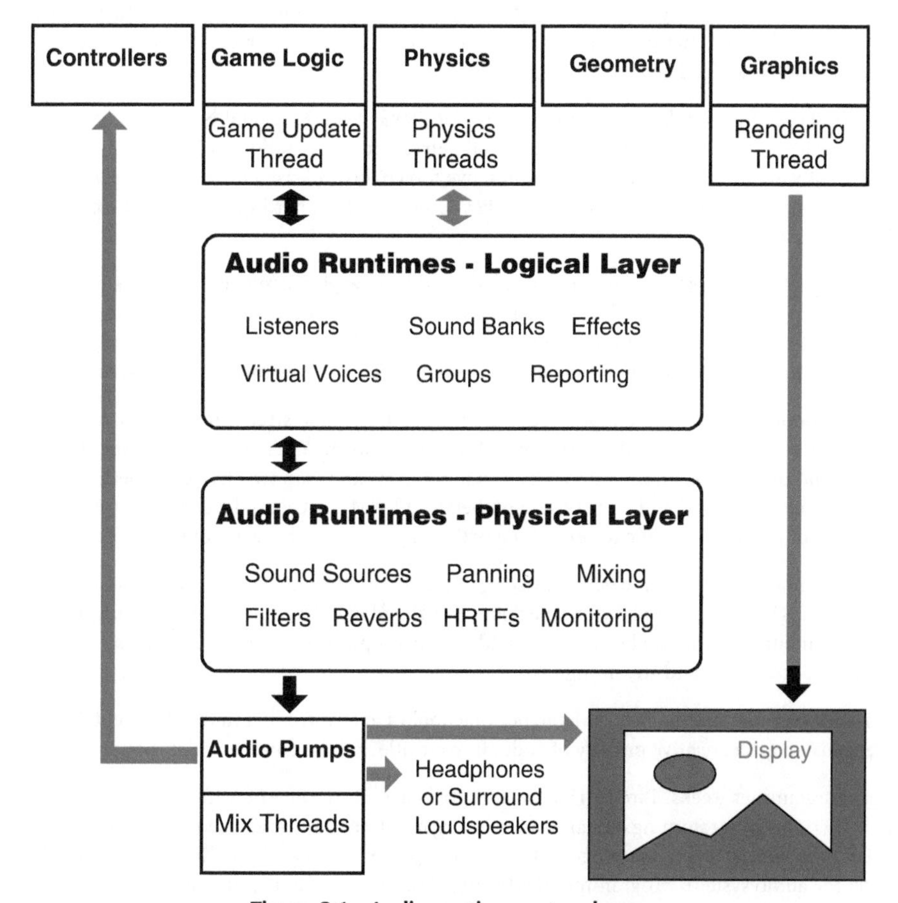

Figure 8.1: Audio runtime system layers

There are eight reasons why this layering is valuable:

- A consistent interface can be maintained even if the internals are radically refactored
- A large-subset implementation can be brought up quickly on a new platform
- The hidden interface between the layers provides a simplified model for maintenance
- Unit tests can focus on either layer, saving regression test time and improving diagnostics
- Platform-specific hardware-accelerated features can be supported transparently
- Multiple audio drivers can be supported on one platform (e.g. XAudio or OpenAL)
- The bulk of the implementation code can be verified and shared between all platforms
- Platform-specific optimisations can be added without cluttering up generic code

It means that new platforms can be added relatively quickly, as the logical layer is common to every platform and will already have been well tested. There may still need to be tweaks, particularly when moving between commercial and non-proprietary compilers.

The differences between Visual Studio and GCC's interpretations of the C++ standards were once wide and problematic, but these have narrowed since Visual Studio 2015 and the switch from GCC to Clang/LLVM on the open-source side. While the compiler options vary in scope and effect, you're likely to find that Clang reports warnings and errors which Visual Studio ignores, with both in their default configuration. It's a good idea to fix these issues, even if you think they're pettifogging—this will make future ports easier and will almost certainly eliminate some problems which might otherwise affect you and your customers. If you must turn off warnings, do it locally in specific files or functions rather than globally so you don't miss errors they might otherwise catch elsewhere or when code is added later.

Over decades I've successfully implemented this layering for PlayStation, Xbox and Nintendo consoles, Android and iOS mobile devices, OpenAL soundcards, Microsoft's DirectSound, XAudio and XAudio2 hardware abstractions and software mixers in four game engines, five toolchains, at least a dozen sets of platform audio libraries and processor architectures: 32- and 64-bit, big and little-endian, hardware DSPs and software-directed graph implementations.

This has enabled me to reliably estimate the time needed to add a new platform to a mature system already supporting several similar ones. I quote three ideal programmer weeks for an initial working implementation capable of loading, decoding, playing, mixing and positioning a substantial number of voices under game control, with position, pitch and volume controls, Doppler and distance effects. Allow four if you're "scrum agile.'"

Multichannel and looping assets will be supported, but reverb, filtering and streaming will probably come later and then in proportion to the needs of the first title on the new platform. You're unlikely to need these for a "proof of concept" port, and they can be mocked up with alternative assets and gain tweaks in the meantime. Likewise, additional voices or codecs can be added in a trade-off between RAM, CPU and development time. These are classic dimensions for platform-specific tuning, along with sample rates and the use of multi-channel assets.

This initial implementation is enough to get an existing build of the title playing audio well enough for the sound designers to start thinking about the platform-specific extras they'd like. Depending on the platform and needs of the title, a similar amount of time will be needed to add platform-optimised DSP effects such as voice and group filters, advanced codecs, reverb and streaming.

Optimisation work on those may be needed later, but that applies to any product in development. The amount of such work depends upon the relative priority of audio in your title and the complexity of the sound design.

By "three ideal programmer weeks" I mean 15 uninterrupted days. Only you know how often these fit around meetings and other non-programming demands in your organisation, but the elapsed time will almost certainly be longer unless you work in an engineering-led firm or (I hope not) a sweat shop. I've also assumed that the clock starts only after the audio systems programmer has become acquainted with the documentation, file and audio code libraries for the platform, building and tweaking some example code that comes with it. This should have been done as part of the initial consideration of the platform.

How long it takes depends not just on the implementer's experience but also on the newness of the platform and the amount of audio tech support available. You can save a day or two if there's enough codec compatibility for the first version to use soundbanks baked for an earlier platform.

If the programmer is not senior or this is the first or second implementation of the physical layer, allow more time—potentially twice as much. There's little point in delivering anything intermediate to the "play and mix, DSP to follow" alpha release; make sure that subset is tested and feature complete, then let people audition and play with that while considering what DSP and streaming extensions the title and platform really needs.

Where's the Glue?

The following chapters include public and private member variables and implementation code for the Virtual Voice, Listener and Group classes and the properties of the playing source objects in the inner physical layer. Member functions are described but only fleshed out in C++ if they include non-trivial audio processing. In particular, the methods used to get and set public member variables are not spelled out, partly to save space for real audio code but mostly because there are many ways to encapsulate and expose those, everyone has their favourite, and when you're augmenting an existing framework, like a game or game engine, it's foolish to add another. Use whatever you've got already—that way you'll benefit from the remote update, logging, mapping, disabling and similar extensions of mature exposure systems, and their usage will be obvious to other team members.

Besides the obvious ways to update and query public members, there should be some way to interpose code when certain values—like m_Rolloff, m_MinDistance, m_MaxDistance—are tweaked, so that private dependencies like m_RolloffFactor keep step. Some properties are read-only, like Voice::m_State; write attempts should return EACCES.

The linkage between classes is similarly simplified. Each Virtual Voice contains a member m_pGroup pointing to its Group, and 3D voice groups find their listener similarly, via m_pListener. Wrapper methods like GetGroup() and GetListener() would provide proper encapsulation but are not essential to these illustrations.

> "Even experts make errors. So we must design our machines on the assumption that people will make errors."
>
> —Don Norman[1]

When error-codes should be returned—which is any time something could go wrong, including Set methods with limited range (ERANGE) or domain (EDOM)—ERRNO refers to a datatype (usually int) to hold standard error codes from the list in errno.h. Don't sweat when trying to match each circumstance to an existing ERRNO code. Follow existing conventions (unless colleagues use EDOOFUS for everything!), and remember that the code is there to allow the caller to sift types of error and handle each appropriately, often in convenient groups which ERRNOs identify. Use the ERRNO to help the caller recover and a separate descriptive message, with an associated priority filter, to describe the context unambiguously.

Fitting the Platforms

Some aspects of the proposed interfaces, layering and object design allow platform-specific trade-offs. For instance, you might be able to replace streams with resident samples on a platform with lots of memory but limited loading bandwidth in game. Remember audio is unlikely to be the only system loading then, even if it's the most likely to show problems if contention occurs.

Soundfield or pre-baked multichannel surround assets can be replaced with mono or stereo samples, perhaps encoded with surround audio hints such as Pro Logic or HRTF (Head-Related Transfer Function, also known as binaural) filtering if it's more important to save space than get the best possible surround. Select these to suit the listening mode, e.g. handheld, headphones, stereo or surround speakers.

Long loops can be replaced with trimmed ones, soundfields with individually positioned loops or one-shot sounds and so on. It's the job of the sound designers to decide such trade-offs and the programmers to give them plenty to choose from, a clear understanding of the pros and cons and a quick—preferably instantaneous—way to audition and compare alternatives.

For example, *Colin McRae Rally 5* continuously ran eight interleaved channels for weather, crowd and local ambiences read in parallel from a single PlayStation DVD file, plus a second independent speech stream also read from that disc. This work, the ADPCM decoding, and the associated memory buffers were allocated to co-processors to limit demands on the main CPU and memory running the rest of the game. Chapter 7 explored the details of that sub-system.

The Xbox version took advantage of the later console's hard drive and extra RAM, loading the eight ambience streams from separate mono and stereo disc files on demand, decoding them from WMA format on the main CPU, and keeping all the co-driver speech for each event in memory, using a simple XADPCM codec and an intermediate sample rate.

The PSP handheld version of *Rally* used shorter memory-resident loops for the ambience, reduced sample rate for the resident speech and was limited to mono or stereo output, with some of the stereo streams downmixed to mono to further save space and battery life.

In each case the streams or banks of samples were customised for that specific stage of the game. This is essential for co-driver speech, detailing each corner and obstacle along the track—in fact over 400 banks of speech samples were needed even on the Europe-only discs, to support speech in French, German, Italian and Spanish as well as the default UK English (with Welsh or Scottish accents, depending upon the co-driver).

The ambience banks were similarly tailored. Their eight sub-streams might be arranged as one or two weather pairs, usually for weather varying from light rain to a Swedish blizzard, plus woodland, city and industrial-area ambiences, plus between four and six mono channels of crowd noise, positioned along the track and triggered or faded up only when the player was in the vicinity. And all the banks of crowd sound were prepared in three levels of intensity or arousal, chosen at the start of the stage depending upon the championship positions and the player's place on the leaderboard. So the selection of the stream or soundbank for a given rally depended upon the location, the stage in the rally and the relative performance of drivers and co-drivers.

This adaptive design was implemented differently on each platform, balancing codecs, loading bandwidth and available memory. But the audio assets and conditions to select them were the same for all platforms, albeit differently edited or compressed. So the sound designer creates a relatively small number of sounds and groupings for those to reflect the context. These are packed together in context-specific subsets, with duplication to ease the trade-off between disc space and loading time. This is worked out by the designer and system programmer together, informed by the game design document.

The game audio programmer implements one set of triggers to select between those banks and individual samples, regardless of how they're loaded. The system programmer offers platform-optimised codecs, loading and streaming strategies to make the best use of the memory, filesystem and audio hardware. That way customers get a game which makes the best use of the capabilities of each device—Xbox's digital surround sound and hard drive, the PlayStation's DVD and co-processors, headphone stereo and UMD on the portable—without forcing the game team to make several different games.

Interactive Hyper-Reality

Realism is subjective in games or VR. First, try to understand what happens in the real world so you have a grasp of the player's initial expectations and which factors could influence audio cues. After that, the game design informs the audio model—are wind and rain decorative or effectual, for example—and you cut down, simplify or emphasise parts of this to support the design. Heed the adage attributed to Pablo Picasso: "*Learn the rules like a pro, so you can break them like an artist*." Rules are not inviolable in creative media, but it helps to know which rules you're breaking, and why.

A realistic sports sim would be unplayable by all but a handful of experts and beyond the experience of almost all players. Motor racing is almost unbelievably loud, but forcing that aspect upon gamers would be unhelpful, if not cruel. Mass-market games, especially those attached to a real-world licence, often model the experience of watching a sport on TV more than that of actually participating. Consider the prevalence of chase cameras, replays, leaderboards and third-person and top-down views. Even so, the benefits of camera control, multi-player interaction and even limited hints to the avatar gives the gamer an interactive edge over the passive TV viewer.

Often this pruning means concentrating on the things that can be modelled most easily, those most obvious to the player or those which can be modulated in a way which provides consistently helpful information. Even if this modulation is not strictly realistic, it is in the player's interest to learn to recognise and interpret its fluctuations.

You may well need to skimp on some other aspects or eliminate realistic but uninformative clutter from the mix for the player to feel fully immersed and simultaneously *in control*. Passive media like cinema and TV can make do with immersion alone and have long employed the pruning techniques of hyper-reality, but interactive audio should go further and faster.

Relating dynamic aspects of the mix to the user—including their speed and orientation as well as relative position and any other factors you consider relevant—has two benefits: it adds variety, so each event sounds different from the last similar one, and it helps the player learn your world.

Find Nearest Sounds

Before leaving the subject of triggers from the game to the audio runtimes and getting deep into the implementation details of audio runtimes, it's worth noting one powerful but rarely used way the audio runtimes can inform the rest of the game rather than vice versa.

How many times have you played a game when a simulated character, controlled by the console or computer, acts unconvincingly deaf? It blithely ignores some crucial event in the vicinity, like a collision or explosion behind them or sometimes in plain view. The same mechanism used to work out distance attenuation and sift out the least important Virtual Voices from the listener's point of view can be exposed almost unaltered to the Artificial Intelligence (AI) programmers or script kiddies on your team.

In its simplest implementation, this FindNearestSound() helper function takes a dummy listener object as an argument, with its matrix and optional polar pattern corresponding to the location and orientation of the AI character. It returns, at a minimum, a direction vector and a loudness in dB. The AI code checks for a certain loudness threshold and turns, ducks or dives taking account of the direction. That's much better.

This is a "dummy listener" because we never actually render a mix for that location. We just do the logical-layer work needed to determine what would be heard there, which is the same scan through all the Virtual Voices we do for the real mix but from a different location. Hardly any new code is needed. The result is derived by re-running the volume and distance attenuation calculations for each Virtual Voice from the position expressed in the dummy listener's matrix.

Implementation and Updates

One consistent and efficient way to do this is to make it part of the logical-layer update which runs every game update cycle. That way, the Virtual Voice data will be already cached, and you can batch together several scans. But it relies on the AI being willing to wait for the result, just as it would for a ray trace from a fast physics system.

We now move into the dangerous area of multi-threading. House rules, caution or prejudice may bring mutexes, semaphores and message-passing into play. The AI programmer might want you to use an existing notification system. If so, and you really want them to use your code, try to use their interface. Otherwise, carefully consider a shared-memory approach.

AI, like physics, often runs on its own threads asynchronously from either update or rendering on modern multicore hosts. It's an ideal process to share out, as it can then be handled much like player input, which has its own independent timing. But it does complicate the interface with the main update thread, where the Virtual Voice system is expected to run.

The quick way to implement this coupling, for a low-level programmer at least, is for the AI client to pass the volatile address of a result vector, which can be atomically updated later, to the audio runtimes when it instantiates the dummy listener. The volume can be patched into the last element of a vector before returning it alongside the X, Y, Z position, normalised or including distance.

This lockless approach relies on only audio code being allowed to write to the vector while the listener exists. If you're not sure the write will be uninterruptible or cannot guarantee alignment of the vector, set the volume momentarily to the noise floor while updating the coordinates and patch the correct volume in last.

This also assumes, like most lockless schemes, that at least one float or pointer can be written to memory in a single transaction, not broken across two cache lines, when another thread might pick up a partial update. Most compilers and memory allocators will enforce this. If so, there's no need for a callback and the associated threading problems.

If the AI needs a result at once, you must accept that you're almost certainly partway through an update cycle, with some voice volumes and positions a frame ahead of others. You also need to make sure the call is in the context of the game update thread . . . But if your AI programmer can't cope with delayed notifications, chances are they're bound to the main thread already.

Which Way Now?

If you only return the loudest sound it's trivial to return the corresponding direction, since you worked it out when moving the source into listener-space—it's the vector associated with the scalar attenuation distance.

You may want to normalise the vector so it just expresses a direction. If it's non-zero, divide by that distance; otherwise return the zero vector. This may mean the AI caused the sound in the first place, as it's co-located or something else has directly struck it. Ultimately the AI system must make that very significant call; but you can help it, and your own profile, by providing a way to filter the Virtual Voices considered. There's no need to sort the voices, just keep track of the loudest yet as you scan through.

This simple implementation has been enough for me so far, but if you want to take account of multiple voices you will need to sort and either return a container load of vectors and volumes—if that's really what your customer wants—or, more likely, tot them up. It's easy enough to compute a combined loudness, though summing the attenuations obviously won't work!

Convert each perceived volume, up to 0 dB, into a gain factor 0..1, sum their squares and convert back to dB to get a non-directional total power. Fast conversion code is presented in Chapter 9. Finding the average direction—

which may be rather nebulous—requires summing all the relative vectors after scaling them in proportion to their power.

In any case, audio programmers can provide their AI colleagues with a flag or set of controls to pass back to select a subset of groups—e.g. weapons, collisions, diegetic speech—which saves time by culling some voices from consideration. It can also help the game sift friendly and enemy actions, as in the case of co-located audio noted earlier. Player sounds and enemy ones are almost certain to be in different groups, and player sounds can often be ignored, as expected, in this context.

It's better not to expose audio-internal group names or indices; bespoke flags can be issued once and transparently remapped if a change inside the audio system requires it. Of course, if the change is not transparent or the mix is dramatically revised, the AI and audio teams need to share. But the designers should be allowed to rename or re-order their domain without breaking an external dependency. This is part of the tragedy of teamwork and good engineering—"the better you do a job, the less anyone thinks you've done."

> The better you do a job, the less anyone thinks you've done.

Unless this subset is likely to vary for different listeners or game modes or you're working in mainline (not a branch) on runtimes for several titles, it's simplest and usually good enough to embed it in the game-specific implementation or a wrapper layer, which you may have already. It's entirely platform-independent code, anyway.

Reference

[1] *The Design of Everyday Things; Donald A Norman, MIT Press 2013, ISBN 978262525671; page 8, see also chapter 5 "Human Error? No, bad design" pages 162–216*

CHAPTER 9

Quick Curves—Transcendental Optimisations

This book assumes that the target platforms have hardware support for floating-point arithmetic. Integer-only mathematics libraries were often used in the 1980s and 1990s, and there's documentation for them and their limitations online and in old DSP manuals, but floating-point hardware has been available even in cheap devices for decades. Its generality and dynamic range make it far more convenient for audio mixing and control than integer or fixed-point approaches which were forced upon programmers of the Amiga or original PlayStation. IEEE-754 is the standard for cross-platform floating-point arithmetic.

While most modern hardware can add and multiply single-precision floating-point values as quickly as it handles integers, standard library functions must work to at least seven decimal places of precision and manage exceptional values like infinity, which have no business in any real-time system. Transcendental mathematical functions are often used in 3D audio runtimes to apply psychoacoustic curves or convert between polar and rectangular representations and can eat up a lot of processing cycles. This chapter includes audio-friendly ways to speed those up while still staying accurate enough for audio control purposes.

Fast Logarithms

The following snippet, inspired by game programmers Dylan Cuthbert and Ian Stephenson, rapidly works out an approximation to the base 10 logarithm plenty good enough for audio mixing. It works by exploiting the fact that the IEEE-754 representation of a floating-point value is roughly proportional to the corresponding base 2 logarithm if the value is treated as an integer. This is because the exponent in the most significant byte of the value has a logarithmic relationship to the value. That is then used as a first approximation for a Newton-Raphson step, which roughly doubles the precision.

This technique made the calculation of distance attenuations about three times faster than they would have been if my PS2 runtimes had relied upon the system log10f() function. It's definitely a hack, only worth using if a realistic profile test tells you that log10f() is a significant bottleneck. It's included here because it's benefited several games, and the accuracy of this version is well matched to the needs of audio, but it might even be counterproductive on other processor architectures, especially those which penalise transfers between integer and floating-point execution units.

The floorf() function is declared in math.h. The magic numbers are artefacts of IEEE-754 representation: invShift23 relates to the 23-bit mantissa and logBodge scales the Newton-Raphson correction amount in y. Scaling by 0.30103 converts the result from a base 2 logarithm to base 10 and can be omitted, saving a little time, if you prefer to use log2 rather than log10.

This code is compiler and platform dependent—it makes assumptions, most notably that the float value is in IEEE-754 format and yet not so very close to zero that it becomes denormalised. Game hardware often skips support for such tiny values and the associated loss of precision, treating them as zero instead, but this should not be assumed—check the documentation and experiment if necessary.

Include a unit test for this sort of optimisation so that compiler updates don't break it or make it counterproductive. Such tests should time the speed as well as the accuracy of the approximation against the system library function across the required input domain and assert if the results are slower or less accurate than necessary.

```
inline float FastLog10(float p)
{
  const float logBodge = 0.346607f;
  const float invShift23 = 1.f / (1<<23);
  float x=(float)*(int32_t *) &p;
  x *= invShift23;
  x = x - 127.f;
  float y = x - floorf(x);
  y = (y - y * y) * logBodge;
  return (x + y) * 0.30103f;
}
```

The logarithm of zero is minus infinity, but anything below -96 dB is effectively silent. This wrapper function traps that case and converts gains into useful decibel attenuations:

```
const float SILENT = -96.3f; // Max 16-bit attenuation in dB
const float MIN_GAIN = ConvertDecibelsToGain(SILENT);

float ConvertGainToDecibelsFast(const float fGain)
{
  if (fGain <= MIN_GAIN)
  {
    return SILENT;
  }
  return 20.f * FastLog10(fGain);
}
```

This computes good-enough approximations to p^b by a similar technique:

```
inline float FastPower(float p, float b)
{
  float a = b * FastLog10(p) * 3.322f;
  const float powerBodge=0.33971f;
  float y = a - floorf(a);
  y = (y - (y*y)) * powerBodge;
  float x = a + 127.f - y;
  x *= (1<<23);
  *(int32_t*)&x = (int32_t) x;
  return x;
}
```

The factor 3.322 converts between base 10 and base 2 logarithms. The factor 0.30103 does the reverse at the end of FastLog10(), so if you merge the functions or implement FastLog2(), both can be eliminated.[1] The test in the complementary conversion avoids the risk of introducing tiny denormalised values:

```
float ConvertDecibelsToGain(const float dB)
{
  if (dB < SILENT)
  {
    return 0.f;
  }
  return ( FastPower(10.f, dB * 0.05f) );
}
```

Fast Trigonometry

FastSine() returns an approximation of sin(x) for x in the range PI to PI*3. It's less accurate than sinf() but quicker on many platforms, depending upon the compiler and library, and accurate within 0.2% in the interval +/−2 * PI. That's close enough for panning and most audio purposes. For cosines, add PI/2 to x.

This assumes that PI is defined somewhere as the ratio of the circumference of a circle to its diameter, around 3.141593. Oddly, neither the C nor C++ standards include this constant, though it's essential for trigonometry and most 3D work, as well as filter configuration. POSIX implementations define M_PI. C++ 20 is expected to standardise on std::math_constants::pi_v<float>.

```
inline float FastSine(float x)
{
  if (x>PI)
  {
    x -= 2.f * PI;
  }
  const float c1 = 4.f / PI;
  const float c2 = -4.f / (PI * PI);
  float y = x * ((fabsf(x) * c2) + c1);
  return 0.225f * (y * fabsf(y) -y) + y;
}
```

A vectorised version, FastSinCos(), computes several sines or cosines in parallel almost as quickly. It's very handy for higher-order Ambisonic encoding. Here it is in Android vector intrinsics, suitable for ARM32 and ARM64; Google's NDK also translates most of these for Intel and MIPS architecture Androids. This snippet computes both the sine and cosine of two angles in radians, supplied in the float array angles.[2] It follows the same sequence as FastSine() but preloads extra constants into c0, c3 and c4. First we initialise six vectors with suitable values. The vdupq_n_f32 intrinsics copy float constants into all four elements of a float32x4_t vector.

```
float32x4_t t0, t1, c0, c1, c2, c3, c4;

c0 = vdupq_n_f32(PI);
c1 = vdupq_n_f32(4.f/PI);
c2 = vdupq_n_f32(-4.f/(PI * PI));
```

```
c3 = vdupq_n_f32(2.f * PI);
c4 = vdupq_n_f32(0.225f);

t0 = (float32x4_t) vdupq_n_f64(*(float64_t*) angles);
t1[0] = t1[1] = 0.f;
t1[2] = t1[3] = PI / 2.f;
t0 = vaddq_f32(t0, t1);
```

The 64-bit vdupq_n_f64 intrinsic duplicates both angles from the first two elements of t0 in the second pair, treating two f32 values as a single f64—this is safe providing 64-bit floats are supported and you don't try to do double-precision arithmetic with the float pair! Otherwise use this scalar equivalent, compatible with Intel's NEON_2_SSE.h translator:

```
t0[2] = t0[0] = angles[0];
t0[3] = t0[1] = angles[1];
```

The final vaddq_f32 advances the second pair of values by PI/2. As cos(x) = sin(x+PI/2), we're then ready to compute two sines and two cosines in parallel. Table 9.1 shows the setup immediately *before* t1 is added to t0 so that lanes 3 and 4 compute cosines. The two input angles are θ^1 and θ^2. I've avoided gratuitous Greek in this book, but this is too apt to miss:

Table 9.1: Data configuration for SIMD trigonometry

Register	lane 1	lane 2	lane 3	lane 4
c0	π	π	π	π
c1	$4/\pi$	$4/\pi$	$4/\pi$	$4/\pi$
c2	$-4/\pi^2$	$-4/\pi^2$	$-4/\pi^2$	$-4/\pi^2$
c3	$2*\pi$	$2*\pi$	$2*\pi$	$2*\pi$
c4	0.225	0.225	0.225	0.225
t0	θ^1	θ^2	θ^1	θ^2
t1	0	0	$\pi/2$	$\pi/2$

Now comes the range reduction, avoiding slow scalar branch instructions.

```
uint32x4_t mask = vcgtq_f32(t0, c0);
t1 = (float32x4_t) vandq_u32((uint32x4_t) c3, mask);
t0 = vsubq_f32(t0, t1);
```

The vcgtq_f32 compares all four angles in t0. If the angle is greater than PI, in c0, it sets all bits in the corresponding element of the mask; otherwise it clears them. Then vandq_u32 applies this mask to the constant c3, so t1 contains 2*PI for all angles less than PI and zero otherwise, and vsubq_f32 subtracts t1 from t0. Similar techniques or vector intrinsic min/max instructions can quickly clamp vector fields or groups of samples to a headroom limit.

Next, form the initial approximation by expressing t0 * (fabsf(t0) * c2 + c1) in simple steps:

```
t1 = vabsq_f32(t0);
t1 = vmulq_f32(t0, t1);
```

```
t0 = vmulq_f32(t0, c1);
t0 = vfmaq_f32(t0, t1, c2);
```

Finally, refine the approximation according to the formula t0+=.225*(t0*fabsf(t0)—t0):

```
t1 = vabsq_f32(t0);
t1 = vmulq_f32(t0, t1);
t1 = vsubq_f32(t1, t0);
t0 = vfmaq_f32(t0, t1, c4);
```

The four floats in t0 return FastSine(angles[0]), FastSine(angles[1]), FastCosine(angles[0]) and FastCosine(angles[1]) in fewer than 20 instructions. The compiler may condense those further. I've used fused-multiply-add instructions, which are not always available; substitute vmlaq_f32 if your compiler objects to vfmaq_f32.

Processor Pipelining

If we have lots of trigonometry to do, this code could be at least four times faster with only simple changes. Once the constants c0..c4 are set up and the four angles are in t0, only eight instructions are needed to generate four sines or cosines. But 16 could be computed almost as quickly, because each of those eight lines depends upon a result from the line before. Even if the processor can submit one vector instruction per cycle it still takes several cycles, known as the "latency," before a result is available for reuse.

Assuming a typical latency of four cycles, it will take 32 cycles for this block to deliver a result, a throughput of eight cycles per sine. If there's no independent work to be done in the meantime, the processor stalls for three cycles of every four. But if we've got sufficient input data we can fill the three slots by computing another dozen sines, and the total time is only three cycles longer, an average of 2.2 cycles each.

For each additional four values, two temporaries equivalent to t0 and t1 are needed. Introducing t2..t7, with 12 more angles in t2, t4, and t6 and results in the same registers, the final line becomes these four:

```
t0 = vfmaq_f32(t0, t1, c4);
t2 = vfmaq_f32(t2, t3, c4);
t4 = vfmaq_f32(t4, t5, c4);
t6 = vfmaq_f32(t6, t7, c4);
```

Corresponding duplication of the previous seven lines, with t0 and t1 replaced likewise, busies most but not all the 16 quad vector registers on ARM32 yet fewer than half of the ARM64 set.[2] The Xbox 360's VMX hardware has a latency of up to 14 cycles, and 128 registers to keep its pipeline busy.[3] It could filter 40 voices in twice the time it would take to do one.

Cartesian-to-Polar Conversion

Graphics use Cartesian X, Y, Z coordinates, but 3D audio panning often favours polar distance, azimuth and elevation representations. Azimuth is the horizontal angle, anti-clockwise from the front, in the range 0 to 360°, 2*PI radians. Elevation has half this range as it's the vertical angle, 0 meaning straight ahead, positive up and negative down.

Standard library routines are slower and more accurate than we need for such conversions. This arctangent approximation computes azimuth from X and Y (or Z) coordinates. It is almost five times faster than the PS3 library atan2() and accurate to within 0.01 radians:

```
inline float FastAtan2(const float y, const float x)
{
  const float c1 = PI * 0.25f;
  const float c2 = PI - c1;
  float abs_y = fabsf(y) + 1e-9f;
  float angle;

  if (x >= 0.f)
  {
    float r = (x - abs_y) / (x + abs_y);
    angle = (0.1963f * r * r * r) - (0.9817f * r) + c1;
  }
  else
  {
    float r = (x + abs_y) / (abs_y - x);
    angle = (0.1963f * r * r * r) - (0.9817f * r) + c2;
  }
  if (y < 0.f)
  {
    return(-angle);
  }
  else
  {
    return(angle);
  }
}
```

The fudge factor 1e-9 prevents division by zero at the asymptote. Exchange the arguments to switch between clockwise and anti-clockwise progression. FastAtan2 builds on Jim Shima's April 1999 posting to Usenet's comp.dsp list. Similar public-domain approximations are collated at dspguru.com.[4]

Floating Points

Bear some general tips in mind. Never divide by a constant if you could multiply by its reciprocal instead. Some compilers may convert (f / 5.f) into (f * 0.2f) for you, but this depends upon how strictly they worry about edge cases and marginal loss of precision. In binary, 0.2 is an infinitely recurring number.

The online Compiler Explorer, by former Argonaut Games programmer Matt Godbolt, interactively shows what code compilers will generate as you fiddle with compiler optimisations and speed-up settings like—fffast-math or Microsoft's /fp:fast.[5] It's wise to make such optimisations explicit in your code. The keyword const and the C++11 feature constexpr expect the compiler to work out subexpressions at compile time, which is why the constant subexpressions are isolated here.

Most floating-point hardware can directly calculate the square root of a number or even the reciprocal of that value, which is handy when working out reflections, but standard library implementations necessarily wrap the fast hardware instruction in domain checks. These are needless overhead if you are sure that all the values you present are in normalised format. David Goldberg's 1991 ACM paper expertly explores the edge cases of IEEE-754 floating-point arithmetic.[6]

On ARM processors with NEON hardware, the GCC/Clang compiler intrinsic __sqrtf() is guaranteed to generate a single instruction rather than a call to a library function. The equivalent for Intel processors, on Microsoft or Intel compilers, is _mm_sqrt_ss().

Similar intrinsics expose other transcendental functions. Their names depend on the compiler and target architecture, and the benefit varies likewise. They're worth using if a profile suggests that the standard library function is slow and confirms a substantial benefit from use of the intrinsic on the hardware you know your customer will own. The latter concern makes them more useful on consoles and embedded devices than generic PCs.

Desktop computers often handle double- and single-precision values with comparable speed, but most embedded systems penalise the use of doubles. Despite the first letter, the fabs() function declared in math.h takes and returns a double-precision value. Use fabsf() rather than fabs() to strip the sign from a single-precision number, or your code will spend more time converting between precisions than it does working out the result. That's also why the single-precision suffix "f" follows all floating-point constants in portable real-time code.

Single precision is ample for almost all interactive audio work and often much faster. Likewise, favour sinf() over sin(), atan2f() over atan2() and powf() over pow(). They're accurate to six or seven decimal places, which is twice as many as you'll need for ear-friendly pitch, volume and positional control. For those four functions the earlier approximations are good enough. Double precision is overkill for audio and disproportionately slow on embedded platforms.

The Four Randoms

> "Any one who considers arithmetical methods of producing random digits is, of course, in a state of sin."
>
> —John von Neuman[7]

The term "random" describes a slippery concept. Von Neumann's point is that pseudorandom values are not strictly random at all, but can often be used, with care, as if they were.

Pseudorandom numbers are used to add variety to games when there's no convenient variable to do the same job more informatively, and sometimes to make "noise" as opposed to pitched audio, as we saw in Chapter 2. When selecting or adjusting sound assets there are four distinct sorts of "random" to consider.

Mathematician Random

Mathematicians assess randomness in terms of the average distribution of long sequences of values. Each value is related to the previous one in a non-obvious way, starting from a "seed" followed by a long fixed sequence of values in a predictable range but superficially unpredictable order.

The simplest way to scramble an unsigned sequence is to multiply the previous value by a well-chosen constant and add another, taking the result modulo the word size. That happens automatically in C as it does no bounds checking. These are called linear congruential pseudorandom-number generators. They're not the best, but they are quick and well understood. These constants are recommended by Donald Knuth and chosen so that the maximum length sequence is obtained.[8] Any value other than 0 can be used as the seed, and the period is 1 << 32–1.

```
uint32_t rand32()
{
  static uint64_t seed = 1;
```

```
    seed *= 69069;
    return seed += 1234567;
}
```

As for most pseudorandom-number generators, the high-order bits are the most random, so a right shift is the best way to extract a limited number of bits: rand32() >> 22 yields well-distributed values in the range 0..1023.

A similarly fast but statistically superior approach, xorshift*, was proposed by George Marsaglia then refined by Sebastiano Vigna.[9,10] This uses three shift and exclusive-or operations to scramble the bits of the seed at each call. Here's a 64-bit version:

```
uint64_t rand64(void)
{
    static uint64_t seed = 1;
    seed ^= seed >> 12;
    seed ^= seed << 25;
    seed ^= seed << 27;
    return seed *= 0x2545f4914f6cdd1d;
}
```

This emulates the polynomial shift generators used in 8-bit sound chips—in fact, it is mathematically equivalent to a linear feedback shift register. Vigna's paper includes alternatives with more state and hence a much longer period. Beware: 64-bit shifts are slow on 32-bit processors.

To derive floating-point values in the range 0..1, multiply the random integer by the reciprocal of the modulus, or mask the random bits into the mantissa of the IEEE-754 float after forcing normalisation:

```
float random1()
{
    union
    {
        float f;
        uint32_t i;
    } t;
    t.i = (0x3f800000 | (0x7fffff & rand32()>>9));
    return t.f - 1.f;
}
```

Bitwise operations are much faster than floating-point division. The subtraction yields values in the range 0..0.99999988.

Designer Random

Sound designers are rarely mathematicians and prefer a subtly different sort of "random," illustrated by the following exchange:

Programmer: Which random would you like?

Designer: (puzzled) Normal random, what d'ya mean, which?

Programmer: Can the same thing play twice in a row?

Designer: (puzzleder) Course not, that's not random is it?

Keep a copy of the previous random number generated in that context. If the next value matches that, after range reduction, call for another. Repeat till it's different. Audio tools programmer Pete Goodwin calls this "Random no repeats."

Gambler Random

To postpone repetition as long as possible, we must shuffle all the possible values and allocate enough memory to record the whole set. This does the shuffling for n possibilities of any type in array a[]:

```
assert(n > 0);
size_t j = n-1;
while (j-- > 1)
{
    size_t k = floorf(n * random1())+1;
    std::swap(a[j],a[k]);
}
```

Use a[0] to a[n-1] in that order. At the end of the set, shuffle again . . . swap the new first and last if the new first happens to be old last to avoid an immediate repeat.

Simulation Random

Sometimes simulations need to be sure that a pseudorandom sequence can be rewound, repeating previous results—for a flashback or replay video, for instance, where the same "random" events should be exactly repeated, lest history fails to repeat itself. You must then derive your pseudorandom values from the generator used in the game update, since its seed will be reset appropriately when required.

It's vital not to call the game random generator at random, or you'll upset the required determinacy, and randomised AI systems might make different choices, leading to chaotic "butterfly effect" outcomes. Always use the game-update random for selecting or positioning sounds if the timing affects subsequent gameplay. This occurs if some event is waiting for a randomly selected sample to finish playing. Try to avoid such circumstances, especially in titles where audio output is optional! For sounds without consequences, such as kickup events, just use local random to select and place them.

Random Curves

Sometimes we prefer random numbers which cluster round a certain value, with outliers progressively less likely. Many real-world probabilities take this form because of combinatorial interactions. Adding results from multiple dice models this, in *Dungeons and Dragons* and similar games: with two dice, a total of 7 is six times more likely than either 12 (double 6) or 2 (snake-eyes), the outliers. Thus 0.5 * (random1()+random1()) maintains the range 0..1 but clusters results round the median, 0.5. The more random numbers you average, the closer the results come to a bell-shaped "normal distribution.'"

Take account of psychoacoustic curves when randomising pitches and volumes. If you randomise the frequency in Hertz of a sample by ±50% you'll end up with sounds that, on average, play a third flat rather than over an evenly distributed range of higher and lower pitches. Fifty-per cent frequency randomisation of a 24,000 Hertz asset yields a range from 12,000 to 36,000 Hertz, where the latter is a fifth sharp and the former a whole octave flat! The solution is to select within an interval expressed on a psychoacoustic scale, such as semitone cents, MIDI note numbers or decibels, before converting the result to a linear pitch or gain.

References

[1] *Fast Powers*: www.dctsystems.co.uk/Software/power.html
[2] *ARM64 Architecture*: http://infocenter.arm.com/help/topic/com.arm.doc.den0024a/DEN0024A_v8_architecture_PG.pdf
[3] *Xbox 360 Architecture*: http://home.deib.polimi.it/silvano/FilePDF/ARC-MULTIMEDIA/XBOX%20360%20Architecture.pdf
[4] *Comp.dsp Tips*: http://dspguru.com/comp.dsp/tricks
[5] *Compiler Explorer*: https://godbolt.org/g/io4Cz9
[6] *What Every Computer Scientist Should Know About Floating-Point Arithmetic; David Goldberg; ACM Computing Surveys, ISSN 0360-0300, Volume 23 Number 1, March 1991*: https://docs.oracle.com/cd/E19957-01/806-3568/ncg_goldberg.html
[7] *Various Techniques Used in Connection with Random Digits; John von Neumann in "Monte Carlo Method," US National Bureau of Standards Applied Mathematics Series, ISSN 1049-4685, Volume 12 Number 6, 1951*
[8] *Seminumerical Algorithms, The Art of Computer Programming, Volume 2; Donald Knuth, Addison-Wesley Professional, 3rd Edition, 1997, ISBN 0785342896848, Chapter 3*
[9] *Xorshift RNGs; George Marsaglia, Journal of Statistical Software, ISSN 1548-7660, Volume 8 Number 14, 2003*
[10] *An Experimental Exploration of Marsaglia's Xorshift Generators, Scrambled; Sebastiano Vigna, ACM Transactions on Mathematical Software, ISSN 0098-3500, Volume 42 Number 4, 2016*

CHAPTER 10

Objects, Voices, Sources and Handles

The interactive audio system comprises a network of objects—Voices, Listeners, Groups, Filters and Reverbs—which work together to create the mix. The maximum number of objects of each type is set when the system is initialised, so that all the memory required can be allocated up front. Each object instance requires a few hundred bytes, at most, so even if you need hundreds of Virtual Voices the RAM required to keep track of all of them is dwarfed by that needed for memory-resident sample data, which usually runs to tens of megabytes, or hundreds on most recent full-price console and PC games.

Figure 10.1 shows how sample voices, groups and listeners are arranged in a simplified circuit-racing game. Sound samples are on the left, feeding into groups. 3D voice mixes go via listeners to the speakers or headphones. AI refers to non-player vehicles, which use a simplified version of the player car audio model and many of the same assets. Split-screen multi-player and effects routing are discussed later.

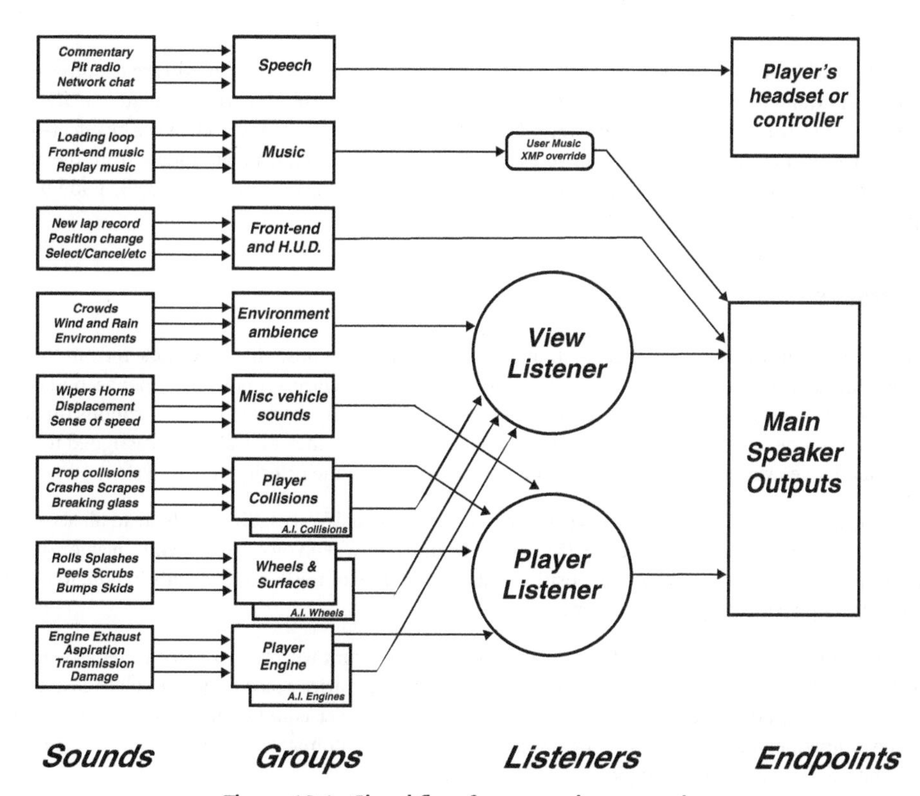

Figure 10.1: Signal flow from samples to speakers

The number of listeners varies depending upon the game design—some need only one, and even elaborate split-screen titles are unlikely to require more than a dozen. Group counts are similarly game dependent—separate controls for menu sound effects, speech, the player's diegetic sounds, AI players (collectively), weather, surfaces, crowds, collisions and weapons are commonplace, and most of them will be duplicated for each active on-screen player. This may add up to a couple of dozen, often but not always associated with a Filter object but never more than one each. There's an example in Chapter 12.

Table 10.1 shows the number of physical voices allocated by eight of the games the author has helped to make, on PC and console platforms.

Table 10.1: Physical voice counts in some Codemasters games

Game	PlayStation	Xbox	PC
DiRT	192	128	60
DiRT2	128	128	192
DiRT3	120	90	192
Showdown	145	134	192
GRID	192	128	100
F1 2010	135	128	131
F1 2011	128	178	128
Bodycount	256	166	N/A

The low figure for *DiRT* on PC is because Microsoft was pushing the PC market from hardware to software mixing in 2007, and we still wanted to support older DSP cards like Creative Labs Audigy series. The later *GRID* required a more powerful X-Fi card, or the "generic software" fallback, but was still limited to 100 voices by effect-routing bottlenecks. *Bodycount* was only released on consoles; like *F1 2011* it benefited from an update from the original XAudio to XAudio2, boosting the number of voices on Xbox 360. Its urban ruins made especially good use of 3D7.1 output on PlayStation 3, but I wish I'd made the sky less echoic!

Virtual Voices

Note that these limits apply to physical voices—simultaneously playing sounds. These are chosen from a much larger pool of "Virtual Voices" according to a flexible priority system which runs many times per second as part of the game's update function. Physical voices, often known as "sources", are rather like asynchronous threads—they may stop at any time, when they reach the end of a one-shot sample. They may also be reassigned to a more important Virtual Voice, with different samples and metadata, as part of the priority update system.

It follows that the game should not access physical voices, or even Virtual Voices, directly. Instead all access is via "smart pointers" or voice handles owned by the game and updated when voices are called to play or stop. Subsequent parameter updates, to change the position, pitch or volume of the sound, use this handle to find the Virtual Voice and its physical counterpart, if any.

Virtual voice systems allow the same sound design and game code to work with two or three times as many voices on some platforms as others. Naturally the amount of detail in the mix varies accordingly, but all the vital voices and most of the important ones were heard in every case. The Virtual Voice count depends on the amount of dynamic culling implemented by the game code—some circuit-racers automatically purge distant vehicle updates—but several hundred Virtual Voices are typical.

Listeners, Effects and Groups are permanently valid after initialisation; references to them can never go out of scope. But after the audio system Update() method is called, existing voice handles may become invalid. Hence any handle played by game code must be validated each subsequent frame, by calling VoiceHandle::IsValid() as the first operation in the update code for each voice.

Just in case this is forgotten, there's a mechanism to soak up attempts to read or update properties of an invalid voice. The Voice and VoiceHandle classes are friends, so they can share code. When a Virtual Voice stops, its handle is remapped to a static dummy voice instance which acts as a punch-bag, accepting but ignoring property updates, with a non-critical warning. This dummy voice is declared in the VoiceHandle class header, along with another static voice instance which contains default properties for newly initialised voices, and a handle for that:

```
static Voice ms_dummyVoice;
static Voice ms_defaultVoice;
static VoiceHandle ms_voiceDefaults;
```

Don't forget to instantiate these in the corresponding .cpp file:

```
Voice VoiceHandle::ms_dummyVoice;
Voice VoiceHandle::ms_defaultVoice;
VoiceHandle VoiceHandle::ms_voiceDefaults;
VoiceHandle::ms_voiceDefaults.m_pVoice = &VoiceHandle::ms_defaultVoice;
```

Voice handles are allocated by and belong to the game, not the audio system, but beware of allocating them on the stack. Looped voices stop when their handle's destructor is called, so if you allocate a voice handle locally the loop shuts down when the function in which the handle was allocated returns to its caller.

The VoiceHandle class contains only one instance member, a private pointer to a Virtual Voice, initialised to NULL by the constructor. IsValid() just checks if it's bound to a voice:

```
inline bool IsValid()
{
  return ( m_pVoice != NULL );
}
```

We access the contained voice by dereferencing its handle with an override of the -> operator:

```
inline Voice* operator->()
{
  if( IsValid() )
  {
    return m_pVoice;
  }
  else
  {
     Moan();
     return &ms_dummyVoice;
  }
}
```

Moan() writes a suitable message about the inadvisability of indirection through an invalid voice and sets an ERRNO which can be checked later. Voice handles must not be copied by construction or assignment, so we disable default constructors in the header:

```
VoiceHandle( const VoiceHandle& ) {}
void operator = ( const VoiceHandle& ) {}
```

When a voice is told to play we start by checking that the handle is not already valid—if so we return EADDRINUSE to avoid having one handle for two voices. Otherwise we get a Virtual Voice object from the previously created pool, set the appropriate sample asset, priority and group pointers from arguments to the Play call and initialise the other voice properties by copying them from VoiceHandle::ms_voiceDefaults. Finally we attach the Virtual Voice to the handle by copying its address into m_pVoice and vice versa. We won't find out if it's actually playing or just virtual till the next full audio update which compares the priorities of all candidate voices and binds the most important or loudest to an output Source.

The Virtual Voice needs access to the handle so that it can know when the voice stops. It does this by calling VoiceHandle::OnVoiceDeath(), which encapsulates m_pVoice:

```
inline void OnVoiceDeath()
{
   m_pVoice = NULL;
}
```

Conversely, the VoiceHandle destructor tells the Virtual Voice to clean up, ready for reuse.

```
VoiceHandle::~VoiceHandle()
{
  if( m_pVoice )
  {
    m_pVoice->OnVoiceHandleDeath();
  }
}
```

Voice::OnVoiceHandleDeath() is explained later, in the section headed "Ways to Stop."

What to Play

Before we can play anything, we need a way to specify the audio data and metadata properties of a sample or a stream. This enumeration lets us identify and correctly play compressed samples:

```
enum CODEC
{
  CODEC_ADPCM  = 0,
  CODEC_MU_LAW = 1,
  CODEC_LPCM16 = 2,
  CODEC_ATRAC3 = 3,
  CODEC_OPUS   = 4,
  CODEC_XMA    = 5,
  CODEC_XADPCM = 6,
  CODEC_VORBIS = 7,
  CODEC_COUNT
};
```

The set of items in CODEC will depend upon the platforms you're supporting. It's possible (though inessential) that there might be more than one ADPCM codec or that the meaning of that enumeration might vary by platform—it could refer to VAG ADPCM on Sony platforms, one of the XADPCM variants on Microsoft ones, Nintendo's original mono ADPCM format, introduced for the Super Nintendo (SNES) and still going strong, or their stereo GCADPCM variant, as implemented on GameCube, Wii and Wii-U.

The following structures contain all the information needed to find and play a sample loaded into memory. Typically these will be packaged together into "banks" for implementation efficiency; the sampleSpec and any special data needed to initialise or seek within a compressed sample is held in main memory while the bank is loaded. The raw audio at sampleData should at least conceptually be stored separately, since many systems have dedicated audio memory from which it can play directly. Muddling up samples and metadata used to play them undermines the efficiency of this arrangement.

Much the same metadata is used for samples in memory and streams from disc. Shared elements of a soundSpec are imported as an anonymous structure into the specialised sampleSpec and streamSpec types. This is strictly a C11 feature but has long been supported by Microsoft and works on GCC if the -fms-extensions compiler switch is used. Clang also requires this, and -Wno-microsoft to suppress related warnings.

```
typedef struct soundSpec
{
  CODEC     soundFormat;
  unsigned  soundChans;
  bool      soundLoops;
  bool      soundMusic;
} soundInfo;
typedef struct
{
  struct    soundSpec;
  char*     sampleName;
  void*     sampleData;
  unsigned  sampleBytes;
  float     startOffset;
} sampleSpec;
```

Here, soundChans is the number of channels, soundLoops is true for a sample that should play as a continuous loop, and soundMusic must be set for samples that contain music so that they can be suppressed if the user already has other music playing from outside the app or game. This is a requirement for Xbox titles and a recommendation for PlayStation ones. I do the same on iOS, so users can replace in-app music with their own, playing in the background from iTunes or similar. However jaunty your front-end song, it will not mix well with users' preferences.

The same structure can be used to play a chain of speech samples loaded from the network or blocks of data generated by granular synthesis. The runtime system doesn't care where it comes from, providing it hangs around long enough to be fully played. Typically the codec and channel count must correspond for chained samples and streams, so replay is continuous. You must not release or re-use sampleData until all voices playing it have stopped.

Since samples normally play from their beginning startOffset is usually zero but it can sound less repetitious if we start loop samples part-way through. If so, startOffset should express the preferred number of seconds of audio to skip. The underlying system may round this onto the nearest block boundary if the data is compressed. Chapter 16 explains how to do this without a conspicuous click.

Streams played from disc, without loading the entire sample first, require different initialisation, specifying the path to the asset rather than sample address and length in memory. Set interleaveOffset if you want this voice to use only some parts of an interleaved stream, numbered from 1, and interleaveChans to the number of channels (e.g. 1 for mono, 2 for stereo) it should use. As soundChans is the total number of interleaved channels, interleaveChans + interleaveOffset – 1 must be less than soundChans. The stream must run while any of the interleaved channels are playing, but only the ones bound to a voice are heard.

It's sometimes useful to be able to play a one-off sample directly from a file, without packing it into a bank.

```
typedef struct
{
  struct   soundSpec;
  char*    streamFile;
  unsigned streamBuffer;
  unsigned interleaveOffset;
  unsigned interleaveChans;
} streamSpec;
```

The streamFile path or suffix can be used instead of sampleName in reports. The streamBuffer is the index number of a pre-allocated loading buffer, specified in the bufferBase and bufferSize initialisation arrays described in Chapter 15. This may be all you need if the format, channel count and looping flag are in the file header, but otherwise you'll need to fill in those three soundSpec fields too.

How to Play

In theory there should only need to be one way to set a sound playing—exchange its address, size and codec type for a Virtual Voice Handle and then use that to track its lifetime and update its properties such as position, volume and pitch.

As I've developed increasingly capable audio runtimes, the number of useful ways to start sample replay has grown. It soon became apparent that looping and non-looping samples required different treatment—one-shot sounds can often be treated in a "fire and forget" way, requiring no further updates after initial triggering, but loops run forever unless you retain a handle with which to stop them.

Occasionally it's useful to replace one playing sample with another, using the same handle to update the later replacement. This technique is needed in the physical layer, as it's the mechanism used to replace a low-priority sample with a more important one. Also, sometimes we might want to "chain" a sample onto the end of another, queuing up the second so that it can automatically take over the output channel and voice handle, without the gap and handle swap associated with waiting for one sound to finish before triggering the next.

Once a voice has been allocated a handle, its update interface is identical regardless of where it's playing from. The physical layer encapsulates all the differences between samples and streams, chains and swaps, loops and one-shots. This means that implementations of a given title can share all their update code without requiring the same data organisation on each platform. The only implementation difference, from the game's point of view, is the initial play call. So here they all are:

```
void PlayOneShot(sampleSpec, group, volume = 0.f, pitch = 1.f);
handle = PlaySample(sampleSpec, group);
handle = ChainSample(handle, sampleSpec);
handle = StealSample(handle, sampleSpec);
```

```
handle = PlayStream(streamSpec, group);
handle = ChainStream(handle, streamSpec);
```

You may get by with fewer or variants of these. Not all titles require Chaining capability. Stealing is rarely needed except at the voice-management level, but since it's vital there and sometimes useful upstream, it makes sense to expose it.

PlayOneShot is deliberately minimal. The volume and pitch are only needed if the defaults—full volume, original pitch—are not required, and the only way to set other properties is via the default voice. PlayOneShot() is most useful for non-diegetic sounds on a Group without a 3D listener; it's great for front-end beeps, background speech and transitional stings.

There's no StealStream(). That would be hard to implement without causing performance spikes, and there's never been a need for it. Techniques for handling interleaved streams and sharing stream buffers are discussed in Chapter 15.

Ways to Stop

Here are the states a voice can be in:

```
enum VOICE_STATE
{
  STATE_WAITING=0,     // Waiting to play
  STATE_PLAYING,       // Playing at last update
  STATE_PAUSING,       // Pause requested
  STATE_PAUSED,        // Paused at last update
  STATE_STOPPING,      // Stop requested
  STATE_STOPPED,       // Stopped when last updated
  STATE_COUNT          // Number of states
};
```

Voices may stop when they reach the end of the current sample unless they're loops or when the underlying Source is hijacked for a more important sound. They can also be stopped deliberately by calling the Voice::Stop() method via the associated VoiceHandle. This kind of synchronous stopping is the simplest to implement. Unsurprisingly StopSource() is the physical layer method which asks a Source to stop playing.

```
void Voice::Stop()
{
  m_State = STATE_STOPPING;

  if ( m_PhysicalSource  != NO_SOURCE )
  {
    StopSource( m_PhysicalSource  );
  }
}
```

NO_SOURCE and m_PhysicalSource are explained below, with the other private class properties. Notice that m_State is set to STATE_STOPPING; it won't change to STATE_STOPPED till the physical layer update has confirmed that the source has actually stopped—remember, sources are asynchronous, like threads. Similarly, the Virtual Voice destructor calls VoiceHandle::OnVoiceDeath() to tidy up when the associated voice dies.

```
Voice::~Voice()
{
  if( m_pVoiceHandle )
  {
    m_pVoiceHandle->OnVoiceDeath();
  }
}
```

It also calls its own method, Voice::OnVoiceHandleDeath(), to ensure the Virtual Voice forgets its previous state and will be re-initialised properly later. This is called from the VoiceHandle destructor and the FadeToStop() method below.

```
const float FLAG = -99999.f;

void Voice::OnVoiceHandleDeath()
{
  m_Velocity.w = m_RelativePositionOld.w = FLAG;
  m_RolloffDirty = true;
}
```

Virtual Voice Priorities

There are thousands of potential sound sources in a complex game or virtual environment. Often there's not enough resources to play all of them, and no need to if all the most important and obvious sounds are prioritised. Every sound that might currently be playing is known as a Virtual Voice. This chapter explains how to tell the difference between Virtual Voices and playing ones and how to make sure that your customers hear everything they need at a consistent rate.

Once you've identified all the things you'd like to play it may be necessary to cull some to fit the capabilities of the runtime system. If there are more voices than it can handle you need to prune out the least important, least obvious sounds from the mix so that more significant ones can play. This is the key to supporting the most capable platforms best, without penalising them to accommodate slower systems or squeezing the latter out of contention. Considerations are:

1. The priority assigned to the voice
2. How loud the sample is
3. Whether the listener is facing the source
4. How far away the source is
5. How many similar sounds are already playing
6. Whether the sample is looping

After experimenting with many priority schemes, a simple set of three has been found more effective than any complicated arrangement with signed or decimal values, where the right choice depends as much upon the ego of the chooser as the needs of the system. The terms VITAL, IMPORTANT and OPTIONAL can be consistently interpreted by all team members, unlike continuous values, and facilitate dynamic sub-categorisation by distance or other analogue measures. The same set is also used to prioritise error and status reports and file loading.

```
enum PRIORITY
{
  PRIORITY_OPTIONAL=0,
  PRIORITY_IMPORTANT,
  PRIORITY_VITAL,
  PRIORITY_COUNT
};
```

Considerations 2, 3 and 4 can all be handled in one step by using the effective volume of the sound after applying distance, group and listener attenuation to its volume. Priority is applied first, and our three-level scheme means that it's easy to group voices that way and sort within each priority group. Vital voices all sort to the top. It's a very serious error if there are not enough physical voices to play all these.

Lower-priority voices come in two sets, important and optional. We play as many important voices as we can, regardless of their effective volume. It should be rare but not impossible for an important voice to be culled. After that we consider voices in the third group, the optional ones. We play as many of those as physical voices permit and don't worry if they're culled, because there must be plenty of more significant sounds playing already.

When Jon Mitchell and I implemented this approach for racing games we found that there were circumstances when important voices in one group, like tyre sounds, were being squeezed out by others, like collisions. Rather than prioritise one group and make the other set of sounds optional, we deferred the decision to the runtime system, informed by the context. The key to this was finding a clear way to express the context which would give predictable results and allow us to uphold the principle that it's a reportable error if there are more vital and important things to play than physical voices to play them.

We introduced the concept of MAX_IMPORTANT_VOICES as a property of each group. Beyond that limit, sounds can still be triggered, but if they have important priority they'll be dropped to optional to give another less-busy group a better chance of having its important sounds played. As important voices finish, this frees up priority slots in the group.

This reassignment of priorities can be handled at several points, with subtly different results. The simplest is to keep a count of the important voices in a group that have already been triggered and apply the cap as soon as the limit is reached. It's safest to recalculate this count as part of each update, so you don't have to adjust the other way as important voices die, either because the end of a one-shot is reached or something more important takes precedence.

For more complicated but potentially better-sounding results, count and compare the volumes of all the important samples in a group that could be playing. If it exceeds the group limit demote the quietest ones to optional. You must do this for all groups before you can be sure which sounds fall into each category. Again, there should be no need to keep history or update counts when voices stop for whatever reason.

The looping consideration might come as a surprise, but it's important for two reasons. If you have a choice between culling a one-shot sample or a loop, you should generally drop the one-shot. It's less obvious to the user and easier for the runtime system. A one-shot sample will stop anyway and typically tails off in volume, unlike most loops. Restarting a loop is relatively complicated, especially if it's a long one and you want it to carry on from somewhere other than the start.

Public Properties of Each Virtual Voice

These are the properties of each Virtual Voice. Any of them can change in any update frame, tens of times a second. Provide a suitable get/set interface to tweak these.

Table 10.2: Virtual voice object, public members

Variable	Type	Initial value
m_State	eState	STATE_STOPPED
m_Volume	float	0.0
m_Pitch	float	1.0
m_Filter	float	0.0
m_Priority	ePriority	PRIORITY_OPTIONAL
m_Velocity	Vector	{0, 0, 0, FLAG}
m_Position	Vector	{0, 0, 0, 0}
m_Size	float	1.0
m_MinDistance	float	1.0
m_MaxDistance	float	100.0
m_Rolloff	float	1.0
m_WetSends	float[]	SILENT
m_EndpointSends	float[]	SILENT
m_LFE	float	SILENT
m_DopplerScale	float	1.0
m_Display	bool	False
m_Name	char*	Voice #

Pitch is a scale factor for the original sample rate. 2.0 is a sensible upper limit, as higher values involve decoding and discarding a lot of redundant data, wasting CPU time and distorting the sound. If you need full-range audio over many octaves, consider cross-fading or granular synthesis rather than relying on one sample. There's no lower limit, but a pitch of 0 is conventionally used to pause a waveform.

Volume is the attenuation in decibels, like WetSends, EndPointSends and low-frequency effects (LFE) which specify attenuations for additional reverb, controller speaker and sub-woofer outputs. You could move some or all of these properties to the Group level, but that may involve creating extra groups which would not be needed if multiple routes were handled on a per-voice basis.

Sends are expensive on some platforms, so there's a workaround when you want more reverbs than you can efficiently send unique mixes. I recommend that the largest reverb—typically the one associated with distant reflections—be allocated first, with local ones—side reflections, room and vehicle interiors—set up later. This is because nearby reverbs can be driven from the same submix as a "dry" output without sacrificing much control, so if you can only afford one true "wet send" it should be reserved for far-away echoes, where the wet-send control allows per-voice distance modelling. This has more than doubled the number of voices I've been able to use on some OpenAL configurations. Individual reverb unit level controls then set the amount of side reflections or cabin reverberation in the mix.

Filters and reverb sends do not need to be updated as often as voice volumes, pitches and positions. If effect updates are expensive, consider implementing a round-robin update that tweaks half the filters each frame and no more than a quarter of sends. This greatly improved X-Fi performance on Windows.

Sources can have sizes, so they get wider close up and narrower with distance. Set **Size** to twice the radius of a volumetric source or zero for a point source. Sizes are more human-friendly measures than radii, though radius is used internally. Use the same units as for distances.

Rolloff and DopplerScale are dimensionless. Rolloff determines how progressively a sound is attenuated between MinDistance and MaxDistance. DopplerScale reduces or eliminates the Doppler effect when otherwise it might seem excessive. These concepts are explained in detail in Chapter 11 and Chapter 13.

The **Filter** property works like a tone or treble control. It's an attenuation expressed in dB, typically in the range 0 (max) to -96.3 (16-bit silence), where 0 denotes a fully open low-pass filter and lower values set the attenuation at 5 kHz for a filter with a slope of -12 dB per octave. This is a second-order low-pass filter, as illustrated in Chapter 20, so -20 gives an attenuation of -32 dB at 2.5 kHz and -8 dB at 10 kHz. This was the definition used in hardware implementations of OpenAL, for Audigy, X-Fi and similar PC soundcards, but beware—the fallback "generic software" implementation, for users who lack DSP hardware, uses a cheaper, shallower first-order filter slope, which falls off much less quickly.

Platform drivers often provide filters with undocumented slopes, which makes cross-platform consistency problematic unless you avoid those and roll your own portable code. There's no correct way to map the control value for a first-order filter to a second-order one or vice versa—the curves are different—but if you can't avoid using a mixture of filter slopes you may get away with converting the expected second-order attenuation in dB into a gain, 0..1, then passing the square of that value (which will be smaller) to a shallow first-order filter control. I've used the square root of the gain similarly, to map a value intended for a shallow software filter to a second-order hardware one. Neither technique is recommended unless it's just a fallback for people with unexpected hardware.

The public interface includes Volume, Pitch and Filter controls for each voice, but if you're implementing all the filters yourself you'll probably wish to expose a larger set of tone controls, replacing the single low-pass Filter with up to five controls—Low, Mid and High band-pass attenuations in dB, plus Bandwidth and Centre frequencies expressed in Hertz.

The bandwidth could be expressed as a range of frequencies in Hertz, or a "Q factor" where a value of 0.5 gives the widest possible midrange band, extending all the way down to 0 Hertz, 1.0 permits the fastest implementation, and higher values give greater resonance increasingly close to the centre frequency and increasing risk of instability. Technically, the reciprocal of the normally quoted Q factor, 1/Q, is the value used at runtime.

An infinite Q (where 1/Q is zero) turns the filter into an oscillator; very high values have a similar effect, though the trapped note will die away eventually. Values in the range 0.5 to 8 are musically useful. Classic mixing desks often used a low pre-set Q, around 0.7, to separate bass and midrange frequencies in an ear-friendly manner. Chapter 20 gives implementation details.

Hertz are the preferred units in which to express the centre frequency. Though a pitch ratio related to the output sample rate could serve the same purpose, a value of 2400 Hertz is easier for designers to relate to than a multiplier of 0.05 (presuming 48-kHz output) and more portable if not all your platforms run at the same rate, as might become the case later even if you start out aiming for the same rate across the board. GameCube hardware mixes at 32 kHz, apart from its single 48-kHz music stream, so filtering anything above 16 kHz is pointless.

Dummies, Stubs and Defaults

The initial values for a Virtual Voice are copied from a special instance of a voice object which is known as the default voice. There's one other special-case voice, the dummy voice, which is automatically used when a real voice stops. The voice handle points to the dummy instead so that belated attempts to update properties (or read them back) don't refer to memory which is no longer bound. Neither default nor dummy voices are associated with sounds that are playing, though they have the same properties as any Virtual Voice.

The stubs and dummy voices comprise some of the "scaffolding" Fred Brooks recommends in the chapter "The Whole and the Parts" in his vital book *The Mythical Man Month*.[1] It's also helpful to have a method to report, to stdout or the debugger log, the internal state of all or any objects in the system. This includes condensed reports on virtual and physical voices, with short codes for enumerated states.

You might prefer reporting scalar speed rather than the velocity vector to save space, but include the full 3D positions, relative to the listener or the origin, and enough information for a designer to identify each playing sample, using bank and sample names, or paths in the case of streams. These parts are the widest and hence hardest to tabulate, so they fit best at the end of the line. Similar reports on DSP effect bindings, Groups and Listener states are helpful during initial development, title configuration, and when investigating unexpected behaviour.

Audio Off—or Is It?

Game programmers will inevitably sometimes want a build of the game with "audio off," often so they can temporarily re-use memory allocated to us for their own experiments. Rather than encourage them to recompile with our code suppressed, which wastes time and defers the discovery of logic errors, it should be possible to reduce the entire audio system to a stub without stripping out all the code. If the interfaces are carefully designed this can be done without disturbing higher-level code for audio or otherwise.

The trick is to use an implied runtime switch based on initialisation state rather than an explicit compile-time one. If the logical-layer hooks for voices, groups, listeners, effects and asset loading are defined as pure virtual interfaces, those can be initially implemented as stubs: short routines that do nothing but refer to the dummy voice. There's no need to allocate memory for samples or effects, and only the container classes are needed to give the game some dummy objects to play with. The physical layer doesn't appear at all, so this allows games to be ported to platforms in parallel with development of the low-level audio driver.

As part of successful initialisation the stubs are replaced by working audio code. The pure virtual interface means that there's no overhead other than replacement of the stub class addresses with those for the full implementations; you'd have to pick up one or other anyway. Make sure others on your team know about this, understand the rationale and use it—once their code is polluted with #if AUDIO switches, your life will get harder. There should only be one conditional test, around the call to initialise the whole audio system.

The VOICE_STATE constants keep track of transitions when a voice is triggered, paused and stopped. The variable m_State is read-only from the point of view of users of the class. The game can safely read it, but changes to the value should only be made from inside the runtime system. There are more states than you might expect, because we need to draw a distinction between asking a voice to pause or stop and knowing it has actually done so.

A voice is only in STATE_PAUSING or STATE_STOPPING for a single update frame, from the time the game requests the state change until the runtime update for all the voices. A voice may also switch from STATE_PLAYING to STATE_STOPPED during the runtime update if the physical layer indicates that a one-shot sample has reached its end.

m_State is often the only part of the voice structure which needs to be checked for every voice at each runtime update. If a voice is STOPPED there's no need to consider the other member variables. This means it's most efficient to put it in a separate array rather than intermingled with all the other members so that a quick scan through the contiguously packed voice states doesn't cause all the other voice properties to be churned through the cache.

When the runtime system is initialised, allocate an array just for the voice states, packed into bytes or words depending upon the processor architecture, and a parallel array of structures or classes for the other members. If you're on a fast platform with plenty of Level 1 cache this complication can be avoided.

Finding the optimal blend between structures of arrays, arrays of structures or—most often—a structure of arrays of structures is a key part of performance optimisation. But it's something that should be left till later, when your prototype is working, and done only if a profile shows the need. Adam Sawkins had a relevant mantra on his whiteboard when we worked together:

> Get it to work,
> Get it to fit,
> Make it fast.

It seems this is a variant of a remark by Kent Beck; the second line is especially relevant to work on embedded systems like phones and consoles; Beck substitutes the vaguer "make it right" there. Donald Knuth famously described premature optimisation as the root of all evil and says we should pass up small efficiencies except in a small proportion of our code.[2]

The order is important; getting things to fit minimum space is often a good step towards making it fast. But if it doesn't work before then, you're definitely being premature. Unless you're short of memory or bandwidth or CPU time, and the profile confirms that the working code is a bottleneck, you're wise to leave it working, and keep things simple. Ninety-seven per cent of the time, Knuth generalised, there's no need to optimise code. But attention to the other 3% may make a disproportionately big difference.

It's tempting but counterproductive to optimise things that look easy to speed up. Profile your code, with a realistic workload, on your most time-critical platforms—not just where it's easiest to get figures; a Windows profile is apocryphal at best. Let that be your guide to what really needs attention.

Private Members of Each Virtual Voice

Table 10.3: Virtual voice object, private members

Variable	Type	Initial value
m_pVoiceHandle	VoiceHandle*	NULL
m_Fading	bool	false
m_StopWhenFaded	bool	false
m_StartFadeVolume	float	0.0
m_EndFadeVolume	float	0.0
m_FadeDuration	float	0.0
m_FadeStartTime	float	0.0
m_RelativeVelocity	Vector	{0, 0, 0, 0}
m_RelativePosition	Vector	{0, 0, 0, 0}
m_RelativePositionOld	Vector	{0, 0, 0, FLAG}
m_DistanceAttenuation	float	0.0
m_ListenerAttenuation	float	0.0
m_pGroup	Group*	NULL
m_PhysicalSource	sourceID	NO_SOURCE
m_FinalWetSend	float	SILENT
m_FinalLFE	float	SILENT
m_pSoundSpec	soundInfo*	NULL
m_RolloffFactor	float	0.0
m_RolloffDirty	bool	True
m_Streamed	bool	False

The sourceID in m_PhysicalSource is an index which binds virtual and physical voices. It has the value NO_SOURCE for voices which are not playing and hence entirely virtual.

```
typedef unsigned sourceID;
const sourceID NO_SOURCE = 0xA000001;
```

The Vectors, Attenuations and "Final" floats are used during voice update. m_pGroup points to the associated Group, so its properties can be factored in. m_pSoundSpec may be cast to sampleSpec or streamSpec depending upon the value of m_Streamed and the audio data location. m_pVoiceHandle points back at the handle, if any, used to play, stop and update the voice. The flag m_RolloffDirty is set after changes to the voice distances or rolloff curve, so that they can be combined during the next update, as Chapter 11 explains, into a new m_RolloffFactor.

Voice Updates and Sorting

The Virtual Voice class static Update() method computes the potential volume of each voice, for the main output and individual effect and LFE sends. It then handles state changes, putting those in STATE_WAITING into a list of candidates to start playing. Voices in STATE_PLAYING or STATE_STOPPING are moved to STATE_STOPPED if the physical layer method IsSourcePlaying() returns false and their Voice Handle set to NO_SOURCE.

Once all the Virtual Voice positions and volumes have been updated they are sorted into three groups by priority. If there are more candidates than physical voices available, the lowest-priority ones are culled. Physical voices are then assigned to the higher-priority ones. A VITAL message is emitted if any VITAL voices cannot play, indicating a serious design error, while an IMPORTANT message denotes the rare but permissible case of that category exhausting the physical voice supply. These messages occur in update, not when the voice is triggered, so they need to identify the asset that failed to play.

Within a priority group, voices are sorted by attenuation on their main output, favouring the loudest. Use a "stable" sort for consistent results when several samples happen to match exactly in volume. Bias looping assets upwards by 6 dB to account for their persistence. When there are no free sources we search for one with lower priority in STATE_PLAYING and call StealSample() to take it over.

Things get hairier if the physical-layer implementation requires different types of source for each channel count. In that case, when there are no free voices in a multi-channel category, you need to carry on looking through the lower-priority voices till you find one with the count you need and steal that, with a warning. Multichannel assets are usually massively outnumbered by mono positional ones and tend to be of higher priority, so steal operations almost always swap one diegetic mono asset for another. You can probably make this issue go away by setting suitable priorities, but it's worth bearing in mind if designers haven't got to grips with 3D yet and want to include more stereo or soundfields than they can realistically play at once. It's not bitten me yet.

Automatic Fading

There are six private members of the voice class associated just with fading up and down. These are not strictly necessary, as all they do is adjust the voice volume progressively, which could be handled in the game's own voice update, but this is such a commonly used process that it's worth building into the runtime system. It can implement arbitrary volume envelopes, including ADSR and supersets, over a seamless chain of samples without intervention except when target levels are reached.

The UpdateFade() method is called at the start of the voice update each frame while m_Fading is true. It adjusts the master volume control of a fading voice in proportion to the amount of time spent so far fading between the start and end levels. It can be used to fade a voice either up or down and optionally to stop the voice automatically when the target volume is reached. Here's the Virtual Voice method that tells a voice to start fading. It takes two parameters, a target volume in dB and a duration in seconds, and uses those to initialise the private members used in subsequent updates while the fade itself takes place.

```
void Fade(float destVolume, float duration)
{
  m_StartFadeVolume = m_Volume;
  m_EndFadeVolume = destVolume;
  m_FadeDuration = std::max(duration, 0.f);
  m_FadeStartTime = GetTime();
  m_Fading = true;
  m_StopWhenFaded = false;
}
```

The call to std::max() copes with the possibility of a negative duration, clamping it to at least zero seconds. There's no unsigned float data type because of the way the sign of an IEEE-754 value is tucked away. The standard template library also provides std::min(). Both may be three times faster than fmax() and fmin() defined in math.h in C99 or later, because they skirt IEEE-754 edge cases irrelevant to well-formed audio. Old-school C programmers who laugh in the face of double evaluation of functions (unlikely but possible and hence risky) use a macro like this:

```
#define MAX(A, B) (((A) > (B)) ? (A) : (B))
```

GetTime() is presumed to return the wall-clock time in fractions of a second, as a float since some arbitrary epoch. Provided you use the same function call in UpdateFade() as you do in Fade(), the epoch start time is irrelevant.

Fading down is often a prelude to stopping a voice playing entirely, so that combination merits special treatment. FadeToStop() calls Fade() as above, then sets a flag to tell the internal update routine to stop and release the associated physical voice once the specified duration is passed. It also releases the associated Virtual Voice handle right away. This forestalls the possibility of a VoiceHandle leak later by turning it into a "fire and forget" voice—subsequent updates will be handled automatically. This extension of the method above completes the public implementation of automatic fading:

```
void FadeToStop(float destVolume, float duration)
{
  Fade(destVolume, duration);
  m_StopWhenFaded = true;
  if (m_pVoiceHandle)
  {
    m_pVoiceHandle->OnVoiceDeath();
    m_pVoiceHandle = NULL;
  }
}
```

Since these are intended as "fire and forget" convenience functions and it'd not make much sense for the game to manually set the volume at the same time as the fade sub-system does so, the implementation clobbers the same Volume control that the game would use for manual level tweaking. If you try to set this from the game it will be overridden while the fade takes place. Insert another private member (e.g. m_FadeVolume) just for this sub-system, and add it to m_Volume each update, if you really need both controls to work at once. As distance-related fading is independent of either, you can probably live without the extra complication.

```
void UpdateFade()
{
  float timePassed = GetTime() - m_FadeStartTime;

  // Find out how far along the fade we are and interpolate
  float fadeCompletion = timePassed / m_FadeDuration;
  m_Volume = ((m_EndFadeVolume - m_StartFadeVolume) *
    fadeCompletion) + m_StartFadeVolume;

  if (fadeCompletion >= 1.f)
  {
    // Stop fading, and perhaps stop the voice too
    m_Volume = m_EndFadeVolume;
    m_Fading = false;
    m_FadeDuration = 0.f;

    if (m_StopWhenFaded)
    {
      OnVoiceHandleDeath();
      Stop();
      m_StopWhenFaded = false;
    }
  }
}
```

References

[1] *The Mythical Man-Month; Frederick Brooks, Addison-Wesley, 1995, ISBN 0201835959*

[2] *Structured Programming with Go to Statements; Donald Knuth, ACM Computing Surveys, ISSN 0360-0300, Volume 6 Number 4, December 1974, page 268*

CHAPTER 11

Modelling Distance

Distant sounds are quieter than nearby ones. This chapter discusses how much quieter and why, and creative ways to exploit the effect. The associated code forms part of the Virtual Voice update, using public and private variables introduced in Chapter 10.

The term "distance attenuation" is quite a mouthful, and the concept is more often referred to as "rolloff" because of the way the audibility of sounds rolls off towards silence with increasing distance. This behaviour and hence the rate of rolloff depends upon several real-world factors, including the composition of the atmosphere, temperature, humidity, time of day and the height of the listener and source above ground. These affect the tone as well as the volume of the distant sound.

These audio runtimes expose a set of controls which can be used either to model the interaction of these factors or directly to create any desired layering of sounds, be that realistic, symbolic or aesthetic. The types and polarities of the factors which influence distance perception are noted, but it's the designer's job to assign their priority for a given model of the world. Strict realism is rarely the best choice in games, but it helps to know what *might* matter before you decide what really *should*.

> Strict realism is rarely the best choice in games.

Atmospheric water vapour—or visible fog, rain, hail or snow—preferentially absorbs high frequencies. This is a function of temperature as well as humidity. Cyril Harris exhaustively tested and documented this for NASA in the 1960s.[1]

When either source or listener is elevated the sound travels along at least two paths, directly through the air and along the ground. Particularly at dawn and dusk, temperature inversions introduce a third longer path and make distant sounds more apparent at that time of day, as layers of the air differ in temperature and density.

These paths differ in length and are also affected by wind and other air movements such as "thermals." These effects vary the timing and phase of each sound, causing comb filtering and some cancelation of high frequencies and muddying of bass ones. You've almost certainly heard this effect when listening to public-address systems at large outdoor venues. The timbre of the sound is alternately muffled and harshened as air moves and paths interact.

It follows that there's no right value for the rate of rolloff. In any case you might want to vary it creatively, either to de-emphasise distant sounds in a cluttered scene or to make distant threats more apparent. The rule of thumb is that sounds fall in volume by about 6 dB for each doubling of distance, but circumstances or aesthetics might make twice or half that factor more appropriate in your context.

Unlike photography, our implementation allows you to vary the property source by source, so birdsong might fade less quickly than the scrabbling of rats. You can even vary it on the fly, if the birds land, for example. This is not strictly realistic, but providing it's consistent and sympathetic with the needs of the game it has a place in real-time audio rendering.

Rather as a camera operator might choose lenses to impose a certain depth of field, literally focussing on objects at a chosen distance and blurring those in front or behind, a sound designer should be able to control the rate of rolloff to suit the game genre, scenario and relative significance of the sounds to the player. A game of exploration might combine occlusion and unnaturally slow rolloff. The opposite works best for a crowded city circuit racer by emphasising the sources closest to the listener. Such design decisions also affect Virtual Voice reuse by delaying or advancing the culling of distant voices, further influencing the mix.

The rolloff associated with a given distance depends upon six factors:

- The distance between source and listener
- The size of the source
- The source's MinDistance
- The source's MaxDistance
- The noise floor (nominal silence)
- The desired rolloff curve

The first factor is the distance from the source to the centre of the listener, which may be slightly affected by the size of the source. We model volumetric sources as spheres in this example, so non-point sources are closer by their radius.

The runtime system computes this distance anew at every update, from the relative 3D positions and the source object's diameter, increasing attenuation with distance. It doesn't matter what units you use here, providing they're consistent—metres are often chosen, but designers are welcome to use cubits or furlongs if that's more natural for them.

The two calibration distances, Min and Max, must use the same units. The first represents the distance within which the sound will not get any louder. This is technically wrong but a necessary simplification, since if a sound gets steadily quieter with each doubling of distance, strictly speaking it should be infinitely loud when the listener is coincident with the source!

The conventional and entirely practical way to get around this snag is to pretend there is a certain MinDistance within which radius the sound volume is a constant maximum, corresponding to no attenuation. Providing this is greater than zero the asymptote is avoided. The value can be tuned to suit the source—a bee would have a smaller MinDistance than a bonfire, for example—but one metre is a good start for most human-scale worlds.

Maxed Distances

MaxDistance sounds similarly simple, but a strange design decision made in Microsoft's DirectSound3D means that audio runtimes interpret it in one of two incompatible ways. When I ask designers what MaxDistance means, they almost invariably say it's the distance at which we can no longer hear a sound. They are generally surprised to hear that DirectSound3D defines it as "*the distance beyond which the volume does not get any lower.*"[2]

Microsoft also made the default MaxDistance a billion metres and supplied a spreadsheet to help mathematically-minded people work out the actual attenuation associated with a set of distances and a dimensionless "rolloff factor" associated with the single listener.[3]

DirectSound3D is still part of Windows but is only emulated by software now. The peculiar DirectSound definition is still used in some audio products, like Firelight's FMOD. But Wwise, XAudio2 and OpenAL all use the other interpretation of MaxDistance, which makes more sense in general, since it defines a consistent "inaudible" volume level reached at that range. In DirectSound3D the maximum attenuation varies whenever the global rolloff factor or either of the per-source distances is changed and only corresponds for sources with the same Min/Max ratio.

Noise Floors

The alternative approach guarantees that every sound will be attenuated to the same low level when it is MaxDistance away from the listener. It therefore requires that this low level be defined when the Virtual Voice system is initialised. It's known as the noise floor and typically set to an attenuation of about -50 dB. It's always a negative number and really an artefact of Virtual Voice management rather than distance attenuation. The correct value for the noise floor depends upon the mix intended.

Any quieter voice is a candidate for substitution. You'd need to be in a very quiet room with speakers turned up loud to notice it stopping, and the assumption is that something will be playing at a higher volume—typically music, weather or location ambience—unless the entire mix is deliberately muted for what James Slavin refers to as "percussive silence," in which case other volume controls will have taken priority.[4]

The noise floor cannot sensibly be changed while sounds are playing, but the minimum and maximum distances can be, as can the rolloff factor which is maintained here for each voice rather than just for DirectSound3D's singleton listener. More often they're set when the voice is played to suit the asset and its place in the mix. Changes between first- and third-person camera views, single and multi-player, or game and replay modes may benefit from distance or rolloff tweaks, just as a photographer might change lenses under such circumstances.

We've defined the volume curve between MinDistance—0 dB of attenuation—and MaxDistance, when it falls to **noiseFloor** decibels. These bounds delimit the vertical axis in Figure 11.1. But for a layered "depth of field" effect we need to change the shape, as well as the slope, of the associated transfer function or "rolloff curve." It may be concave to emphasise close sounds or convex to give a greater audio depth of field, as shown.

Values less than 1.0 roll off more slowly than normal with distance, making distant sounds more audible, whereas greater values emphasise nearby sources and rapidly attenuate them as they move away. Rally games I've worked on use a low gamma, while circuit racers use a higher value to emphasise the immediate environment. In either case the sound fades smoothly to the noise floor as the maximum distance is approached and is loudest at and within the minimum distance.

The calculation of attenuation is performed in two steps. The private member m_RolloffDirty is true only when either the curve m_Rolloff or the min or max distances have changed; if so, we must recalculate m_RolloffFactor,

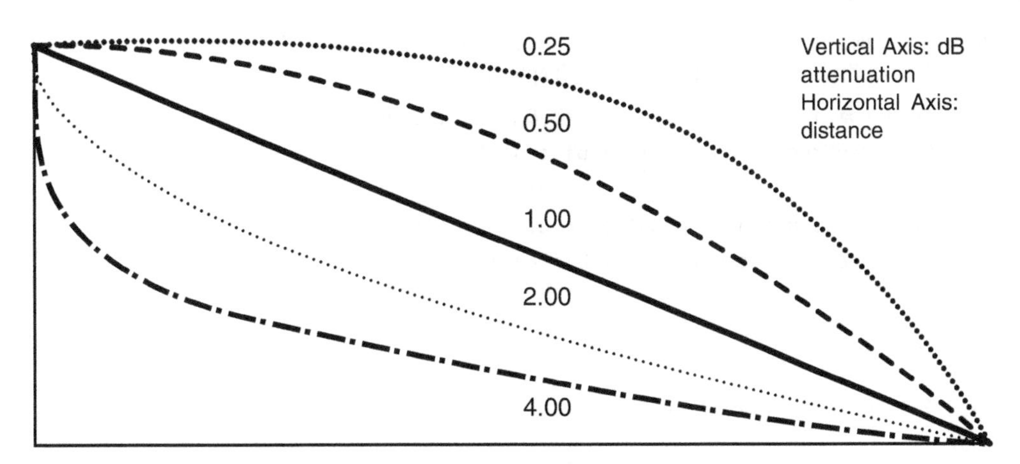

Figure 11.1: Distance-related rolloff curves

which is used later to convert a distance into a suitable attenuation. The dirty flag is set by the methods which set m_Rolloff and voice distance properties.

It's rare for these distances to change once a voice is playing, so to keep the runtime work to a minimum we cache this value while the distances and curve remain unchanged. This eliminates the need for costly floating-point divisions and calls to the powf() library function when rolling off to a new noise floor at MaxDistance.

The Virtual Voice member m_Rolloff set the curve shape of attenuation with distance. The per-voice rolloff factor is then a function of the noise floor (that being the range in dB between the levels at minimum and maximum distance) divided by the difference between minimum and maximum distances, raised to the power of the chosen m_Rolloff.

```
if (m_RolloffDirty)
{
  m_RolloffFactor = AudioConfig.noiseFloor /
    powf(m_MaxDistance - m_MinDistance, m_Rolloff );
  m_RolloffDirty = false;
}
```

After this the "dirty" flag is set to false, freezing the calculated value of m_RolloffFactor until the associated distances or curve are deliberately changed. Dirty flags are generally useful as long as the work they seek to avoid is much slower than simply testing the flag. It's worth using them to avoid a library function call but not to skip a multiplication or addition. Bear in mind that they only speed up the typical case, when a voice has not been totally updated in the current frame—this means that they don't benefit the worst case, when everything everywhere changes, and that's the most important path to optimise.

Most of the public members of the Virtual Voice class can be updated directly, though it's good practice to provide inline functions to Get and Set their values, as that improves encapsulation, enables some properties, like m_State, to be read only—and makes debug logging easier. An optional stack trace on these methods helps track down game-code references by identifying the calling sequence.

The interactions between m_Rolloff, m_MinDistance and m_MaxDistance mean that whenever any of those three properties is changed, m_RolloffDirty must also be set to true. The same applies whenever the voice is re-initialised, before it is played for the first time, to ensure that m_RolloffFactor is correct.

This calculates m_DistanceAttenuation in decibels, using the posInListenerSpace Vector computed by the TransformToListenerSpace() method fleshed out in Chapter 13:

```
m_DistanceFromListener = VectorLength(&posInListenerSpace);
if (m_DistanceFromListener > m_MinDistance)
{
  m_DistanceAttenuation = m_RolloffFactor * powf(
    m_DistanceFromListener - m_MinDistance, m_Rolloff);
}
else
{
  m_DistanceAttenuation = 0.f;
}
```

Tonal Fluctuations

That is the minimum needed to make sounds fade up and down smoothly as they move around the listener. Several other manipulations can improve the perception of specific distances or simply add variety to the mix. The amount you will use those depends upon the extent to which you're already using filters creatively, the size of the game world and how much processing time you have spare.

Distance affects tone as well. This can be implemented at the game level via the voice property m_Filter or internally by using the distance attenuation factor to progressively attenuate high frequencies. In this case the same Min and Max Distances can be used, but you may find that a faster rolloff curve is desirable; this can use the same algorithm as the volume calculator, but with a distinct and lower control value for the curve, controlled via an additional m_RolloffHF value.

The varying tone associated with hearing distant sounds is caused by air movement and multi-path interactions. Modelling this adds variety to open world environments, providing the results are not unrealistically predictable. The periodicity of fluctuations and the amount of the effect should increase with distance, while the variability is a function of wind turbulence.

Figure 11.2 shows the control of panning, cross-fading and asset volume based on the player's orientation in an FMOD middleware project. This recreates the feeling that the wind is directional and the effect of facing into the wind.

Air velocity and turbulence modulate weather effects like rain, wind and snow. They also affect Doppler shift associated with movement of the source and listener, providing the air movement is modelled as a directional vector rather than a scalar speed.

Fluids like air and water flow in two modes, laminar or turbulent, with characteristic sounds. It's useful to know that the power associated with smooth laminar flow is proportional to velocity, whereas the power and hence volume associated with turbulent chaotic flow is proportional to the square of its velocity.[5] The interaction between these two curves, as speed increases and flow becomes chaotic, gives an audible insight into game physics.

The phasing of sounds from similar directions should be correlated, as this helps the listener associate them together. A few low-frequency oscillators with relatively prime periods of a few seconds can be ganged together to wobble the voice filter settings, with their period and amplitude proportional to distance and relative phases determined by their absolute source position. This approach ensures predictable maximum deviation and a long irregular period before the wave pattern repeats, as well as making adjacent sounds vary sympathetically.

Distant Reflections

Rolloff affects only the direct or "dry" sound from the source—the level of reverberation, or indirect "wet" sound, is deliberately not affected, so that as sources recede the echoes predominate, in accordance with the theory that the ratio of reverberation to direct sound is a primary distance cue. This is the "Craven hypothesis" due to surround-sound pioneer Peter Craven, explored in a paper by Eoghan Tyrrell.[6]

The amount and timbre of reverberation associated with a playing voice depends upon the settings of the voice, its group and associated reverb units—longer delays indicate a larger space, and lower attenuations make it sound more reverberant. As with direct sounds, both the voice and the associated group set the level—the m_WetSends array stores reverb input attenuations in decibels.

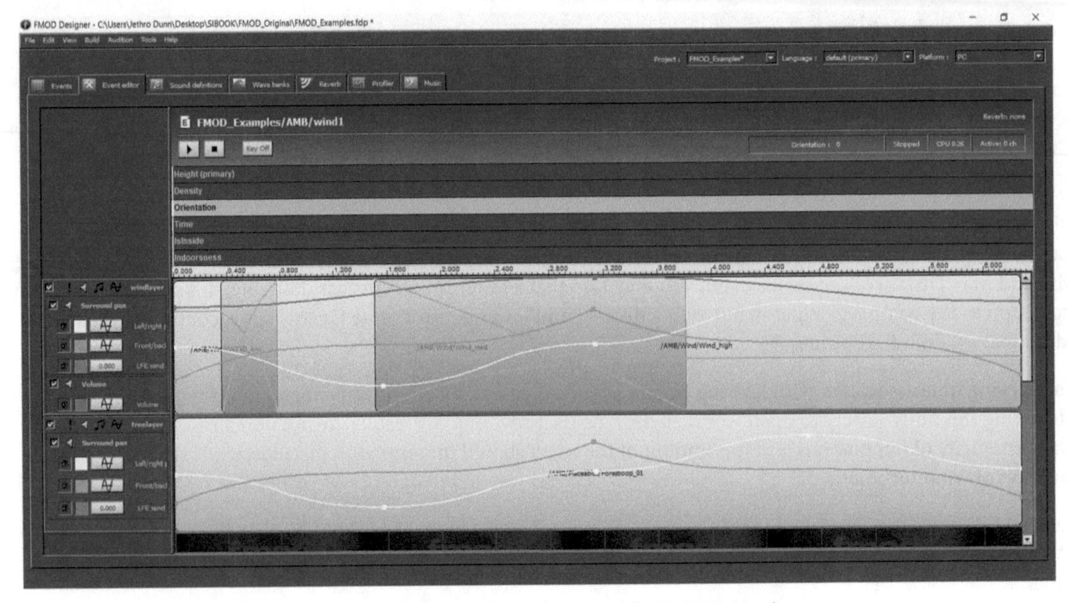

Figure 11.2: Directional wind control in FMOD Designer

The final output volume for the voice is computed by adding all the component attenuations, m_Volume, m_ListenerAttenuation and m_DistanceAttenuation, with that from the associated Group, m_pGroup->m_Volume; this total will be used to work out which voices to purge in a heavily loaded system.

Double Jeopardy

If you're building a cross-platform abstraction over existing middleware you need to make sure that it's not doing its own, often sparsely documented, distance modelling in parallel with yours. Otherwise the effect is likely to be excessive and different between platforms. Disabling theirs also speeds things up.

The method to do this depends on the host. OpenAL users ensure this by passing AL_NONE to the alDistanceModel() function. Unity 5 requires us to zero the Audio Source's Spatial Blend setting. XAudio needs appropriate X3DAudioCalculate() flags. 0x10001 denotes X3DAUDIO_CALCULATE_ZEROCENTER | X3DAUDIO_CALCULATE_MATRIX, so that the panning coefficients are computed but Doppler, Reverb and Distance filtering are not, and the front centre speaker is left free for non-diegetic content.

Propagation Delays

There are times (sic) when you might want to deliberately delay a sound. Sound travels much more slowly than light, so the synchronisation between the display and sound of a muzzle flash, lightning strike or thermonuclear explosion depends upon the distance between the source and the listening observer. When asked by designers how to calculate this delay, pedants reply, "In what medium?" It's usually air, but this—and even wind-speed and direction—makes a significant difference.

In air, on earth, at sea level, sound takes about 5 seconds to propagate 1 mile. But if the sound is passing through water it's more than four times faster, and it's 15 times faster through solid material like steel. On alien planets it goes at whatever speed you like, though if you're game is sci-fi rather than fantasy you may want to base the figure on physical properties of known materials. It's well known, since Ridley Scott's film *Alien*, that "in space no one can hear you scream," but strictly speaking no one can hear *anything* in a vacuum—there are no molecules to carry the vibration.

As Principal Programmer on the binaural rendering system for the space exploration title *Elite: Dangerous VR*, this gave me pause for thought. Luckily the designers at Frontier had a neat backstory to explain being able to hear remote spaceships and their weapons as well as local sounds—in essence, the ship adds back audio cues you need to survive—so the player's sonic skills and our own are not wasted.

But of course it's a game, and designers can do whatever they like, preferably with some sort of rationale. It's still important that the distance delay model should be consistent throughout the game world, even if it acts differently from mundane reality. And you should think hard before introducing a delay that might seem unfair to your customers, even if it's strictly realistic.

Bullets travel faster than sound, so the at-times-painfully realistic *Operation Flashpoint* games delay both the arrival of the slug and the associated sound, when a sniper picks you off from a distance. This means you may see the muzzle flash if you happen to be facing the right way, but you're liable to be dead before the associated audio reaches you. Unless the first shot misses, audio will not help you locate and avoid the shooter in this sort of game.

This exemplifies processes that can make a simulation more realistic but less fun. Brutal realism is a selling point in *Operation Flashpoint—Dragon Rising* but unlikely to impress casual players of *Call of Duty*. Even "friendly fire"—the accidental or deliberate targeting of players on the same side—is typically disabled there, except in hard-core multi-player modes. The needs of simulations and games and expectations of their players necessarily vary by genre and often by game mode too.

A recurrent question relates to the time delay between seeing and hearing fireworks or between lightning and thunder. A delay of one millisecond per foot of distance from the firing and detonation of a rocket works well. Thunder is a volumetric sound without parallel, emitted from all along a lightning bolt and its forks, so while you can work out a typical delay, in practice this is hardly necessary. Providing the sound follows the image, players are unlikely to be distracted by inexact timing. But the closer—and hence bigger and louder—the bolt is to the player, the shorter the delay; this coincidence is a great way to ramp up tension through interactive audio.

The very rare case when lightning—or a mortar shell—strikes right next to the player is important enough to deserve special coding. Imagine a sudden full-screen flash of brightness with accompanying vital-priority top-volume reverberant volumetric sound. This works best if the game runtimes anticipate it and create space a fraction of a second in advance. Fade down inessential groups, ambient reverberation or perhaps the entire mix to create a gap the thunderclap can fill.

Momentary darkening of the ambient light a frame or two before does the same for the eye. The combination is more than the sum of its parts, even if most of the praise attaches to the graphics people. Audio folk are used to that!

References

[1] *Absorption of Sound in Air Versus Humidity and Temperature; Cyril M Harris, Journal of the Acoustical Society of America, ISSN 0001-4966, Volume 40, 1966, pages 148–159*: https://ntrs.nasa.gov/archive/nasa/casi.ntrs.nasa.gov/19670007333.pdf

[2] *DirectSound3D Distances*: https://msdn.microsoft.com/en-us/library/windows/desktop/bb318697.aspx

[3] *DirectSound3D Listeners*: https://msdn.microsoft.com/en-us/library/windows/desktop/ms804993.aspx

[4] *Percussive Silence: James Slavin, Panel Discussion at the "Special Event: Discussions into the future of audio in games" at the AES Audio for Games Conference 2009*

[5] *Physics for Game Programmers; Daniel M Bourg, O'Reilly 2002, ISBN 0596000065, pages 61, 109–111*

[6] *Craven Hypothesis*: http://eoghantyrrell.com/wp-content/uploads/2016/10/TheDistancePanpot_EoghanTyrrell_2016.pdf

CHAPTER 12

Implementing Voice Groups

There may be hundreds of voices in a game but fewer sub-systems (e.g. speech, engine sounds, front end, environment, collisions) sharing those voices. We need a way to manipulate sets of playing sounds that does not force us to keep lists and send loads of messages to voice subsets when all we want to do is change the balance to make the game more realistic or exciting. This is handy not just to let the player tweak things with the usual front-end sliders—music, effects and speech volume—but also for balance and emphasis as the game is played, fading stuff up and down to emphasise key interactions and changes in context.

Groups give designers and programmers convenient ways to control, switch on and off, and route sets of voices tailored to the game. Groups makes game coding a lot easier and simplify tuning of the game audio for situations and platforms. In replays, with a zoom-lens effect, you could dedicate a group to the sounds close to the point of focus of the camera so they don't get lost in a long shot, as well as the usual group for other environmental sounds.

Figure 12.1 shows how Groups are interposed between Voices and Listeners or Endpoints. The first couple of voices are routed via a non-diegetic Group to the first Endpoint, a controller, as well as to the main speakers, the second Endpoint. Most of the Voices are routed there through three more 3D Groups and two Listeners. The last Voice and its Group are only routed to the third Endpoint.

Every Voice belongs to exactly one Group. Each 3D group has one Listener. Off-screen speech and music has a Group but no Listener. Multiple Groups may share a Listener, and Groups may target multiple Endpoints. Within these constraints, the routing is arbitrary. Groups are also used to share filter and reverb effects between Voices.

Groups are obviously useful in multi-player situations, where you need at least one group for each view and another for non-diegetic sounds. They are also a natural way to share DSP effects like filters and echoes between associated voices. There's much more about DSP in Chapter 16, Chapter 17 and Chapter 20.

Table 12.1 shows the properties of a group and the "private" internal state used to maintain it. You set the maximum number of groups you may need when you initialise the system. Groups are cheap to process and in memory—less than 100 bytes per instance.

Group Properties

The ReverbsMask keeps track of the reverb units bound to a group, using one bit for each, from a set returned during initialisation. EndPointsMask is similar but used later to determine which endpoints receive sounds in this group. Both use bit fields so that multiple reverbs or endpoints can be selected by setting more than one bit in the value. The GroupFilter value, in contrast, is an index into the set of filters requested at initialisation, and each group allows only one of these, in addition to the per-voice low-pass filters mainly used to model occlusion and distance. Group filters are counted from one, so zero denotes no group filter.

MaxImportantVoices is used to cap the number of voices a group uses when voices are in short supply, so that each group can be guaranteed a minimum number without running the risk that it takes more than its share

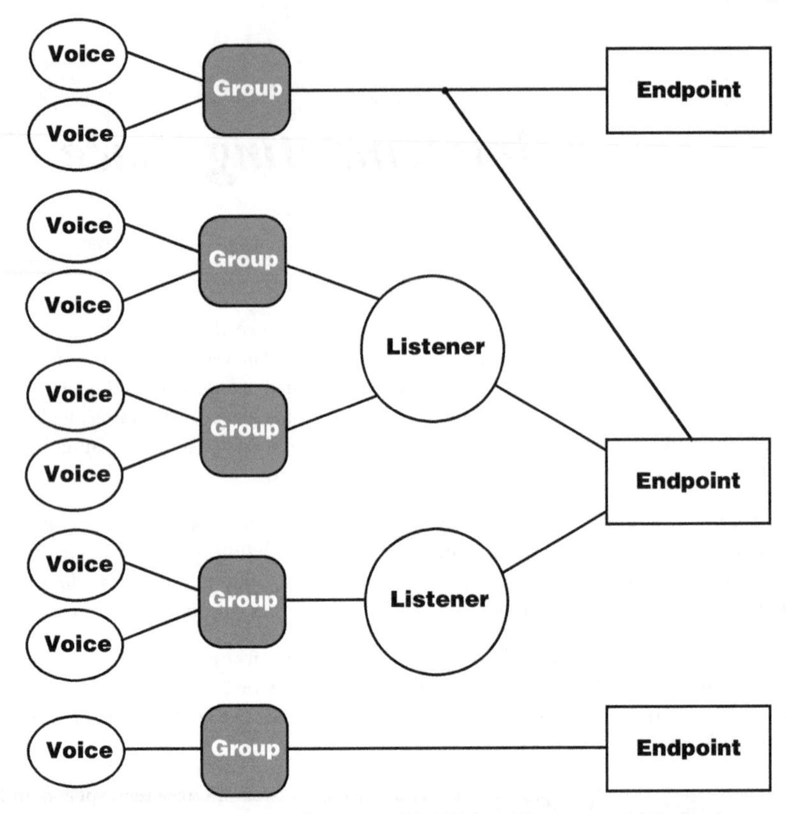

Figure 12.1: How groups interact with Voices, Listeners and Endpoints

Table 12.1: Group class members

Variable	Type	Initial value
m_Volume	float	0.f
m_ReverbsMask	unsigned	0
m_EndPointsMask	unsigned	1
m_GroupFilter	unsigned	0
m_pListener	Listener*	NULL
m_Paused	bool	false
m_Display	bool	false
m_Name	char*	Group #
m_MaxImportantVoices	unsigned	4
m_ProximityBoost	float	0.f
m_PriorPause	bool	false
m_PauseOn	bool	false
m_PauseOff	bool	false
m_PlayingVoices[PRIORITIES]	unsigned	0
m_WaitingVoices[PRIORITIES]	unsigned	0

from other groups playing at the same time. Once more than this value of IMPORTANT voices is playing, any further voices are reduced in priority to OPTIONAL—this allows for capping of the number of voices that a group actually plays without limiting it in situations when the total voice count is no more than the hardware can handle.

ProximityBoost provides an optional low-frequency proximity effect, discussed in Chapter 19. The Voice update passes this down to the platform layer for all voices in the group, after adjustment for other the other attenuations applicable to that voice: fading, distance, orientation and the shared group and individual voice volumes.

The three later Pause members are private—check **m_Paused** to find out if a group (and all the voices in it) is currently paused. The voice-count arrays are read only, updated once per update frame.

Group Static Methods

The Group class needs static methods to allocate, update and keep track of group instances. The constructor is called during initialisation and sets up an array of groups and a method to look them up by index number.

Groups are mainly used as containers for information needed during voice updates, so the only things you need to do with each of them every frame, before voices are updated, is clear the PlayingVoices and WaitingVoices counts and update the pause flags.

First compare **m_Paused** with **m_PriorPause**. If they match, nothing has changed, so set both state-change flags **m_PauseOn** and **m_PauseOff** to false; otherwise if **m_Paused** is true, set **m_PauseOn** and if false set **m_PauseOff**. Finally copy **m_Paused** to **m_PriorPause**. Thus the On and Off flags are only set in the update frame when the Pause control has changed state.

The voice update checks those flags and pauses or un-pauses associated voices accordingly. It also checks if the voice's Group has a Listener and applies distance attenuation and shifts the coordinates into listener space if so. These steps are only needed for 3D voices, but the group volume adjustment is always applied to voices in that group.

Groups have no position in space, but it's still useful to have a Render method to visualise their state. This loops over all the groups once the voice update is complete. If m_Display is true it overlays a summary line on the screen showing the group name, volume and number of playing voices at each priority. There should also be a method to write this to the log file with a time stamp.

Group Manipulations

One powerful aspect of Groups is the way they define the balance of a game in terms of sound categories. At the game level, accessible to the sound designers without coding, there should be a mechanism to load a "group mode volume" setup file. This can be binary to save space or CSV or XML for ease of tweaking.

The Group Mode Volumes file contains a list of group names, one per line, alongside several named columns of m_Volume values, corresponding to different game modes, such as front end, practice, single-player race, replay, two-player race—plus additional debug modes like engine test, environment check and suchlike.

The game stores this data internally, allowing instant switching between group-mode columns by name. Fast A:B comparison is very useful when comparing mixes. The Group pause mechanism should also be exposed to designers so they can readily mute or solo groups—this will also inhibit associated voice updates, speeding up debug builds. There should also be a debug mechanism to reload the data file after editing without re-starting the game.

Groups of Groups

This design does not directly include support for groups containing groups, or further nesting. I did consider this and offer it to customers while developing similar systems over the years, but it was either not considered necessary or a lower priority than other features. We may have dodged a bullet. One of the risks of such complication is that as volume and other controls are cascaded it becomes more likely that "false neutrals" will crop up—mysteriously muted configurations where sounds get lost between the layers—or that subtle bugs will be introduced.

The term "false neutral" comes from manual gearbox design. A succession of overdrives, transfer boxes and high and low ratios produces not only a large set of useful—if sometimes duplicated—combinations but a plethora of useless settings when one break in the chain makes all the other selections moot. Anyone who has been forced to daisy-chain recording studios or mixing desks is likely to have run into similar frustrations, especially if working as part of a group (sic) with differing objectives.

Nested hierarchical organisation is fundamental to programming language and data-structure design, so programmers soon tend to take it for granted—if not, they'd struggle to learn their trade—but it's hardly intuitive to others. As user-centric design guru Alan Cooper points out, we rarely encounter filing cabinets inside filing cabinets or drawers within drawers in the real world.[1]

> "Hierarchies are generally very difficult for users to understand and use."
>
> —Alan Cooper

Confusion naturally arises when containers of the same type are cascaded—is it the Group or the Group of Groups we want to tweak? The wrong choice might yield superficially suitable results and be hard to unpick later when you realise that more or less of the mix is affected than was intended. Nesting of DSP effects is particularly unpredictable—and potentially expensive at runtime.

There's also some runtime cost, though it's a minor consideration. You don't even need to convert several linear gains from attenuations or multiply gains hierarchically if you sum all the attenuations in a signal path before doing the logarithmic to linear translation. The difficulties come not from the implementation but from the muddling of the mental model.

The closest we came to implementing nested groups was when two-player split-screen features were added to existing brands. This led to a situation where some of the player-one groups were duplicated for a second player. Adam Sawkins chose to implement this grouping on the game side, building on top of the Listener/Group/Voice hierarchy. The meta-groups were therefore Players, with groups therein, and no need to nest filters or effect sends. Groups can be subdivided into sets by sequential allocation or odd/even tests. It's not really an audio system runtimes problem.

The game controls the assignment of names and purposes to Groups when it initialises the runtime system. Adam implemented a natural super-group structure at this point by splitting the groups sequentially and conceptually into two sets—sounds independent of the per-player view and those specific to each player. The player-specific groups also share distinct listeners. Local Ambience may include industry, wildlife, watercourses or anything else that comes and goes with locations.

Figure 12.2 shows a simplified representation of this hierarchy. Non-diegetic sounds like music and menu confirmations are allocated and used first, regardless of game mode. In single-player mode the subsequent location-specific groups are used, but only the first set is needed. Two-player mode uses the first or second set as required by context, and the same scheme can be extended for an arbitrary number of players. Each group costs only a few bytes, and the maximum number of players is known in advance, so the overhead of asking in advance for as many as might ever be needed is insignificant.

- **Shared non-diegetic Groups**
  - User interface sounds
  - Background music

- **Player 1 Groups**
  - Vehicle engine/exhaust
  - Surfaces
  - Collisions
  - Damage
  - Local ambience

- **Player 2 Groups**
  - Engine/exhaust
  - Surfaces
  - Collisions
  - Damage
  - Local ambience

- **Non-player Groups**
  - AI/network exhausts
  - AI/network collisions
  - AI/network surfaces

Figure 12.2: Per-player voice groups

A few compile-time constants are then enough to identify the super-groups—a groupCount of the number of per-player-groups and an array pointing at the first group of this type for each player. Operations intended for just one player were performed by picking up the first group index for that player and iterating over the appropriate count of groups.

Operations on a class of sounds—such as music or the intensity of all surface sounds—go directly to the non-diegetic group in the first case and through the player-specific groups in the second, stepping by groupCount from Player One's Surfaces group until the current number of players have all been serviced. The non-player groups are for competitors controlled by artificial intelligence or remote systems. They use a merged engine/exhaust model and simplified surfaces—often modelling a single central wheel for audio, though not of course physics—to reduce their demands on the system.

Reference

[1] *Alan Cooper, About Face, IDG Books 1995, ISBN 1568843224, pages 276–277*

References

CHAPTER 13

Implementing Listeners

A listener works like an idealised microphone placed in the game world. Many audio systems support only a single listener, but it can be convenient and creatively enabling to allow more than one and vital for split-screen games where there's more than one window into the game world.

Listeners are statically allocated when the runtime system is initialised—you specify the maximum number you may want in advance so they can be stored together in an array or pool and there's no need to dynamically allocate memory for them later. Each listener instance consumes about 150 bytes of memory. The constructor is called during initialisation and sets up an array of listeners and a static method to look them up later by index number.

Figure 13.1 shows how 3D voices share listeners via their groups. Sounds that are not positional are routed directly from their group to one or more endpoints, ignoring listeners.

There is no limit on the number of listeners, but as they're associated with groups it's pointless having more listeners than you have groups. Some groups share listeners, while others are associated with non-positional sounds, so they don't have a listener.

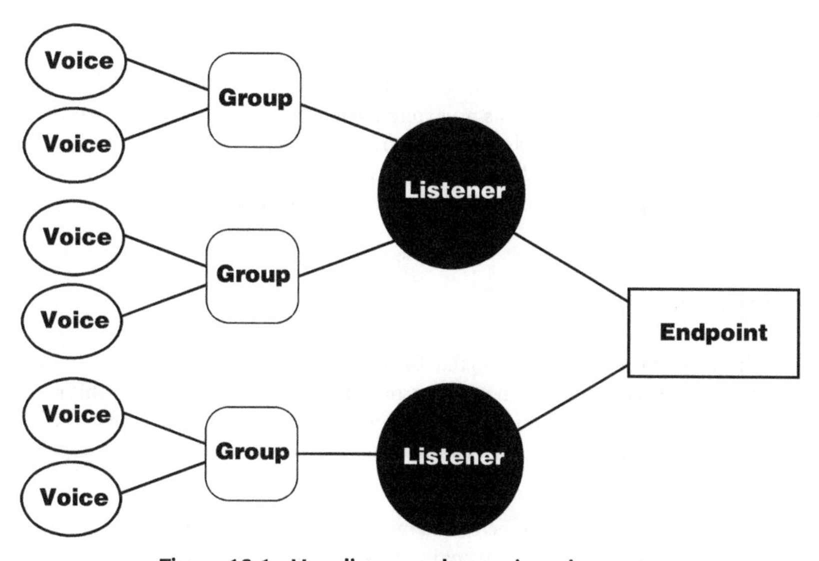

Figure 13.1: How listeners share voices via groups

Table 13.1: Listener class members

Variable	Type	Initial value
m_Orientation	Matrix	Identity matrix
m_Velocity	Vector	{0, 0, 0, FLAG}
m_Scale	Vector	{1, 1, 1, 0}
m_Offset	Vector	{0, 0, 0, 0}
m_PolarPattern	POLAR_PATTERN	OMNI_PATTERN
m_Display	bool	False
m_Name	char*	Listener #
m_OldPosition	Vector	{0, 0, 0, FLAG}

The positions and velocities of each voice and its listener combine to determine volume attenuation with distance and Doppler pitch shifts with relative speed. The listener position, in world coordinates, is held in the last row of the orientation matrix. Storing it this way means there's no risk of the position and orientation getting out of step and that a single parallel matrix operation can adjust for the position and direction in one easily optimised process.

Listener Static Methods

The Listener class is primarily a container for information used in Virtual Voice processing but still needs a few static methods to allocate, update and keep track of listener instances. Like everything the audio runtime system places in the game world, the listener name and its orientation can be overlaid on the screen for testing purposes.

The Render() method displays the listener's axes to visualise its position and orientation when **m_Display** is true. I draw the horizontal axis in red, vertical in green and the depth (Z) axis in blue, all labelled in white with the name of the listener.

The listener Update() method catches one special case detailed later under the heading "Doppler Variations." If **m_Velocity.w** still contains **FLAG** this means no explicit velocity has been set. If **m_OldPosition.w** also holds **FLAG** this is the first update and we clear **m_Velocity**. Otherwise we extract the current position from **m_Orientation** matrix and subtract **m_OldPosition** from that to derive the velocity from the change in position then copy the current position into **m_OldPosition** for use at the next update, implicitly clearing **FLAG**.

Listener Transformation

Multiple listeners are handled in the logical layer, so that the physical layer can just assume that the listener is at the origin and doesn't need to worry which listener any particular voice is associated with. This means we must convert each source coordinate, which the game will express in coordinates consistent with everything else placed in the world, into listener-relative coordinates, offset against the position of each listener in the world and rotated to match the listener's orientation.

This transformation requires some generic 3D vector and matrix functions. In some products I've used standard library routines, such as those in Codemasters' Ego engine, to do this, but if your audio runtimes need to be portable between engines it makes sense to minimise dependencies by bundling in the handful of 3D arithmetic routines required.

This is not a mathematics book so only the minimum code required to do the job is presented here. The aim is to explain what we do and why, exploring the trade-offs and potential optimisations without burying the key concepts in inessential specialisations. This is the most mathematical part of the book. You don't need to understand in detail how these functions work as long as you're clear about what effects they have and why they're needed.

One implementation along these lines was incorporated into three different game engines and came with its own little mathematics library, so the only adaptation needed was to rotate the listener matrices between row and column major order before passing them down—assuming the game uses Microsoft-style row-major matrices rather than OpenAL/OpenGL column-major ones. The data is the same either way, but the indices of corresponding matrix elements differ. Rotating the matrix, exchanging rows and columns, is a trivial operation as it only involves re-ordering elements, and it takes little time as it only needs to be done for the handful of matrices, one per listener, updated every frame.

The example code suits Unity and OpenAL column-major matrices. It uses two very simple representations of four-element vectors and 4 × 4 element matrices. These wrap four or sixteen float values into suitable structures. Optimised code would use arrays or parallel vector types to store these, but vector representations and operations tend to be platform dependent, so the example has deliberately been left in scalar form.

If live profiling on real data shows that performance is an issue, consider using corresponding platform-specific datatypes and functions. These functions are typically only called once or twice per active voice update, so you may find you can live quite happily without optimising them.

It may seem odd that the Vector type has four elements when only three are needed for X, Y, Z 3D coordinates. The fourth element W is sometimes used as an update flag in this system, but it's mainly present to finesse the Matrix implementation.

```
typedef struct
{
  float x, y, z, w;
} Vector;
```

This type of matrix has four rows and four columns, though it strictly needs just three of each to express a rotation and three more scalars for the position. A few systems use two unit vectors, typically pointing up and to the front, to express orientation, but those can readily be upgraded to rotation matrices by introducing their cross product as a third orthogonal vector. Four-by-four matrices are preferred, as they're easier to optimise for binary parallel hardware which processes chunks of two, four or eight values at a time rather than odd groups of three or nine. They can also be easily transposed without changing shape, as they have an extra row (or column).

```
typedef struct
{
  float _11, _12, _13, _14;
  float _21, _22, _23, _24;
  float _31, _32, _33, _34;
  float _41, _42, _43, _44;
} Matrix;
```

The final reason to prefer quad rather than three-wide structures is that they pack naturally onto cache line boundaries, allowing a "Single-Instruction, Multiple Data" (SIMD) architecture processor to fetch and store four elements at once. Depending upon the compiler, additional directives like __declspec for Visual Studio or __attribute__ for GCC may be needed to facilitate this alignment, but they're not needed for unoptimised scalar code.

Mathematical Assistance

The VectorSub() function takes pointers to two vectors to be compared and returns the difference between them in a third vector. The asserts just check that the pointer arguments are not null:

```
void VectorSub(Vector* pDest, const Vector*
  pSrcVector1, const Vector* pSrcVector2)
{
  assert(pDest);
  assert(pSrcVector1);
  assert(pSrcVector2);

  pDest->x = pSrcVector1->x - pSrcVector2->x;
  pDest->y = pSrcVector1->y - pSrcVector2->y;
  pDest->z = pSrcVector1->z - pSrcVector2->z;
  pDest->w = pSrcVector1->w - pSrcVector2->w;
}
```

VectorLength() uses Pythagoras's Theorem, extended to three dimensions, as follows:

```
inline float VectorLength(Vector* pSrc)
{
  assert(pSrc);

  return sqrtf((pSrc->x * pSrc->x) +
    (pSrc->y * pSrc->y) + (pSrc ->z * pSrc->z));
}
```

This function is used later to compute distance attenuation and Doppler shifts for 3D voices. Notice that pSrc->w is not used in this calculation; three dimensions are enough. The assert() catches the problematic possibility that the address pSrc passed in is null.

The DotProduct() helper compares two vectors and returns a scalar value which tells you what amount of one vector goes in the direction of the other. If the two vectors are opposed, the result is negative. This will also help us work out Doppler shifts:

```
inline float DotProduct(const Vector* pSrcVector1,
                        const Vector* pSrcVector2)
{
  assert(pSrcVector1);
  assert(pSrcVector2);

  return (pSrcVector1->x * pSrcVector2->x) +
         (pSrcVector1->y * pSrcVector2->y) +
         (pSrcVector1->z * pSrcVector2->z);
}
```

This scalar arithmetic implementation is easily optimised to do all the multiplications in parallel if you've got a vector co-processor to hand. Systems designed for gaming often have built-in primitive instructions to compute the dot product, sometimes known as the scalar product, in a single step. It's much quicker to do this than to scale the cosine of the angle between the vectors, and the result is equivalent.

The dot product is also useful when you need to know if a sound source is pointing directly at the listener. Driving games use it to adjust the volume of rearward exhaust sounds—if the dot product of the car direction and the listener direction is negative, we know the exhaust is facing away from the listener and can attenuate it accordingly.

I'm assured that gunshots sound very different when the barrel is pointing directly at you, as happens when the dot product of barrel and listener directions is most negative. Cross-fading between oblique and direct sounds tells the player if a shot is bang on target or just off. This is useful information for gamers, especially in the second case.

The MatrixInverse() function takes pointers to two matrices, makes sure they're not null, and assigns the inverse of the second matrix to the first. This is what we need to convert the rotation properties of the listener matrix, passed from the game, into an inverse matrix which reorientates a position in world space so that it's in the required direction relative to the listener.

```
void MatrixInverse (Matrix* pDestMatrix,
              const Matrix* pSrcMatrix)
{
  assert(pSrcMatrix);
  assert(pDestMatrix);

  pDestMatrix->_11 = pSrcMatrix->_11;
  pDestMatrix->_22 = pSrcMatrix->_22;
  pDestMatrix->_33 = pSrcMatrix->_33;
  pDestMatrix->_44 = pSrcMatrix->_44;

  float temp1      = pSrcMatrix->_12;
  float temp2      = pSrcMatrix->_21;
  pDestMatrix->_12 = temp2;
  pDestMatrix->_21 = temp1;

  temp1            = pSrcMatrix->_13;
  temp2            = pSrcMatrix->_31;
  pDestMatrix->_13 = temp2;
  pDestMatrix->_31 = temp1;

  float temp3      = pSrcMatrix->_32;
  temp2            = pSrcMatrix->_23;
  pDestMatrix->_32 = temp2;
  pDestMatrix->_23 = temp3;

  temp1 = -(pDestMatrix->_11 * pSrcMatrix->_41 +
          pDestMatrix->_21 * pSrcMatrix->_42 +
          pDestMatrix->_31 * pSrcMatrix->_43);
```

```
    temp2 = -(pDestMatrix->_12 * pSrcMatrix->_41 +
              pDestMatrix->_22 * pSrcMatrix->_42 +
              pDestMatrix->_32 * pSrcMatrix->_43);
    temp3 = -(pDestMatrix->_13 * pSrcMatrix->_41 +
              pDestMatrix->_23 * pSrcMatrix->_42 +
              pDestMatrix->_33 * pSrcMatrix->_43);

    pDestMatrix->_41 = temp1;
    pDestMatrix->_42 = temp2;
    pDestMatrix->_43 = temp3;

    pDestMatrix->_14 = 0.f;
    pDestMatrix->_24 = 0.f;
    pDestMatrix->_34 = 0.f;
}
```

ApplyMatrix() uses a matrix to transform a 3D coordinate vector to a new position and orientation. We use this to re-express world-relative positions and velocities relative to a particular Listener.

```
void ApplyMatrix(Vector* pDest,
        const Vector* pSrc, Matrix* pMatrix)
{
    assert(pDest);
    assert(pSrc);
    assert(pMatrix);

    Vector v;

    v.x = (pSrc->x * pMatrix->_11) + (pSrc->y * pMatrix->_21) +
          (pSrc->z * pMatrix->_31) + (pSrc->w * pMatrix->_41);

    v.y = (pSrc->x * pMatrix->_12) + (pSrc->y * pMatrix->_22) +
          (pSrc->z * pMatrix->_32) + (pSrc->w * pMatrix->_42);

    v.z = (pSrc->x * pMatrix->_13) + (pSrc->y * pMatrix->_23) +
          (pSrc->z * pMatrix->_33) + (pSrc->w * pMatrix->_43);

    v.w = (pSrc->x * pMatrix->_14) + (pSrc->y * pMatrix->_24) +
          (pSrc->z * pMatrix->_34) + (pSrc->w * pMatrix->_44);

    *pDest = v;
}
```

Listener Adaptations

The TransformToListenerSpace() method uses those functions and the listener matrix to transform the position and velocity members of a voice into listener-relative values. It starts by copying the listener matrix and clearing the elements that encode the position of the listener. This converts it into a rotation matrix. Inverting that gives us a matrix which can be applied to convert a coordinate or velocity vector into one relative to the orientation of the listener.

Subtracting the listener position from the world-space position of the source then rotating it with the inverse matrix gives us the vector from the listener to the source. This is **m_RelativePosition**.

The second part of the method computes **m_RelativeVelocity**, which will be used for speed-related Doppler shifts. This calculation is implemented in two alternative ways; the slowest one requires the source velocity to be computed and updated by the calling code, while the second infers the velocity from changes to the position.

This is where the **FLAG** value in **m_Velocity.w** comes in handy. This is only set while the source velocity has never been explicitly set, since updating **m_Velocity** vector would normally clear it. If we've been passed an explicit source velocity we work out the relative velocity by taking the difference between the source and listener velocities, both in world space. The magnitude of that vector gives us the relative speed—if they both have the same velocity, it'll be zero—but to get the correct direction as well we need to push that vector through the inverse matrix we computed previously.

The faster alternative is used only if **m_Velocity.w** has not been changed since the voice was initialised. In this case we need to know the previous position to work out how far the voice has moved in the current update frame. This hack relies upon the game updates coming at a regular pace, though the occasional dropped frame can be tolerated.

At the first update, when a voice has just been told to play, there will be no previous position. We detect that much like the unset velocity, by looking for the **FLAG** initialisation value in **m_RelativePositionOld.w**. If this is present, we clear **m_RelativeVelocity**—it'll catch up next frame.

Otherwise the difference between the current position and the old one, stored last frame, gives us the velocity of the source. We use the relative positions previously computed in listener space, so there's no need for any more matrix manipulation, as changes in the listener position and orientation have already been incorporated.

The last step is to take a copy of the current relative position of the source. We stash this in **m_RelativePositionOld**, which clears the **FLAG** if necessary, so subsequent updates can compare this with updated relative positions and infer the velocity.

There are two things to remember if you choose to skip explicit velocity updates. When a voice is initialised or re-used in a new position, you must put the **FLAG** value back in both **m_Velocity.w** and **m_RelativePositionOld.w** to re-enable velocity optimisation on the re-used Voice.

```
void TransformToListenerSpace()
{
  Matrix invListenerMatrix = m_pGroup->m_pListener->m_Orientation;
  invListenerMatrix._41 = invListenerMatrix._42 =
    invListenerMatrix._43 = 0.f;
  MatrixInverse( &invListenerMatrix, &invListenerMatrix );

  Vector PosInListenerSpace;
  VectorSub( &PosInListenerSpace, &m_Position,
```

```
    &m_pGroup->m_pListener->m_Position);
  ApplyMatrix(&m_RelativePosition, &PosInListenerSpace,
    &invListenerMatrix);

  if (m_Velocity.w != FLAG)
  {
    Vector listenerVelocity = m_pGroup->m_pListener->m_Velocity;

    VectorSub(&m_RelativeVelocity, &m_Velocity, &listenerVelocity);

    m_RelativeVelocity.w = 0.f;
    ApplyMatrix(&m_RelativeVelocity, &m_RelativeVelocity,
      &invListenerMatrix);
  }
  else
  {
    if (m_RelativePositionOld.w == FLAG)
    {
      m_RelativeVelocity.w = m_RelativeVelocity.x =
        m_RelativeVelocity.w = m_RelativeVelocity.w = 0.f;
    }
    else
    {
      VectorSub(&m_RelativeVelocity,
        &m_RelativePosition, &m_RelativePositionOld);
    }
    m_RelativePositionOld = m_RelativePosition;
  }
}
```

Listener Orientation

Our listeners have orientation—they can be made to point in a particular direction, most often corresponding to that of the camera—and a "polar pattern" (like a microphone) which determines their sensitivity to sounds from particular directions. Some commercial systems use the concept of "sound cones" to suppress sounds from a particular direction; this is a concept appropriate to lighting but tends to yield unrealistic effects, with sounds popping in and out of the mix, in audio—sound goes more readily around corners than light does.

The reference implementation only uses volume effects, but the same technique, perhaps with a different polar pattern, can be applied to add low-pass filtering for sounds outside the preferred direction. These are the polar patterns implemented here; additional ones can easily be added:

```
enum POLAR_PATTERN
{
```

```
    POLAR_OMNI = 0,       // Listener orientation makes no difference
    POLAR_CARDIOID,       // Sensitive to front, insensitive to rear
    POLAR_FIGURE8X,       // Sensitive only to sound from either side
    POLAR_FIGURE8Y,       // Sensitive only to sounds above or below
    POLAR_FIGURE8Z,       // Sensitive to front and rear but not sides
    POLAR_HRTF,           // Mimic pinnae response by rear attenuation
    POLAR_COUNT           // Number of polar patterns supported
};
```

A switch statement in the voice update selects appropriate code to implement each type of POLAR_PATTERN. An omnidirectional pattern picks up sound from all around.

A cardioid pattern, like a unidirectional microphone, emphasises sounds in front of the listener and is often more appropriate for replay listeners that track a camera's position and orientation, as it filters out voices that are behind the camera and hence invisible to the viewer. You can orientate it however you wish, but most commonly it'll be pointing in the same direction as player cameras, so they can reorientate both the visual and audio field at the same time.

```
float angle = atan2f (PosInListenerSpace.x,
                      PosInListenerSpace.z );
```

Angle is the direction in radians from the listener to the source. For simplicity this ignores height, so only the horizontal (x) and front/back (z) coordinates are addressed; later notes extend this to handle elevation.

Depending upon the coordinate convention you've chosen, you may need to flip the sign of the z coordinate—the example assumes the right-handed coordinate system customarily used by OpenAL, so negate the z term if working in the left-handed Unreal Engine or Direct3D-style. It's possible to instruct either system to switch handedness, but in practice Microsoft-based systems are almost always configured so that positive z values extend towards the viewer, while mobile devices and Japanese consoles favour the opposite convention.

This formula gives a subtle attenuation of sounds behind the listener. It falls far short of the full HRTF implementations in Chapter 18 but is enough to give players a sense of when a source is right behind them, even if they're only listening in mono or on a slow platform:

```
case POLAR_HRTF:
{
  m_ListenerAttenuation = (1.f + cosf(angle)) - 2.f;
  break;
}
```

This subtle cardioid, with only one or two dB of attenuation to the rear, gives a sufficient directional cue by making sounds passing behind the player appear to be quieter, as they would be behind the ears, enabling even a mono listener to identify by experimental listening which way to turn to keep the sound centrally behind the player and block the source of the sound from passing on either side, even though the source is not visible on screen.

A stronger effect, mimicking a unidirectional microphone, can be obtained by scaling the result. The following code increases the rearward attenuation to a maximum set by the constant rejectionRatio, which is -20 dB in this case:

```
const float rejectionRatio = -20.f;
```

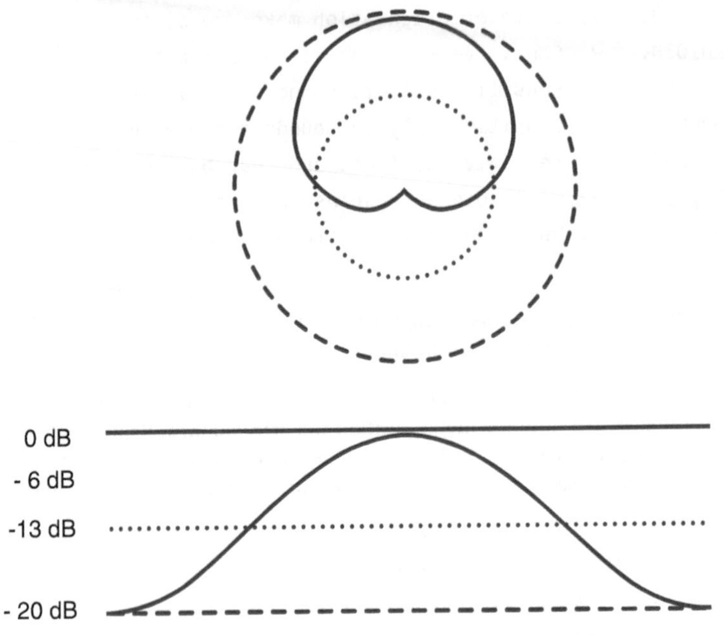

Figure 13.2: Polar pattern of the cardioid directional listener

```
case POLAR_CARDIOID:
{
  m_ListenerAttenuation = (1.f + cosf(angle)) - 2.f;
  m_ListenerAttenuation *= (rejectionRatio / -2.f);
  break;
}
```

This type of listener was useful in a replay when the camera happened to be placed on a corner right in front of a waterfall but facing away from it. During races, when the player was approaching the corner and the waterfall was clearly visible, an omnidirectional polar pattern was used.

In replays the cardioid listener was selected, so that the sound of the approaching player was not drowned out by the now-invisible waterfall. This listener-centred approach was smoother and less error-prone than ad-hoc code to adjust the source volume depending upon the game mode.

The omnidirectional listener is trivial to implement:

```
m_ListenerAttenuation = 0.f;
```

Other polar patterns can be defined to meet the needs of the game—for instance, listeners with a figure-of-eight polar pattern, like ribbon microphones, can pick out sounds from two opposite sides and attenuate those from other directions; depending upon the orientation of this pattern, they might emphasise sounds from in front and behind.

```
case POLAR_FIGURE8X:
{
  m_ListenerAttenuation = rejectionRatio * fabsf(sinf(angle));
  break;
}
case POLAR_FIGURE8Z:
{
  m_ListenerAttenuation = rejectionRatio * fabsf(cosf(angle));
  break;
}
```

The only difference between the code for X and Z orientations is the use of Sine or Cosine functions. The elevation angle, derived from the relative magnitude of the Y direction component, should be used to simulate a microphone oriented to emphasise sounds above or below the listener if you need the additional case POLAR_FIGURE8Y.

How Voices Bind to Listeners

Every 3D voice is associated with a single listener via its group. Listeners belong to groups, not vice versa! One listener may be used by several groups, e.g. for sets of vehicle sounds: engine, exhaust and collision sounds might all use the same listener, so they're heard from the same place but various groups to make balancing easier.

As each voice is assigned to a single group—and hence to not more than one listener when you play it—if you want to "hear" it through several listeners, you must play it multiple times. Only the listener associated with that group "hears" that voice.

Separate voices are needed because the pitch of each voice may vary depending on Doppler shift caused by relative motion of the listeners, so they won't necessarily all sound the same or finish at the same time. Doppler shift is the property of sound which causes perceived pitch to rise when source and listener are moving closer and fall when they move apart. The pitch at the source is unchanged, but the relative motion causes waves to arrive at the listener either closer together or spread out, affecting both the pitch and how long a one-shot sound seems to last.

Doppler Variations

The private member **m_OldPosition** is only used internally as part of the pitch update for Doppler shift. The runtime system allows velocity to be set explicitly, via the Velocity vectors of the listener and voice, or inferred from changes in their position. Unless you want this flexibility, you can standardise on one approach or another, saving 16 bytes per listener object, but the voice update code shows how to support both in a single runtime system. The public Velocity property is not needed unless you use it; contrariwise, the private member is only needed if you just want to set voice positions and let the runtime system infer how fast they're moving.

The latter approach reduces the number of property update calls, as only the position needs to be set rather than position and voice vectors for each 3D source, but may introduce audible glitches if the update is not run at a regular pace or the listener position is suddenly teleported across the world, as might happen if the camera is moved from first- to third-person perspective or between replay and in-game positions.

That sort of change can result in a disturbing momentary pitch change for all affected voices. It can be hidden by momentarily muting the associated 3D voice groups during such transitions, under cover of a suitable sound effect or by reinitialising the Velocity of the associated listener. Before the listener is first used, a dummy non-zero value known as FLAG is stashed in the fourth row of its position vector. The runtime system uses this as a sign that this is not a real position but an indication that the true position has yet to be set.

Every update extracts the position from the listener matrix and copies it to the private variable **m_OldPosition**. Doppler shift is only applied when a new true position has been both set and updated once, signified by the absence of the **FLAG** marker in the private member.

Similarly, the listener update uses the Velocity properties directly to compute Doppler shift unless it is still in its initialised state, with **FLAG** set. If so, the actual velocity is computed from the change in relative position. Since voices and listeners have independent positions and optional velocities, the pitch shift may be computed from the implied position delta of the listener and the explicit velocity of the voice or vice versa.

This works fine because there's a giveaway **FLAG** value in each object. It's most common for an app to use velocity either everywhere or nowhere, but you can use one technique for voices and another for listeners if you wish. As there are more voices than listeners, implicit updating of voice velocity is the most common optimisation.

Doppler Implementation

Doppler shift modifies the pitch of a playing voice in proportion to its speed towards or away from the listener. Approaching sounds rise in pitch, departing ones fall. To work out the amount of modification we need to know the velocity of the source and the direction to the source from the listener. The angle between those two vectors tells us the amount to shift the pitch and the direction, up or down. The DotProduct() helper function gives us just the value we need.

To scale the speeds correctly we need one configuration constant, the speed of sound, expressed in the same units as velocity. On Earth, the formula is 331 + 0.6 * T where T is temperature in Celsius: 340 m/s for dry air at 15°C.

We use the emitter's velocity (relative to that of the listener) and the position of the source, relative to the listener at the origin, to tweak the pitch of the voice up and down depending on how fast it is moving towards or away from the listener. It may seem that we should use the direction of the source rather than its position, but as the listener is at the origin that's the same thing, and we can reuse the **m_RelativePosition** and **m_RelativeVelocity** vectors computed by TransformToListenerSpace().

```
float dopplerFactor = DotProduct(&m_RelativePosition,
                                 &m_RelativeVelocity);
float distance  = VectorLength(&m_RelativePosition);
dopplerFactor /= distance;
dopplerFactor *= m_DopplerScale;

if (dopplerFactor > -AudioConfig.speedOfSound)
{
  dopplerFactor = AudioConfig.speedOfSound /
    (AudioConfig.speedOfSound + dopplerFactor);
}
else
```

```
{
    m_ListenerAttenuation = SILENT;
    dopplerFactor = 1.f;
}
```

The conditional test catches the special case when the source is moving towards the listener faster than the speed of sound. In this case **dopplerFactor** will be negative, so we negate **speedOfSound** before comparing. Providing the source is not moving that fast, we end up with a **dopplerFactor** which can be multiplied by **m_Pitch**; then the combination is used directly to adjust the source's replay pitch.

If the source is travelling faster than sound, you will not be able to hear it. This happens if the source and listener are converging at more than 767 miles per hour (1235 KPH) for the constant given here. How you handle this case is up to you. You could trigger a one-off "sonic boom," clamp the pitch shift at some high value or mute the voice entirely.

This example clobbers the previously computed **m_ListenerAttenuation**, silencing the sample at such high speeds, and resets **dopplerFactor** to 1 so the sample continues to play, albeit silently; we could pause voices by setting **dopplerFactor** to 0, implicitly zeroing the pitch, but that usually leaves a static DC offset. Many early PlayStation games had sporadically distorted audio while paused. This happened unpredictably if loud voices were frozen by zeroing their pitch, as recommended, while the mix was momentarily near peak level. Further sounds riding on that peak were asymmetrically clamped within the output range. The Fade() and de-click routines in Chapter 16 are a better solution.

There are two function calls and a couple of slow floating-point divisions in this routine, so it should only be used for 3D voices. If the sound is non-diegetic you can skip the whole thing and just pass **m_Pitch** unmodified to the mixer.

CHAPTER 14

Split-Screen Multi-Player Audio

Console games often support simultaneous two-player action using a single split screen. This is a challenge for the entire game engine, not just the audio, as it means running the entire game twice, from two viewpoints, without sacrificing much of the single-player experience. It's not strictly double the workload, as the physics system delivers contact and collision data for both views, but there's almost twice as much to be drawn on-screen and potentially twice as many sounds and effects to mix. There's also a lot more fun to be had, sharing the experience in the same room, providing the audio mix is not so confusing that people are misled by sounds meant for the other player.

Compared with lone-player circuit-racer audio routing, the main changes double up the player-one vehicle systems and add a second player listener. Some simplifications limit disc-access. Pit Radio is dropped, and the network chat (from remote players) and local commentary systems are routed identically to both headsets or controllers. Ambience must also take account of the distinct "environments" in each view.

Three techniques can be used in combination to overlay soundfields for multiple players: directional listeners, panner transformations and DSP effect tweaks. Depending upon the platform, you may also be able to take advantage of tiny speakers introduced into each player's input controller.

Nintendo pioneered those in the Wii, and Sony followed in their PS4 controller update, adding a headphone socket capable of full-bandwidth stereo, as well as a little speaker. Some Microsoft controllers support optional headphone connections, but these are more often used as an alternative to the full speaker output than as an extra endpoint for player-specific sound, because not all controllers have the audio capability built in.

If you know your customers will have controllers with stereo headphone outputs and that they're willing to wear headphones while playing, the ideal way to deliver split-screen audio is to create two separate mixes and route them direct to the controller audio outputs. You can then do all the headphone-specific tricks, like front/rear and height-related filtering. But if people are playing together in the same room, they may prefer to be able to hear one another clearly and share a set of full-sized speakers rendering music and other sounds for them both. So it's still useful to understand the soundfield overlaying techniques discussed later in this section.

The in-controller speakers are too harsh and tinny for a full mix, especially where bass-heavy engines and explosions are concerned, but their characteristics are ideal for short speech snippets. If your game features real-time advice from a squad commander or pit engineer, this is best routed directly to controller speakers, as there are no directional cues for such non-diegetic sounds, and it is vitally important that players should know whom they are intended for.

Nintendo games tend to use the controller speaker to confirm player actions, like the release of an arrow from Zelda's bow, but these have little informational content as they directly follow finger movements on the buttons. It's more helpful, and hence more immersive, if controller speakers are used to announce things that happen TO the associated player, such as being hit.

A suitably percussive high-pass filtered crash or crack sound can be routed to the controller, backed up by a beefier non-directional boom played through the main speakers. The routing of the initial sound to one player's controller means they know they've been hit, and other players know they've escaped.

One of the characteristics of the console update cycle is that split-screen games tend to arrive partway through the market life of the latest game machines. Launch titles on new hardware concentrate on the single-player experience, and only after that's well-tuned do game teams try to double it up—or sometimes, as in games like N64 classic *Golden Eye*, render four views at once.

The associated graphics challenges are for other books, but the display format affects the mix even if the audio team may have little say in decisions about it. Fitting two views onto one screen may require changes to the aspect ratio, camera angle and depth of field, with implications for audio rolloff, but the biggest implication of this choice for us is the way we must render two soundfields at once, through a single set of speakers, without confusing the players.

The simplest possible way to handle a vertical split is to assign left and right stereo speakers to the corresponding view. But this means each mix is mono. An alternative is to offset and scale each mix, so that sounds for the left view pan from hard left to centre, and those corresponding to the right pane pan from the middle to the right edge. That way you still get some directional cues, but they're spatially compressed to reduce overlap. There's no need to hard-wire the offset or scaling—depending on the game, some overlap might be permissible, or you might even want a "guard band" in the middle to reduce confusion there.

Surround systems give more flexibility, though the audio configuration must be designed to take advantage of those without making them essential. You might render "dual stereo" on a surround system, with sounds for one view, probably the top slice, panned to the front pair and those for the lower view panned to the rear. This gives you excellent separation at the expense of most of the front–back cues.

Directional listeners reintroduce those even for a mono mix, especially if the listener orientation is under the player's free control. Scanning the scene will emphasise or mask sounds right in front or behind, respectively. Subtle though those differences may be, humans are accustomed to using them to refine their idea of where a sound behind them is coming from with small head movements. Similar listener-turning thumb nudges that sweep the sound across a notch of rearward attenuation help players decide if they're turning the right way to dodge or confront others chasing them.

Field of View and Aspect Ratio

The shape of the screen is a critical factor in the design of local multi-player games and affects audio runtimes too. Early television screens were designed for radar, so the display tube was round; a curve-cornered frame masked this to a photo-shaped rectangle, almost as tall as it was wide. The display area was 25% wider than it was tall, described as a 5:4 aspect ratio. Later, as tubes developed, this was stretched to 4:3, a better match for content made for mid-20th-century cinema. Decades later analogue then digital widescreen systems pulled this out to the current near-standard of 16:9, with a difficult transition period while some countries stuck with 4:3. It seems unlikely that 16:9 will last forever, and tablets, phones and laptops, let alone cinemas, do not necessarily follow the TV fashion.

Early arcade games like *Pac-Man* and *Space Invaders* turned the tube on its edge, for a tall narrow 3:4 aspect, but that's only an option if you're building your own cabinets—few games followed the Amiga's *Side Pac-Man* in expecting players to rotate their TV display, even though Dell and similar office monitors support portrait as well as landscape orientations, following the example of classic Mac *Radius Pivot* displays.

The 4:3 aspect ratio of analogue TVs and the way the picture was drawn one pixel line at a time meant that 2D split-screen games on older systems invariably positioned two foreshortened views one above another. The pause between lines, originally intended to allow old valve sets to "fly back" their beam from right to left, allowed time for the display to be reset to a different view.

Modern hardware developments, like the necessary process of clipping 3D displays to fit a rectangular frustum view, and development of wider digital displays with 16:9 aspect ratio and no beam delays, have made side-by-side split-screen views equally practical. The choice now depends on the genre—a wide horizontal split makes sense for

Figure 14.1: Black borders preserve split-screen view aspect ratios

a flat circuit-racer, but side-by-side works best for game types with more looking up and down or rhythm-action games where height denotes time.

If the game supports remote network play as well as local split screen, it would be unfair for some players to have a different aspect ratio—they'd be able to see farther round corners or above and below the scene visible to their competitors. This has led to widescreen splits, which preserve the usual aspect ratio by offsetting two smaller 16:9 views on a single screen. Each view is the same shape as usual, but they have blank areas either side. The positioning of these shape-matching blanks has implications for the audio mix—we prefer the two views flush to opposite sides so we can take advantage of stereo panning to mix distinctly for each player.

Distinct audio mix techniques are appropriate to horizontal or vertical format splits; in writing future-proof cross-platform game audio runtimes we should make it straightforward for the designer to tailor any game to any layout—perhaps even multiple layouts in a single game, for one, two or four players. Stereo speakers suit a side-by-side split, but the conventional display split into top and bottom sections, most likely to be adopted by predominately 2D racing titles, is less appropriate for full-3D games and wide-aspect screens.

In this configuration left/right per-player panning would be confusing, and we have the option of using front/rear offsets in surround or using other techniques, like filters and reverb. Playing sounds for one player dry, the other gently bedded down by reverb, helps to disambiguate overlaid sounds for each player. These are all imprecise, but using them in combination works pretty well.

Scales and Offsets

Each view has at least one dedicated listener, associated with the camera orientation and position. To bias a soundfield towards particular speakers the direction of a sound on each axis is normalised to the range -1 to +1, with 0 in the centre. This is a normal part of the panner. Distance-related aspects such as volume attenuation and filtering are applied to the mono sound, then the normalised vector is used to set the ratio of its level in each available speaker, depending on its relative direction.

Two vector extensions to the listener properties enable view-specific scaling and offsetting. These are implemented in the logical layer of the runtime system, where listener, group and distance attenuation are already handled, so they work identically on all platforms.

The Listener class OFFSET is a vector allowing 3D offsets to be applied to the pan position of sounds played via that listener. The range is +/−1, matching constants in Chapter 18. Alongside OFFSET lives SCALE, a vector of length 0..1, where 0 collapses panning in that dimension to the offset (centred with no panning if the offset is 0) and 1 gives full panning range (requiring zero offset). Thus to make everything mono, but retaining distance volume effects, zero OFFSET and SCALE. All this can be tweaked to allow *Resi5*-style vertical and horizontal offsets, vertical or horizontal splits, front/rear and left/right pans of varying proportions.

So SCALE is used to shrink the panning range and OFFSET to position it off centre. Here's a simple stereo example for a game with two views side by side. Set SCALE.x to 0.5 for both listeners, halving the pan image width, and set OFFSET.x to -0.5 for the left view and +0.5 for the right. Now sound positions that would use the full -1 to +1 range in single-screen single-player mode occupy the left and right halves of the panning interval. The SCALE.x settings reduce the range from -0.5 to +0.5, and the different OFFSET.x values mean that the view on the left pans from ((-1..+1)/2–0.5), i.e. -1..0, and the other side uses ((-1..+1)/2 + 0.5), which boils down to 0..+1.

If you consider that a bit of overlap is worth tolerating in return for a wider range each side, try a SCALE.x of 0.7, reducing the range to -0.7..+0.7 and offsets of -0.3 and +0.3, respectively. Now sounds associated with the left view can range from -1.0 to +0.4 and those on the right occupy -0.4 to +1.0.

Whatever you do, make sure that the final output range does not exceed -1..+1 in either direction; this should be verified in the code to set the offset and scale. The scale should generally be in the range 0..1. Expensively, 0 forces the mix from stereo to mono.

Negative values would swap left and right, which only seems useful for upside-down players or screens, and magnitudes over 1 will hit clamps in the panner, making it non-linear. The order in which you set the values before verification is critical and depends on whether you are scaling up or down relative to the previous setting.

A neater approach, given the close coupling between these two values, is to provide one listener-class method to set both at once. You can then bury the required order of application and checking inside the method.

Since sounds can nowadays be convincingly placed behind as well as in front of and above and below the listener, these values are 3D vectors, equally capable of suppressing the soundfield in any direction. If you know the players share a 3D speaker system you can apply differential vertical OFFSET.y values, directly reflecting the position of two views one above the other. If you only know that they have rear speakers you can offset the panning for each view between front and rear, using OFFSET.z. These can be combined, so you might put the top slice elevated and the bottom one set back, like this:

Player 1 (top slice)

```
OFFSET = {+0.f, +0.5f, -0.25f};
SCALE  = {+1.f, +0.5f, +0.75f};
```

Player 2 (bottom slice)

```
OFFSET = {+0.f, -0.5f, +0.25f};
SCALE  = {+1.f, +0.5f, +0.75f};
```

This example assumes a right-handed coordinate system, like OpenGL and standard mathematics and physics texts, where Z coordinates in front of the listener are negative. Microsoft's Direct3D is driven the other way, with left-handed coordinates. To adjust for this, flip the signs of both 0.25 terms.

Split-Screen Effects

One of the aspects of the mix that will certainly need to be pared down is the use of expensive digital effects like reverberation. The single-player experience may include left- and right-side reflections, cockpit reverb, and more distant reverb appropriate to the environment—forest, town, valley, plain, tunnel, ridge or cutting. Rendering all those well is often the most processor-intensive part of a single-player audio mix, and just doubling them up—eight reverbs—is unlikely to be an option, especially as other game sub-systems will be working harder and quite likely calling on the audio team to surrender some of their RAM and processing time, especially in the painful late stages of game profiling and optimisation.

The settings of the reverb unit—such as pre-delay, damping and decay time—depend upon the player's location in the game world, so you can't rely on sharing one reverb unit between two scenes; if they happen to require the same pre-set, that's great, but when one player leaves town and the other dives into a tunnel you must pick one effect, and the only controls you have over the processing of each view are the effect-send levels.

Thankfully there is a synergy to be exploited. Even if you had the processor resources, you probably don't want to run two copies of all the effects in your game in the split-screen scenario, because the varying extra layers are more likely to muddle the mix than to help the player tell which sound belongs to which view. Just as lead and backing vocals and instruments are set apart by mixing foreground music tracks relatively dry, a horizontal split-screen mix can be layered by deliberately stripping out most or all of the reverb from one slice. There's no convention as to which layer should be driest, but it's obvious to players which assignment you've made. Consistency, rather than swapping scarce effects to the most-needy view, helps their ears stay trained on the relevant mix.

Split-Screen Voice Management

The voice group system, in conjunction with multiple listeners and the "max important voices" cap on groups, can be used to route, offset and differentially process sounds for each player. There will be at least two listeners for each view, bound to appropriate groups as each voice is played. The polar and positional characteristics of these listeners can be adjusted on the fly to allow 3D sounds for each player to be grouped into (probably) two main categories, typically allowing offset and image width to differ for some sounds, and experimentation to ensure that a compromise between mono and full surround can be selected for each major context. Proper configuration of these listeners requires the game to know the user-selected speaker mode (e.g. stereo or surround).

The platform listener is pre-set to be at the origin, facing forward, and the logical layer uses the appropriate listener matrix to convert each voice position into one relative to the single central listener. This way you can take advantage of platform-specific Doppler, panning and distance attenuation code or do it in a generic way for all platforms and disable the lower-level interaction. This code turns off distance and Doppler processing in OpenAL—similar calls are available in other platform abstractions:

```
alDistanceModel(AL_NONE);
alDopplerFactor(0.f);
```

The game must either play sounds to both listeners or compute the loudest and bind and position it there—which gets tricky as it moves between listeners. This needs game-specific tuning, informed by the voice budget, but the Group, Listener and Virtual Voice systems, with adjusted MAX_IMPORTANT_VOICES group caps for split-screen play, can accommodate many approaches.

Force-Feedback and Rumble

Haptic feedback—subsonic vibrations, known as force-feedback or controller-rumble—can similarly be used to steer information to specific players. Sega's 1976 arcade motocross game *Fonz* introduced hardware to make the handlebars shake in a collision. Later arcade racers give force-feedback so the steering wheel becomes realistically easier or harder to move depending upon speed and cornering, adding jolts associated with changes of road surface, collisions or sudden braking.

While not strictly an audio feature, it makes sense for rumble to be controlled by the same game systems that regulate surface and collision sounds and low-frequency audio effects. *RaceDriver GRID* game audio programmer Adam Sawkins popularised the idea of 5.1.1 sound—the second .1 referring to subsonic rumble—and demonstrated the advantages of controlling all those outputs together rather than leave force-feedback to physics and the rest to audio, which can lead to jarring (or not) inconsistencies. The audio programmer necessarily has all the physics data, including acceleration, rolling surface and collision cues, and responsibility for scaling those to select appropriate sounds and their levels. Adding rumble to the same systems is an efficient use of coding resources and gives players more coherent and nuanced feedback.

Rumble is extra-valuable in a split-screen game, because it provides extra player-specific feedback channels. But you can't rely on it, as to reach the maximum possible audience games rarely require a specific add-on controller, let alone a pair of them, and even handheld controllers sometimes have rumble motors, sometimes not. It's also normal for the feature to be optional, even if the controller has the capability. Rumble can adversely affect targeting, battery life and ease of use for some players, so there are typically options in the console menus or each game to disable it.

Nintendo introduced the Rumble Pak as an optional add-on for the N64 console in 1997. This required separate batteries and had a single motor with a simple on-off control. Timed pulses were needed to give proportional feedback. Sega's Vibration Pack (Jump Pack in USA) was a similar plug-in for the Dreamcast controller. Sony went further with their DualShock controller, released at the end of the original PlayStation's life and standard on the PS2. This includes two independent rumble motors, allowing for a sort of subsonic stereo—Konami's original *Metal Gear Solid* game made great use of this in its helicopter scenes.

Patent litigation forced Sony to pay high retrospective royalties on that aspect of the DualShock design, so the PS3's palindromic trademark SIXAXIS controller had no force-feedback at launch; the feature was restored a year later.[1] Microsoft paid up from the start, so all Xbox controllers have twin rumble motors, one in each handle, though they're not strictly symmetrical—the offset weight in the right handle is substantially heavier than the one on the left, complicating their use for directional cues.

Reference

[1] *SIXAXIS Rumble*: www.gamasutra.com/php-bin/news_index.php?story=4325

CHAPTER 15

Runtime System Physical Layers

The physical layer of the audio runtime system is where all the platform-specific aspects are implemented. It's simpler than the logical layer, because Groups, Listeners, Velocities and Virtual Voices have been interpreted upstream, leaving it with just samples, offsets, DSP effects and output channels to juggle. The logical layer of the audio runtime system is identical for all platforms and sits on top of a simpler Physical Layer interface which encapsulates platform-specific implementations. These can be entirely different, or some platforms may share common code to exploit low-level similarities, such as a dependency upon OpenAL or XAudio2.

The Physical Layer should not be accessible to game code—the only way to call its methods is via the cross-platform Logical Layer abstraction. This means there's no need to stub it when audio is disabled, and more importantly, it can be refactored inside the runtime system without requiring any changes to client code. This makes it easier to repeatedly switch implementations, for instance from DirectSound3D to OpenAL and then to XAudio2, as new features became available on Xboxes. There may be associated data changes higher up, to expose new codecs or endpoints, but often these will take advantage of features already available on other platforms.

Physical Layer Methods

Methods that implement the Physical Layer are tabulated in two sets on page 149. The first set deals with the system as a whole. The second set handles starting or replacing sounds and returns an index or "sourceID" which will be used in subsequent updates of that sound. These sourceIDs are re-used as voices stop, so their range is limited by the maximum number of physical voices the system can play. As with other classes, there should be generic methods to update each source's properties while it's playing—these will be called most often.

Most of these functions return quickly, as they just set up information for the subsequent call to UpdateMix(), which flushes fresh data and parameter changes to the mixer, but StopCapture() and ShutDown() may perform synchronous file operations which take an arbitrary time to complete. These should not be called mid-game except while testing. When they return you are guaranteed that files are closed and associated memory can be re-used.

Physical Layer Configuration

Initialise() and ShutDown() set up and release the physical layer. The AudioConfig structure is passed to Initialise() from the logical layer, which interprets cross-platform settings like the number of Virtual Voices, Effects, Listeners and Groups. It includes the maximum number of physical voices to be tracked and mixed and the number and type of codecs to be pre-allocated to meet that requirement, potentially including special consideration for multi-channel samples, as those might need extra decoding resources.

The aim is to pre-allocate all the resources which will be needed later, to avoid memory churn at runtime. For this reason the maximum number of endpoints and associated channel counts and sample rates for each should be pre-ordained.

Table 15.1: Runtime system configuration values, AudioConfig structure

Variable	Type	Usage
verbosity	PRIORITY	Suppress reports below this priority
numVoices	unsigned	Maximum number of Virtual Voices
numSources (IN/OUT)	unsigned	Maximum number of physical voices
numGroups	unsigned	Maximum number of groups
numListeners	unsigned	Maximum number of listeners
numFilters	unsigned	Maximum number of group filters
numReverbs (IN/OUT)	unsigned	Maximum number of active reverb units
reverbMask (OUT)	unsigned	One set bit for each available reverb
samplePool	unsigned	Bytes to allocate for samples and streams
numStreams	unsigned	Number of stream buffers to allocate
bufferSize[]	unsigned	Byte size for each of numStreams buffers
bufferBase[] (OUT)	void*	Start address for each stream buffer
noiseFloor	float	Lowest audible attenuation in dB, e.g. -48
headroomScale (OUT/IN)	float	Multiplier to mix all without clipping
rolloff	float	Default rolloff curve, typically 0.25..4
speedOfSound	float	Metres per second, mainly for Doppler
pathPrefix	char*	Prefix prepended to audio filenames
numEndPoints	unsigned	Number of output endpoints supported
endPoints[] (IN/OUT)	outputSpec	Associates endpoints and channel counts

Unless otherwise shown, all variables are inputs to the system. Hardware limitations might mean you get fewer physical voices or reverb units than you'd like, especially when using DSP audio expansion cards, so check those values after initialisation. One bit is set in reverbMask for each available unit, ready to be plugged into the Group configurations. Groups, Filters and Listeners are cheap, so you should get all you ask for.

HeadroomScale is marked as an output and an input. This is the gain factor to be applied when mixing each source to ensure that when they're all playing the output dynamic range is not exceeded. It's marked as (OUT/IN) since the configuration system computes a plausible default based on the formula in Chapter 3, but you can subsequently tweak it up or down if you know that groups are faded down or multichannel assets predominate. The default is:

```
headroomScale = sqrtf(numSources)/numSources;
```

Count each B-Format soundfield or stereo voice as two mono sources, and if the output has more bits of resolution than each source, scale accordingly: for an 18-bit mix of 16-bit sources multiply headroomScale by four, that being $2^{18}/2^{16}$.

There are usually some specialisations of the AudioConfig structure, to take advantage of or compensate for platform differences. Particularly problematic is numSources. Some drivers only count mono or stereo voices, and to use multi-channel ones you may need to specify how many of each channel count, or the number of additional codec instances needed to support the worst-case multi-channel load.

Wii-U supports a mixture of hardware-accelerated and additional software voices, 32- or 48-kHz mixing, and a simple compressor with headroom and release-time parameters, so a specialisation for that platform adds options for those. The OpenAL router allows a choice of hardware or software drivers; it's sometimes useful to be able to over-ride the default on Windows to allow quick comparisons without renaming files or shuffling hardware.

Such parameters are passed on through the logical layer to the physical layer, so that one platform-configuration structure meets both needs.

The **endPoints** array specifies the maximum number of channels each output can support and the type of mix to send to them, which may not automatically correspond. For instance, stereo could be mixed for speakers or headphones, 5.1 surround could support game or cinema layouts, 7.1 could be horizontal or 3D. Defaults for these should be determined by asking the user (via a setup menu and saved configuration) or by interrogating the platform hardware. WASAPI is the safest way to do this on Windows PCs; use AVAudioSession on Apple and AudioDeviceInfo on Androids.

The consoles have custom methods to query the number of channels their HDMI, analogue and optical outputs support, but these are secret, so I can't list them here. It's wise to allow for as many channels as you ever intend to support, as the system or HDMI configuration might change on the fly. Register callbacks for notification when endpoint properties change, then call SetEndPoint(), documented below, to adjust accordingly. Console products must follow strict rules to cope with peripheral disconnection and reconnection, so try to leverage the game input code which handles controller state changes, and test this carefully before product submission.

```
enum ENDPOINT_TYPE
{
  EP_MAIN   = 0,
  EP_SECONDARY = 1,
  EP_CONTROLLER_1,
  EP_CONTROLLER_2,
  EP_CONTROLLER_3,
  EP_CONTROLLER_4,
  EP_COUNT
};
```

The channel count is typically six or eight for EP_MAIN, the default output, to allow 5.1 or 7.1 channel surround on a console or PC, two for a secondary stereo mix such as Sony's PS3 multi-out or the WiiU GamePad and one or two for each controller speaker or headphone socket. The number of controllers connected may vary during a session; this structure allows you to allocate support for as many as you'd like, including none, and makes sure that output buffers are allocated for all you may later need.

The system allocates **numEndPoints** bits to denote each endpoint and updates the mask value in each **outputSpec** with the associated set bit. So a Group routed to just the first two endpoints sets m_EndPointsMask to endPoints[0].mask | endPoints[1].mask. The PAN_MODE determines the number of channels sent to each and how their contents are spatialised, as detailed in Chapter 18.

```
struct outputSpec
{
  unsigned maxChannels;
  unsigned mask;
  ENDPOINT_TYPE endpoint;
  PAN_MODE pan;
};
```

For simplicity I've assumed that multiple filter and reverb units are all equivalent on a given platform. To support several different types of filter or reverb, add an array like the **endPoint** and **bufferSize** ones with a suitable enumeration or sub-configuration for each variant. Buffers for stream loading are statically allocated from within the **samplePool** into which sound banks and individual samples are loaded. This keeps things simple but can waste memory under some circumstances; if necessary, consider allocating appropriately aligned stream buffers from one end of the sample pool, and adjust the **bufferSize** and **bufferBase** arrays on the fly—but be very careful not to play anything from the affected areas as you do it!

Config Callbacks

The **AudioConfig** structure may include callback functions which allow platform-specific operations to be written one consistent way throughout the runtime system. A callback is a function interface exposed by the system and implemented by the client, allowing them to share code and data in a managed way.

Most of these are functions which a full game engine or app framework will supply, and it'd be mad to ignore or duplicate them. One reads the system clock, returning a fine-grained value since a consistent epoch, ideally in microseconds. The exact granularity may vary, as it's only used for relative timing.

Another callback writes reports to a log file or scrolling console. Each message has a priority, and those below the "verbosity" threshold are suppressed. Audio programmers might use PRIORITY_OPTIONAL to track everything while debugging; sound designers will prefer PRIORITY_IMPORTANT so they know about things that the runtime has had to work around, while PRIORITY_VITAL is enough for most game team members who only need to know about serious, unrecoverable problems in the audio system and are liable to disable audio entirely (as is their right) if there's too much spam to their tired eyes. By default, if the callback is not set, this should fallback to printf() or outputDebugString().

Another handy callback asynchronously loads blocks of data from a specified file or disc offset to main or audio memory. Completion can be notified by a further callback with an identifying tag or address, or by passing the address of a volatile ERRNO variable which goes to EBUSY while the loading is underway then EOK or EIO in the event of success or failure. A very basic synchronous fallback can use fopen(), but it's useful to abstract this so it can leverage data-finding methods used in the rest of the game.

During development files are often read piecemeal from a host PC or local hard drive, but the associated path or drive changes once assets are packaged for release—often in a large indexed and compressed file to reduce host file-system overheads. Wrapping up the loading of file-blocks allows the audio system to adjust automatically to such configuration changes and provides a hook for introspection into the audio system's loading pattern.

Compressed sample data should be marked so that it is not recompressed; this also makes it practical to load it en bloc by DMA, and if a load-request priority scheme is implemented the load manager can re-order load requests according to their position on the disc to minimise seeking time as well as to prioritise the most urgent requests. The top priority should only be used for stream updates—initial loading can be at a lower priority, since it's much worse for a stream to starve once it's started than it is for it to start a little late.

User music (Microsoft's XMP feature) must automatically over-ride all voices with the m_Music flag set. Sony and Microsoft provide functions and callbacks which tell you when user music starts and stops. Use these to mute (but still play) flagged voices when they'd otherwise clash with the user's tunes.

Two more callbacks allocate and release aligned blocks of memory and should ideally be used only during initialisation and shutdown. Other than local stack allocations, which do not persist, all memory should be managed via these, so you can track all audio memory usage in one place. Pass these functions down to XAudio, Wwise, Sony's Scream or similar underlying sub-systems, and monitor their use—middleware can churn thousands of blocks, especially during initialisation, and that's much the best time for it. If any of your platforms support dedicated audio memory, there should be separate callbacks for this type of allocation so that compressed, aligned sample data and anything else with custom hardware support goes to its own space via suitable DMA and cache configuration, and general processor data is kept separately in main memory.

Arithmetic helpers and platform-independent debug graphics functions to draw lines, spheres and text labels, can also be included among these callbacks. They're used in the Logical Layer when visualising listener orientations, source positions and ranges, using the game coordinate system. It may also be useful to include volatile flags or upstream callbacks into the game from the audio system, to demark the critical regions while Virtual Voices are mapped to physical channels, and while the actual mixing of channels to the endpoints take place.

UpdateMix() runs as part of the game update and defers the actual mixing of sample data to the output audio pump or another asynchronous thread, either of which may be embedded in the host system. It iterates over all the channels and either passes new settings to them or detects that they've stopped and can be re-used. It also services streaming and decompression to ensure that associated voices don't run out of sample data.

The SetEffect() parameter **pFX** may be the address of a block of I3DL2 reverb configuration data, filter characteristics, or some custom combination of these—it all depends upon the shared effects which you choose to implement, as described in Chapter 20 and Chapter 21. Not all the parameters will be used in all configurations, and cut-down implementations might ignore some or all of them. The idea is to include all the information which might be useful and let the physical layer interpret them as best it can. Unlike per-voice filters, this data is shared between all sources routed to that effect and all groups sharing a filter.

```
ERRNO Initialise(AudioConfig* config);
ERRNO SetEndPoint(PAN_MODE pan, ENDPOINT_TYPE index=0);
ERRNO SetEffect(unsigned index, effectParams* const pFX);
ERRNO UpdateMix();
ERRNO StartCapture(ENDPOINT_TYPE index=0);
ERRNO StopCapture(ENDPOINT_TYPE index=0);
ERRNO ShutDown();
```

StartCapture() and StopCapture() determine when a copy of the audio output to the specified endpoint is to be "captured" to a file for subsequent analysis. The system supports multiple endpoints, including controller speakers and headphones, as well as the main multi-channel mix, designated endpoint 0. A specific capture file path could also be passed here, but I prefer to pass this as part of the AudioConfig since file paths are platform specific, and capture files might end up on the host or a controlling PC, depending upon the development environment.

If space is scarce the capture file can be implemented as a circular buffer in RAM or other temporary storage, holding the last few seconds of output, wrapping back to the beginning when the limit is reached; in this case StopCapture() should record the point reached in a file header before closing the file, alongside the sample rate and channel count, so the two parts can be spliced together properly before auditioning. If there's lots of space each StartCapture() call may open a new file of a suitable format, incrementing the filename accordingly.

Playing Samples and Streams

The term "source" refers to a block of samples controlled and output together and might be a mono or multi-channel sample or stream. If so, the number of channels of any replacement assets set later with StealSample() or ChainSample() must correspond to the original sampleSpec.

```
typedef unsigned sourceID;

sourceID PlaySample(int route, &sampleSpec);
ERRNO StealSample(sourceID, &sampleSpec);
ERRNO ChainSample(sourceID, &sampleSpec);
unsigned SamplesQueued(sourceID);
```

```
sourceID PlayStream(int route, &streamSpec);
ERRNO ChainStream(sourceID, &streamSpec);

bool IsSourcePlaying(sourceID ch);
void StopSource(sourceID ch);
```

The argument **route** is a bit-field derived from the voice group specification. Packed into the integer is an index number to identify the associated group filter, if any, and one bit for each possible reverb unit which the voice could be using. This enables routing a voice to multiple reverbs at once. This is only provided for newly opened sources—if a new stream or sample is swapped or chained onto an already-playing source, it re-uses the routing of the original. This follows logically, as it must also be in the same group. Further bits indicate which endpoints the voice should output to, in accordance with its group EndPointsMask. One more bit corresponds to m_Music, to support Microsoft's XMP and similar user-music overrides.

StealSample() replaces the current sample playing on a source with the replacement specified. ChainSample() is similar but waits for the current sample to finish first. ChainStream() can be used to seamlessly link two one-shot streams or to transfer from a memory-resident pre-roll stub, as described in Chapter 7.

It's also useful to have a method to find out the duration of a sound, at its default rate, from a **sampleSpec**. This could be implemented in either the logical or physical layer, depending upon the proprietariness of the codecs you support.

Try Not to Copy

If you're writing your own mixer, the **sampleSpec** tells it where to find the sample data, and you can read it directly from there, with no need for intermediate copying. But if you're layering onto a lower-level system like XAudio, OpenAL or (if you must!) DirectSound3D, take pains to make sure it doesn't copy all the sample data, or there could be big spikes in RAM and CPU usage when a long sample such as an engine sweep or loading music is referenced or when the same sample is playing several times, as often happens with common player-equipment or surface sounds.

One of the flaws of early PC audio stemmed from the need to support sound cards with dedicated sample memory. One of these, the AWE32, could only play one instance of each sample at a time. If you wanted more than one, you needed to load extra copies rather than play the data several times by reference. DirectSound3D confounded this flaw; even 3D and non-positional sounds (sometimes confusingly called 2D, though usually 0D mono or 1D stereo) with identical sample data must be packaged and played separately!

This legacy affected OpenAL, so the default alBufferData() function used to present sample data for replay internally copied the samples to private memory, whether or not this was dedicated to the audio system. This helped with DMA replay and support for sound card X-RAM, but it was generally a waste of time and space, especially on memory-tight embedded systems. The fix, assuming the sample data is persistent, is to pass a reference to it with alBufferDataStatic(), which is supported by most OpenAL drivers. Unless you're streaming or decoding a sample on the fly, in which case the copying of short sections might be convenient, seek out such functions and use them. At least the separate-buffer approach means that you no longer need duplicate data for each playing instance.

The **streamBuffer** index in the **streamSpec** structure indicates which of the pre-allocated stream file-loading buffers the stream playing methods should use. Interleaved streams require additional data, since one file may deliver data for several independent output voices with their own position and level (but not pitch!) settings. In this case it's simplest to separate the methods used to control the file transfer from those which decode and play channels within the file. At least three extra methods are needed: a pair to start and stop the file transfer, taking a **streamSpec** to specify the source and destination of the data, and another specialised Play call which extracts **interleaveCount**-specific channels from a **streamBuffer interleaveOffset** and binds those to a physical voice as explained in Chapter 10, on page 100.

Source Updates

Methods should be provided to update and read the following properties of playing sources. These could use the same Get/Set framework as the logical layer, or simple C functions if you prefer, in this internal interface. Each takes a sourceID argument to identify the voice concerned. Originally I wrote separate trivial functions to set each of the float parameters of a source, but as these were typically all called in succession to update each 3D voice it's more practical to have a single UpdateSource() call which passes in a job lot of floats liable to change most frames.

The position and radius update call should be separate since it only applies to 3D voices. Some of these will be stationary and few will vary in radius, though the panner supports this, for situations like spreading fires and dam breaches.

Table 15.2: Playing Source properties

Variable	Type	Usage
m_position	Vector	Relative to the origin (0,0,0)
m_volume	float	Attenuation in decibels
m_LFE	float	Low Frequency Effect send in dB
m_filter	float	dB attenuation at reference frequency
m_wetSends	float[]	Reverb send attenuation levels in dB
m_EndPointSends	float[]	Attenuations for non-default endpoints
m_radius	float	distance from centre to edge of source
m_pitch	float	pitch modifier ratio, typically 0.5..2
m_gain	float	m_volume converted from dB to gain
m_previousGain	float	Internal click-elimination history
m_loop	bool	true when the current sample is a loop
m_paused	bool	true while channel output is paused
m_playing	bool	true while channel is in use (or paused)

m_loop is read only, derived from the **soundSpec**. IsSourcePlaying() simply checks the **m_playing** flag. This must be updated for each playing source every frame, either via a callback or by polling if you're not doing the mixing yourself. In OpenAL, read the AL_SOURCE_STATE with alGetSourcei() and check for AL_STOPPED. On Nintendo hardware, check for AX_PB_STATE_STOP.

In XAudio2, call GET_XAUDIO2_VOICE_STATE and check for state.BuffersQueued ==0. When streaming or chaining samples, poll this or the similar OpenAL property AL_BUFFERS_QUEUED to find out when a queued buffer has been consumed, so that a fresh buffer of data can be appended before the current one is exhausted. Chapter 6 explains more about double buffering.

AoS and SoA

Since there may be hundreds of sources each with lots of properties, at least two parallel arrays should be used, one with just status flags for playing and paused, the other holding the active-channel properties, so that doesn't need to be churned through the processor cache at each update unless it refers to an active voice. This is known as the SoA/

AoS decision—Arrays of Structures are easiest to design and read, but Structures of Arrays work more efficiently if you don't generally need all the information in each structure element. The same approach can speed up virtual-voice processing on platforms with limited cache.

Private Members Only

Each implementation will store private platform-specific information alongside the properties tabulated above. For instance, an OpenAL wrapper will keep track of one alSource for each voice, an alFilter and a couple of alBuffers to allow streaming or seamless chaining of samples, plus data for decompression, such as OggVorbis_File structures and associated decoding buffers. You'll also need an alFilter for each group filter and an alEffect and alAuxiliaryEffectSlot for each reverb send. These resources should be allocated during initialisation, according to the voice and effect configuration, so that they are readily available later.

The names vary, but corresponding objects occur in most low-level audio drivers. For instance, each XAudio2 channel needs an IXAudio2SourceVoice, Nintendo channels need an AXVPB for each mono sample they play, AXFX_REVERBHI structures for reverb, AXPBLPF filters, and so on. If you're writing your own mixer you still need pointers into the sample data, filter state variables, plus buffers for delays and impulse response reverb.

CHAPTER 16

Mixing and Resampling Systems

After DSP effects, the mixing system is the most processor-intensive part of most audio runtime systems. It usually runs on a separate thread from the main game and audio updates, synchronised by the output routine that pulls blocks of mixed samples from the game to the endpoints. This chapter explains how mixing works, including techniques for resampling, metering and the avoidance of "clicks" in the output as voices start and stop.

The mixing process is simplified in floating point, as there's scant risk of overflow, though you should take pains to make sure that non-arithmetic symbols like NaNs and Infinities don't reach the mix buffers. Even though they're part of the IEEE-754 specification, they have no place in sample data. Underflow is another problem.

Pause Audio Strangles Console

One multi-million seller ran into a weird system bug late in development, and you can still hear—or measure—the workaround if you pay close attention. The Xbox 360 version of *GRID* behaved bizarrely when the game was paused! After a few seconds the XAudio CPU load would shoot up till it was dominating the system scheduler and inter-processor communication. This strangled the game update, which fell to a jerky fraction of its usual rate. But only after pausing for a while. Splendid, a silent class 1 must-fix audio bug.

It turned out not quite silent and ended up that way. The problem was a philosophical clash between the Microsoft-optimised Eventide reverb and IBM's strict implementation of the floating-point standard. A paused game would leave sounds tailing away in the reverb. DSPs silently cast infinitesimal values to zero before loss of precision, but the IEEE-754 standard says they should persist, with dwindling precision, till both the scaling exponent and fractional mantissa parts of the value are zero.

The rules for these very tiny "denorm[alised]" values are complicated, and the RISC (Reduced Instruction Set Computer) processors in most consoles use software emulation rather than expensive silicon to handle them. It takes hundreds of times as long as normal arithmetic, but it's a rare case.

Not in a filter with feedback, it isn't. Audio building bricks like delays, reverbs, shelf filters and parametric equalisation all churn state values inexorably and geometrically downwards to silence, absolute 0, -∞ dB. When the Xbox reverb got close, it triggered IBM's supervisor-level denorm handler and the console croaked.

We fixed this hastily by playing a very deep, very quiet tone slowly on a spare voice. This kept the state delays throbbing at –47 dB, magnitudes from near-zero. Microsoft fixed it properly the year after. Beware of zero: computer pioneer Konrad Zuse presciently treated zeroes and infinities as equivalent exceptions in his World War II vintage electromechanical floating-point contraptions.

Mixing Sensible Values

Excluding NaNs, denorms and infinities, each output or submix channel is associated with a float buffer, which is cleared before mixing starts and large enough to contain one multi-channel block, typically 256, 512 or 1024 samples long. Samples from each playing voice are filtered, converted to the output rate and added into these float buffers. They're attenuated according to headroomScale, the voice and group volume controls and distributed by multi-channel panning for 3D sounds, or directly to the associated output channels for non-diegetic sounds.

You can get by with just one multi-channel mix buffer for each endpoint, but more advanced 3D techniques benefit from an intermediate buffer which separates out the 3D components of a soundfield mix into a speaker-agnostic representation of its spatial distribution. Chapter 19 explains this in detail, using Ambisonics as the intermediate representation.

Figure 16.1 illustrates the signal flow through the Physical Layer to a 5.1 channel endpoint, for individual 3D sources, pre-rendered 3D soundfields and non-positional sounds like stereo music, front centre (non-diegetic) speech and sub-bass sweeteners intended for the sub-woofer. Voices are on the left and output speakers on the right. The same flow, with minor tweaks, caters for stereo speakers or 7.1 or more, in 2D or 3D.

The SRC (sample rate conversion) units up-sample every voice to the output rate. The anti-aliasing filter is only needed if the endpoint can't support the final mix rate. Skip it—or rather, leave it to the receiver—when mixing to HDMI at 96 kHz with input samples at up to 48 kHz and a maximum pitch multiplier of 2. The non-diegetic sounds—most speech, stereo music and LFE output—do not need anti-aliasing unless they're recorded or up-shifted above the output rate, which would be wasteful or unlikely, so here they're passed through unfiltered to the output.

To avoid a tangle, reverbs and filters are not shown but easily explained. Per-voice filters appear in the SRC block for each voice. Group filters and reverb sends are implemented by suitably routing sub-mixes of the group outputs. Filters are applied in line with each group output, even if they're shared between groups.

To filter an Ambisonic submix, just apply identical processing to each channel, in parallel. If the channels are aligned and interleaved this can readily be optimised with parallel vector operations. Sub-mix outputs are routed to the reverb units according to the **WetSend** levels and bits set in **m_ReverbsMask**, extracting mono or stereo as described in Chapter 18 if the reverb unit lacks surround inputs. Spatialised reverb outputs can be further panned or mixed directly to the speaker outputs.

You can safely rely upon the LFE output being low-pass-filtered downstream, but just in case someone makes a receiver that can handle 8.0 channels—ideal for cube or octagonal rigs—I treat it as potentially a full-bandwidth channel. The line connecting 3D voices and the LFE allows each source to drive the sub-woofer, if any, for additional rumble and proximity effects. It uses the same input signals as the Ambisonic encoders but bypasses panning, as LFEs are non-directional.

Higher-order Ambisonics just uses more channels between the encoder and decoder—nine for second-order, eight for hybrid third-order, 16 for third. The decoder, shared between all input channels and a given endpoint,

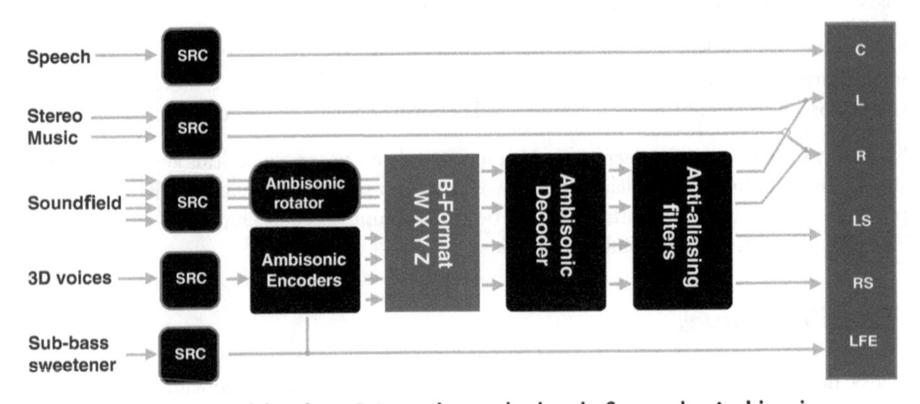

Figure 16.1: Mixing for a 5.1 speaker endpoint via first-order Ambisonics

supports any number of output channels, in any 3D position. Chapter 19 explains the Ambisonic blocks in detail.

Ambisonic to Binaural Mixing

A stereo endpoint is handled more simply; the Ambisonic encoders, intermediate B-Format buffer and decoder can be replaced with a pairwise panner direct to the speakers for mono sources, and code to extract stereo components from the soundfield channels W and Y after rotation. Left and Right speaker signals, with an image width controlled by the factor S, can be directly derived as W+Y*S and W-Y*S, where S is zero for mono.

Michael Gerzon's Ambisonic UHJ encoding makes better use of two channels by hinting at depth and height, even without a matrix decoder, using a 90° phase-shift filter as well as some B-Format munging magic. But be wary of his four-amplifier six speaker proposal in the same fecund paper[1]—at a glance it's elegantly symmetrical, but the W amplifier will surely overload!

Binaural faux-3D could just convolve left and right stereo channels into HRTF-related speaker positions, with ambient reverb for envelopment, but preferably it replaces the Ambisonic decoding of 3D voices with convolution of personalised left and right HRTF data for the position and sound of each source. Pre-mixed B-Format and stereo tracks may be spatialised to virtual speakers, or stereo may be routed directly to the ears, as it would be for normal headphone listening.

Unless the sample rate and channel counts correspond, each endpoint requires its own custom mix and multi-channel mix buffer. Since some sounds may be routed to particular endpoints, such as controller speakers or individual players' headphones, it's unwise to try to share output buffers even if their properties correspond.

Except in the special case when stereo music, speech or soundfields happen to already be at the ideal rate for output, every voice must pass through a resampling step. This involves a very significant trade-off between speed and sound quality which is the focus of this chapter.

Resampling

Before we can mix sources for output they must all be at the same sample rate. Sound designers supply samples at any rate sufficient to preserve their sonic character. The lower that rate, the more assets can fit into memory. Relative motion alters the effective rate, according to Doppler shift, and samples like impacts, scrapes and mechanical sounds may be deliberately played faster or slower to convey information or add variety.

It follows that *it's the exception rather than the rule for a memory-resident sample to play at its original pitch.* Dynamic rate changes are problematic for streams, as they complicate update timing, but there's still no reason that a stream should necessarily be authored at the final output rate, and in a cross-platform project this is rarely known in advance, for all platforms and hardware configurations.

Remember that each output is at a fixed rate, typically 48 kHz. A few systems use the older 44.1-kHz CD rate and HDMI supports rates up to 192 kHz. There are advantages, explained later, of mixing everything up to 96 kHz if your endpoints can handle it.

Interpolation is the process of computing accurate values for points in between those recorded in the original sample data. This is an essential part of sample rate conversion, so we'll explore three ways to do it, of increasing sophistication. The simplest, and noisiest, is known as linear interpolation.

Linear Interpolation

```
inline float LinearInterpolator(
  const float s0, const float s1, const float t)
{
  assert(t >= 0.f);
  assert(t <= 1.f);
  const float c1 = 1.f - t;
  return (s0 * t) + (s1 * c1);
}
```

Pass adjacent input samples **s0** and **s1** and the interpolation factor **t**, where t=0 to return only **s0**, t=1 for **s1**, and intermediate values deliver a linearly weighted average, e.g. t=0.5 for half of each, t=0.75 for one-quarter **s0** and three-quarters **s1**. The assert statements ensure that **t** is not beyond the pair of samples being considered and will be stripped out in release builds when the NDEBUG pre-processor symbol is defined.

About all that can be said for this approach is that it's fast, requiring only two multiplications, one addition and one subtraction to do its work. It's low quality because the straight-line segments between sample values introduce noise, especially at high frequencies, which is audible and liable to "pump," intrusively coming and going as the resampling offset moves between the samples to be interpolated and the error term increases.

That said, a lot of titles do still use linear interpolation, especially on PC platforms. At the time of writing linear interpolation is the only resampling technique implemented in XAudio2 for Windows. It's also the technique used in the Creative Labs generic software OpenAL driver, though that may be partly because it puts into a good light their E-MU hardware resamplers, which take account of eight samples around that to be interpolated.

Whole books have been written about interpolation and resampling. The general principle is that the more samples either side of the interpolated one you take account of, the more accurate the result. For instance, cubic spline interpolation fits a curve to the three closest samples and then traces that curve to find the intermediate level. This works a little better than linear interpolation but hardly enough to justify the extra effort. We'll consider two more techniques, of increasing cost and quality, before referring you to the literature.

Hermite Interpolation

Several fast game runtimes use a four-point Hermite interpolator, which takes a weighted average of the values of two pairs of samples straddling the point of interest:

```
inline float HermiteInterpolator(const float s0,
  const float s1, const float s2, const float s3, const float t)
{
  assert(t >= 0.f);
  assert(t <= 1.f);
  const float c1 = (s2 - s0) * 0.5f;
  const float c2 = s0 - (s1 * 2.5f) + (s2 + s2) - (s3 * 0.5f);
  const float c3 = ((s1 - s2) * 1.5f) + ((s3 - s0) * 0.5f);
  return s1 + (((c3 * t + c2) * t + c1) * t);
}
```

To use this, pass four adjacent input samples **s0** to **s3** and the interpolation factor **t** between **s1** and **s2**, where t == 0 at **s1** and t==1 at **s2**. It's quick and sounds better than theoretical analysis would suggest. For a comparison of this and similar ways of weighting and averaging nearby samples, see Olli Niemitalo's paper.[2]

Sinc Interpolation

For better quality we must take account of more measurements around the point we want. Samples before and after that point are decreasingly influential with distance, and we can't afford to consider every sample before and after—perfect results would require scanning infinitely far in both directions. Microprocessor audio pioneer Hal Chamberlin has recommended considering the nearest sample to the point and six either side of that sample, using a cleverly smoothed impulse pattern stored in a table.[3]

The principle is that any single impulse in a balanced dynamic system creates a symmetrical ripple before and after it, called a "Sinc" wave. This ripple has peak amplitude at the time of the impulse, and then oscillates alternately down and up for sample pairs either side, so those have decreasing effect with distance, but big spikes some distance from the interpolated point still have the proper influence there.

Once you know the shape of this wave and have "tailed" it to a practical size, you can create offset versions of it corresponding to positions in between measured samples. The offset affects the weights given to nearby samples, and the consistent Sinc pattern means the offset versions can be used to accurately predict the level at points in between.

For instance, a Sinc ripple delayed by half of one sample interval can be used to predict the interpolated level between any two measurements, by scaling its coefficients against the actual sample values either side. This process can generate intermediate points all along a wave, effectively doubling the sample rate. Interpolate two intermediate shifts either side of that one, and you can generate an output wave with four times as many samples in it, at four times the frequency of the original but with the same shape and sound.

Understanding Sinc

You may find it hard to believe that a wiggly wave extending infinitely either side of an impulse spike has anything uniquely special to do with audio filtering. I certainly did, until Robert Bantin explained, as follows.

> *When we look at an array of samples, we consider them to be a stream of spikes that blur together to give us the auditory illusion of uncoloured continuity. The reality is not that simple. A spike is a theoretical concept, with infinite energy as it rises vertically, but zero width in time. In practice that continuity requires that each spike bear some relation to ones around it.*
>
> *Consider a typical real-world recording setup. If you were to record such a spike, or the closest approximation you could make, with a microphone and digitise it, the signal would necessarily pass through an anti-aliasing filter before analogue to digital conversion could occur, to limit its bandwidth to that which the sampler could support. Therefore, what would be captured in the digital domain would not be a single spike, but the impulse response of that anti-aliasing filter: a steep analogue low-pass filter with a corner frequency set just short of the Nyquist limit at which frequency aliasing starts. So any audio digitised at a constant sample rate is actually a series of these impulse responses, overlapped and added together. This overlap and interaction captures how each single sample would be related to the next.*
>
> *With that in mind, when we interpolate between single samples to create a new digital signal at a higher rate, we should model an anti-aliasing filter to maintain that colourless continuity. A low-pass brickwall filter is an ideal anti-aliasing filter. The characteristic wiggle of a Sinc function comes from defining the required rectangular filter in the frequency domain and then converting that to the time domain. If we model our anti-aliasing filter using time-domain Sinc waves, the process should be acoustically transparent.*

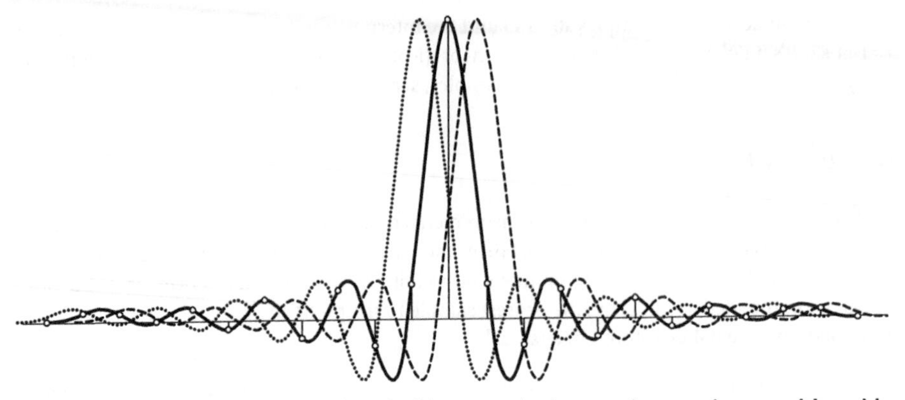

Figure 16.2: The Sinc wave associated with a sample time, and two points on either side

The practical snag is that an infinitely sharp low-pass filter is equivalent to an infinitely long Sinc wave, so we must truncate the wave to a finite length to make this technique practical. The more we truncate, the less transparent the anti-aliasing is. As with most engineering problems there is a trade-off between quality and computational complexity, so the question arises "how much truncation can you live with?"

How much truncation can you live with?

Using Sinc

Chamberlin presents a table of Sinc coefficients for 31 positions in between pairs of measured samples representing interpolated levels at intervals of 1/32 to 31/32 of the sample period. The 32nd case is trivial as an interval of 0/32 exactly corresponds to a measured sample, so involves no interpolation. By using each table offset in turn this allows a wave to be oversampled from its original rate to 32 times that—from 48 kHz to around 1.5 MHz—while preserving its original 24 kHz bandwidth.

To play that sample at any other rate which is an exact fraction of 1,536 kHz (48000 * 32) we just step through the oversampled wave pulling out samples at the desired interval. For instance, we could resample to 64 kHz by using one in 24 values of the oversampled wave. It will sound a perfect fifth flat (ratio 3/4, 48000/64000) if replayed at the original 48 kHz, or identical to the original if played at the new 64-kHz rate.

Thus the Sinc table can solve two problems, either converting samples from a lower rate to be output at a higher one—such as when we save space by storing speech at a lower rate than the maximum—or for arbitrary pitch shifting, to account for Doppler shift, designer preference or real-world fluctuations like gear wobble: the tendency of gearbox speed to flutter after a gear-change, as the mechanism adjusts to the new rate. This wobbling varies between gearboxes, but manual shifts sound wrong unless you model it.

I used this approach, with a table much larger than Chamberlin's, to resample CD audio from 44.1 kHz to 48 kHz for output by the music game *Dance Factory*. The table had to be large because of the lack of common factors between 44,100 and 48,000. The ratio simplifies, somewhat, to 147/160, which means that for every 147 samples coming from CD I needed to generate 160 for the PlayStation's 48-kHz output. High fidelity was essential, as the chosen music is the dancer's focus of attention; they'll be familiar with the CD original, as the unique selling point of *Dance Factory* is its ability to work with the customer's favourite songs.

The table consists of entries for 160 interpolated positions between each sample of a 44.1-kHz wave. As each 48-kHz sample is needed the offset in the table is advanced by 147 positions, modulo 160 to stay within the table and advancing by a whole input sample at each overflow. The resultant stream is at 48 kHz, as 44.1 / 147 * 160 = 48. Since CDs are stereo and the interpolation offset is the same for both left and right samples, the same coefficients apply to both channels.

Chamberlin shrewdly observes that since the Sinc wave is symmetrical, the table only needs six entries per offset, applied to six samples either side of the nearest. But that's still 960 coefficient values, each a four-byte float, even if we use each four times over in the stereo case. The table is almost 4K in size and takes a while to be pulled into the cache; after that, the time needed to sum the result of 13 multiplications (24 additions and 26 multiplications, for stereo) predominates. The floating-point fused-multiply-accumulate (FMA) operation is tailor-made for this purpose, doing the addition and multiplication in parallel. Vector instructions, like the AltiVec, VMX and ARM64 sets, can do two, four or eight of these operations in parallel.

Even for stereo this means only 13 instructions are needed to perform 50 FLOPS (Floating-Point OperationS), and those instructions have few enough input dependencies that an optimising compiler can do a good job of scheduling them to minimise pipeline delays. It's no more code to interpolate an interleaved (stride 16) four-channel B-Format soundfield, on a four-way parallel vector unit (PS2 VU, PS3 SPE, AltiVec/VMX, SSE etc). AMD Jaguar hardware in the PS4 and Xbox One consoles can crunch eight floating-point values in parallel, but the AVX hardware splits them up into pairs of four with separate multiply and add steps, so it's not necessarily an improvement over earlier console hardware, particularly Sony's SPEs.

However big your Sinc table, if you use it directly it will limit the number of intermediate steps you can predict per sample interval—to 32 in Hal's example or 160 steps for the CD up-sampler. That limits the number of sample rates you can convert to, as the ratio between the old and new rate must be a multiple of the reciprocal of that count. We learned in Chapter 2 that a reciprocal rule does not yield chromatic frequency interval steps, so the interval between table entries is neither ideally distributed for display nor listening.

But there's nothing, bar a doubling of CPU cost, to stop us interpolating between the two closest table entries to get any ratio and hence any frequency or chromatic interval, to 32-bit float precision. The bigger the table the less noise the final interpolation step will generate, and even if it's only a few lines of data the oversampling it implies will push the frequency of that noise out of the audible spectrum.

Interpolating New Interpolators

At this point I can present a concise example Sinc table, with only five entries, yet good enough to oversample the input by a factor of six times. This can be used to resample input from 16, 24 or 32 kHz to 96 kHz, for example. Since the Sinc itself is a wave, you can use the Sinc interpolator to resample its own coefficients to generate new Sinc interpolators for any pitch interval! That's how I made the data for the *Dance Factory* resampler. Here's the condensed version:

```
const float FullImpulse[] =
{
  0.f,  0.f, 0.f,
  0.f, -0.0011597f, -0.0022889f, -0.0032350f, -0.0035401f, -0.0026246f,
  0.f,  0.0041810f,  0.0089724f,  0.0127567f,  0.0134281f,  0.0093081f,
  0.f, -0.0131840f, -0.0267342f, -0.0359813f, -0.0361034f, -0.0239875f,
```

```
    0.f,  0.0312815f,  0.0623798f,  0.0820032f,  0.0808130f,  0.0530717f,
    0.f, -0.0690329f, -0.1370281f, -0.1823481f, -0.1834773f, -0.1246681f,
    0.f,  0.1832636f,  0.4024171f,  0.6262398f,  0.8208258f,  0.9531846f,
    1.f,  0.9531846f,  0.8208258f,  0.6262398f,  0.4024171f,  0.1832636f,
    0.f, -0.1246681f, -0.1834773f, -0.1823481f, -0.1370281f, -0.0690329f,
    0.f,  0.0530717f,  0.0808130f,  0.0820032f,  0.0623798f,  0.0312815f,
    0.f, -0.0239875f, -0.0361034f, -0.0359813f, -0.0267342f, -0.0131840f,
    0.f,  0.0093081f,  0.0134281f,  0.0127567f,  0.0089724f,  0.0041810f,
    0.f, -0.0026246f, -0.0035401f, -0.0032350f, -0.0022889f, -0.0011597f,
    0.f,  0.f, 0.f
};
```

This totals only 312 bytes, even though half the data is redundant—note the symmetry between the bottom right corner, reading up and reading down from the top left. I've presented it this way to keep the example code simpler.

TableStride is the number of impulse response instances in the table. As this is a six-times oversampler:

```
const unsigned TableStride = 6;
```

The SincInterpolate() function reads samples from **pIn** and stores **countOut** resampled ones at **pOut**; **countIn** is the number of input samples at **pIn**, and **step** is the fractional interval between input samples for each interpolated one output, so to convert input from 8/16/24/32/40 to 48 kHz, pass a **step** of 1, 2, 3, 4 or 5 respectively. In other words, the SincInterpolate() function generates (6 / **step**) output samples for every one input.

pIn, **pOut**, and **frac** are updated on return from the function; **frac** is the fractional step to use when computing the next sample from that slice of the table and is incremented modulo **TableStride** after each sample. Before each wave is processed, but not between subsequent blocks, this needs to be centred as follows:

```
static unsigned frac = TableStride >> 1;
```

To avoid a glitch between successive calls to the function, we need to preserve the value of **frac** across calls for a given input buffer. The code is re-entrant, so it can be used for several waves and sets of input and output buffers, providing we keep track of **frac** for each. It re-uses the output buffer automatically, in a circular fashion—you will need to implement the function WriteOut(), with this prototype, to copy each output buffer out wherever you need it.

```
void writeOut(float* source, unsigned count);
```

When that destination is full, you'll need to keep the unused part of the output buffer for use in the next frame. After that's been copied away, the entire output buffer is free for reuse. The ideal size of the output buffer depends upon the subsequent way the data will be used. We'll use an array of 512 floats to collect output, in this case; this will become a member variable in an object-oriented implementation:

```
#define outSize (512)
float buffer[outSize];
```

Assuming we have at least "chunk" samples at **pWave**, we can start to interpolate them to a new sample rate of 6 * original rate/**step** like this:

```
float* pNextIn = pWave;
const unsigned step = 2; // For 3:1 interpolation


SincInterpolate(pNextIn, buffer, chunk, outSize, step, &frac);
```

This will call WriteOut() as necessary, given the step size and input and output buffer sizes and return after processing "chunk" input samples with **pNextIn** pointing past those. **pOut** and **index** should similarly be updated and preserved.

To process the next chunk of **pWave**, adjust "chunk" to make sure it does not pass the end of the input and then call SincInterpolate() again till all the input has been consumed. At that point, either wrap the input back to the start, for a loop, or submit a dozen zero samples to flush out the tail of the input, then pad the buffer from **pOut** to **&buffer[outSize-1]** with zeros to complete the final block and pass it to writeOut().

```
void SincInterpolate(float* pIn, float* pOut,
  const unsigned countIn, const unsigned countOut,
  const unsigned step, unsigned* pFrac)
{
  assert(step<TableStride);
  assert(step>0);

  unsigned frac = *pFrac;
  float* pStart = pIn;
  float s;

  while (pIn < pStart + countIn)
  {
    float* pFilter = pIn;
    s  = *pFilter++ * FullImpulse[ 1 * TableStride - frac];
    s += *pFilter++ * FullImpulse[ 2 * TableStride - frac];
    s += *pFilter++ * FullImpulse[ 3 * TableStride - frac];
    s += *pFilter++ * FullImpulse[ 4 * TableStride - frac];
    s += *pFilter++ * FullImpulse[ 5 * TableStride - frac];
    s += *pFilter++ * FullImpulse[ 6 * TableStride - frac];
    s += *pFilter++ * FullImpulse[ 7 * TableStride - frac];
    s += *pFilter++ * FullImpulse[ 8 * TableStride - frac];
    s += *pFilter++ * FullImpulse[ 9 * TableStride - frac];
    s += *pFilter++ * FullImpulse[10 * TableStride - frac];
    s += *pFilter++ * FullImpulse[11 * TableStride - frac];
    s += *pFilter++ * FullImpulse[12 * TableStride - frac];
    s += *pFilter   * FullImpulse[13 * TableStride - frac];

    *pOut++ = s; // Clamp here if desired

    if (pOut >= &pStart[countOut])
    {
      writeOut(pStart, countOut);
      pOut = pStart; // Start again
    }
```

```
            // Advance window
            frac += step;
            if (frac >= TableStride)
            {
              frac -=TableStride;
              pIn++;
            }
        }
        *pFrac = frac;
    }
```

Wrapping Up

Since the process uses a sliding window of 13 samples around the point at which a sample value is to be interpolated, odd (but smooth) things may happen to the samples at either end of the processed stream unless we do some extra work. There's a delay of six input samples between input and output. To preserve the initial transient of one-shot waves, at the expense of slightly more delay, prefix them with 12 zero samples. Likewise, append a dozen zero samples to wring out the final wiggles of the wave. This is not necessary for a continuous stream or looped sample.

It's possible, for instance, when interpolating across the peak of a wave, that some of the output values will marginally exceed the range of the input. If we were interpolating from and to short integers a clamp to the range −32768..+32767 would be required before storing the output sample "s" in the output buffer, but we can live without this when working entirely in floating-point arithmetic.

That's enough to get decent results. I recommend you refer to Chamberlin's book[3] and Nigel Redmon's blog[4] for programmer-friendly explanations of resampling.

Clicks and Interpolation

The mixer combines the decoded, resampled, panned buffers of floating-point samples for each voice into final output blocks for the output hardware. As the name suggests, it does this by adding corresponding sample values. The resampling means that each channel's panned output is already at the same sample rate as the required output mix.

The only complication is the need to adjust the level at which the samples are added to match the expected gain for that voice. In theory, this could be done at the same time as the volumes are adjusted by the panner to distribute mono input across multiple output channels. In practice, we do this during mixing, and after decoding and panning, to hide clicks which would otherwise occur if (as is likely) the game has adjusted the voice volume since the last block was mixed.

Consider what would happen if we used the final voice gain, set by the logical layer, to scale samples during mixing. Since the mixer runs every time the output hardware needs a new block of data, and the logical layer adjusts the voice gains once per game update, if we used the voice gain directly while mixing there would be a sudden jump in output level, up or down, when the voice gain value was modified. This may be audible and is probably unwelcome.

Depending upon the amplitude of the sample being mixed at that instant, there might be a big "click" as the gain adjustment was suddenly applied. The size of the click is hard to predict—it depends upon the amount of the gain change and the magnitude of that sample in the wave. If the wave happens to be at a zero-crossing there will be no click, whereas if it's near the top or bottom of its range the click will be larger, directly proportional to the gain change.

The solution is to distribute the change in channel gain over several samples, so it is applied smoothly rather than suddenly all at once. The easiest way to do this, which yields good results, is to take a snapshot of the gain for each voice after mixing its samples into the output and compare that with the latest value from the game update just before mixing the next block.

If they're both the same we just scale the entire block of voice data by the unchanged gain and add it into the output channel buffers. Since gain doesn't always change, and the logical updates are less frequent than the generation of output blocks, the case where the gain has not changed is most common, and we use a separate simpler mixing loop then.

If the gain *has* changed since the last block was mixed, we need to do a bit more work. It's important that this necessary overhead should be efficiently implemented, since it will be done for every output sample for each voice that's changed in gain—since a simple movement of the listener is enough to vary the volume of all associated 3D voices, the extra work may apply to most of the samples mixed into the output block following the next logical layer update.

If the output blocks are very long-lasting or the game update is very fast, you could theoretically get a situation where game updates run more than once as a single output block are mixed. Applying the changes only across block boundaries might mean that intermediate gain updates get ignored, as they will have been superseded by the time the block mixing is complete.

In practice, this doesn't happen in interactive audio systems, because the output block size is set to minimise latency and the update rate, synchronised with the rest of the game, is relatively slow—typically 192 Hertz for output blocks (512 samples at 48 kHz, a block of 5.3 milliseconds) and between 25 and 60 Hertz for the game update. In VR applications, the quoted game update may be faster—perhaps 100 Hertz—but still less frequent than the audio spits out mixed blocks.

If your endpoint requires such large blocks or low sample rates, you may need to mix half blocks so that all the gain updates are honoured. I've never seen that in practice, and it would slightly complicate the final output to the endpoint because of the need to accumulate blocks before pushing them to the hardware, but it cures the problem. All is well so long as you choose the mixing block size to ensure output chunks are generated more frequently than game updates arrive. Accepting gain changes while in the middle of mixing a block is much more complicated and problematic, so we tune the system to remove that need.

The implementation depends upon three extra "gain" control variables for each channel, one of which is persistent—and so needs to be in the low-level voice structure—and two which are temporary and can be local to the mixing loop. We'll call these **m_previousGain**, **targetGain** (for the gain we want to end up with) and **currentGain** for the smoothed intermediate values. The required gain, set when the voice is updated, is **m_gain**, part of the per-voice structure. The output buffer is **mix[]** and the resampled voice data is presumed to be in **voiceOut**[]. All these are float values. BLOCK_SIZE is the number of samples in each chunk of output, and hence the upper bound of both arrays.

```
float targetGain = m_gain; // Snapshot value from latest update
if (fabsf(m_previousGain - targetGain) < 0.1e-5)
{
  for (size_t sample = 0; sample<BLOCK_SIZE; ++sample)
  {
    mix[sample] += voiceOut[sample] * targetGain;
  }
```

```
}
else // Gain changed significantly since last block was mixed
{
  float currentGain = m_previousGain;
  float gainStep = (targetGain - currentGain) * (1.f / BLOCK_SIZE);
  for (size_t sample = 0; sample<BLOCK_SIZE; ++sample)
  {
    currentGain += gainStep;
    mix[sample] += voiceOut[sample] * currentGain;
  }
  m_previousGain = targetGain;}
}
```

When the voice starts to play, after the initial level has been set but before mixing, set **m_previousGain** to **m_gain**, rather than the obvious default gain of 0, unless it's starting from an offset. This preserves the full dynamic attack of the sample for a voice which is meant to be heard right away. If every newly triggered voice started with **m_previousGain** = 0 it would be faded up as the first block is mixed. Notice that **m_gain** is only read once, then a local copy is used thereafter. This avoids glitches when mixing has its own thread, so **m_gain** might change asynchronously.

The volume ramp is linear rather than logarithmic, so the perceived size of the steps falls as the gain values increase. This means it doesn't sound as smooth as it might. Progressive scaling of **gainStep** could be done inside the loop or by reference to a table of BLOCK_SIZE values populated with a logarithmic sequence. Thus deltas became larger as the gain increased, or vice versa, giving a smoother curve, but the blocks are so short that this complication is not worthwhile in practice. The higher-level Fade() method works logarithmically, up or down, but between mixer updates not samples so it's much cheaper—and still smooth as long as updates are regular.

This is a fairly efficient implementation—you might cavil at the multiplication by 1, but the parenthesised sub-expression (1.f/BLOCK_SIZE) hints that the fraction can be computed at compile time; depending upon optimisations simply dividing the required gain difference by BLOCK_SIZE might otherwise lead a dim compiler to use a slow floating-point division operation rather than multiply by a constant, which is many times faster in hardware.

You might also wonder why we compare the magnitude of the difference with a "magic number" 0.1e-5 (one part in one hundred thousand). This factor corresponds to a difference less than -100 dB. This happens for two reasons. Firstly, if the two gains are almost identical there's no click to suppress, so we can skip the extra work.

Secondly, **gainStep**, the difference in gain applied to adjacent samples, might otherwise end up so tiny, being the difference of two similar values divided by the block size, that loss of precision might occur, distorting the ramp gradient. Given the limited 23-bit mantissa of the single-precision float value and the possibility of distributing a very small gain difference over a long block, strict IEEE-754 implementation might end up using a denormalised value for **gainStep**.

These exceptional values are bad news in audio code, illustrated by the earlier tale about game performance collapsing while paused. They may be handled hundreds of times more slowly than normal values, and need filtering out. Sometimes these come from timing or physics code, especially during start-up or changes in game mode.

A robust audio system will check all floating-point input and catch denormalised values and other horrors like infinities and NaNs (special values meaning "not a number") before allowing them to propagate. These are the kind of checks worth having in an otherwise optimised development build of the product. Only in the final release, after extensive testing, is it safe to strip those to wring out the last possible bit of extra speed. Indeed, you may need to do so when using speed-up options like—ffastmath with optimising compilers like GCC and Clang.

All float buffers and control values should be checked when first set to save duplicated effort later. Warn and substitute zero for any special values found, then investigate the cause. It's usually uninitialised data or a zero Physics time-step but needs fixing wherever it came from. The following function returns false if an IEEE-754 float value is infinite or not a number:

```
inline bool finiteFloat(float f)
{
  return f - f == 0.f;
}
```

Vector Mixing

This loop is crying out for SIMD vector optimisation. It could be several times faster if consecutive groups of four samples—for ARM NEON, SSE, AltiVec and similar vector units—or eight (for AVX) were processed in parallel.

Here is the first loop re-expressed using ARM intrinsic instructions. GCC's **vfmaq_f32** (Vector Fused MuLtiply Accumulate) NEON intrinsic picks up three clumps of float32x4_t vectors, each comprising four consecutive float values, and adds the first to the product of the second and third, combining eight arithmetic operations into one instruction:

```
float32x4_t gain = vdupq_n_f32(targetGain);
float32x4_t* pvMix = (float32x4_t*) mix;
float32x4_t* pvMixEnd = (float32x4_t*) &mix[BLOCK_SIZE];
float32x4_t* pvVoiceOut = (float32x4_t*) voiceOut;

while (pvMix < pvMixEnd)
{
  *pvMix = vfmaq_f32(*pvMix, *pvVoiceOut++, gain);
  pvMix++;
}
```

Notice how the for loop and "sample" counter have been eliminated in favour of a check if the output pointer **pvMix** has reached the end of the mix buffer. The loop needs to increment the input and output pointers as it works through the entire block anyway, but the counter is redundant, as the address just past the end of the block is pre-computed outside the loop in **pvMixEnd**.

BLOCK_SIZE must be a multiple of four, or eight if using AVX, since the loop now processes that many samples at a time. It's vital to check this, or you might run off the end of the buffer—if block size was nine, say, the last iteration would read and clobber three unintended values. Block sizes are generally powers of two, for all the reasons implied by binary computing, and we can make certain of this by a compile-time assert, or a single runtime test if variable block sizes are permitted by your engine.

You should also ensure, by design or at runtime, that both arrays are vector aligned in memory. This means they should start and end at an address which is an exact multiple of the vector size, minimising the number of memory bus transactions needed to fetch each one. Use this for 16-byte vectors:

```
assert((((size_t) pvMix) && 15) == 0);
```

On some complex processor architectures misaligned data still works, slowly, as each vector fetch invokes multiple memory accesses and masking operations to combine the misaligned values. On others the misalignment will be ignored, and you'll read and write different addresses from the scalar version—disaster!

Vector De-Clicking

That snippet only handles the fastest case when no scaling is needed. We need a couple more variables to vectorise the gain ramp. **vGainStep** is just a four-wide equivalent of **gainStep**, used to advance the step value for four adjacent samples at once. Since the loop processes four at a time, the step is four times as large. Since we still want a smooth ramp, a second vector **vTargetGains** contains the first four **currentGain** values which the scalar loop would use.

```
float gs4 = gainStep * 4.f;
float32x4_t vGainStep = {gs4, gs4, gs4, gs4};
float32x4_t vTargetGains =
  { currentGain + gainStep, currentGain + (gainStep * 2.f),
    currentGain + (gainStep * 3.f), currentGain + gs4};

while (pvMix < pvMixEnd)
{
  *pvMix = vmlaq_f32(*pvVoiceOut++, *vTargetGains, *pvMix);
  pvMix++;
  vTargetGains = vaddq_f32(vTargetGains, vGainStep);
}
```

Merge in the if/else test from the first example and the vector variable declarations and simple-case loop from the second. Don't forget to update **m_previousGain** from **targetGain** after smoothing a block.

This example has been written with ARM-style vector intrinsics. Equivalent datatypes and instructions exist for Intel/AMD SSE processors, AltiVec (PowerPC, not supported on Wii/Wii U) and AVX (PS4, Xbox One), alongside many more eccentric ones.

If you are writing portable C++ code but all your platforms are built with GCC, Clang or compatible compilers, there's an even more elegant way to write vector code.[5] Vector Extensions use operator overloading and optimised code-generation, so you can use conventional operators like + and * on suitable datatypes and trust the compiler to use parallel instructions when they're available and fallback to scalars otherwise. Providing you check the code generation, especially to make sure fused multiply-add instructions are being used, these are the best way to vector-optimise code for the free compilers.

Fourier Transforms

In 1822 Joseph Fourier determined that any continuous wave can be reinterpreted as the sum of a set of sine waves (or any consistent shape) of increasing frequency. The amplitude of each sine wave depends upon the proportion of that frequency in the input. Since the component sine waves don't necessarily all start at the same time, their phase must also be taken into account to determine where the corresponding peaks and troughs occur in the input. The Fourier Transform converts a signal from the "time domain"—samples measured at regular intervals—to the "frequency domain": component frequencies which can be combined to recreate the original wave.

Some filtering and frequency-manipulation problems are most easily done in the frequency domain—time-invariant pitch shifting is an obvious example. Fourier Transforms are often used in audio systems to perform convolution efficiently to apply impulse-response reverb or long FIR filters.

The following minimal code converts BLOCK_SIZE samples of the array **in**[] into the same number of pairs of values X and Y which represent the phase and magnitude of frequency components ("bins") of the wave. The code performs a discrete Fourier transform, or DFT, which is only suitable for small values of BLOCK_SIZE as the time it takes is proportional to BLOCK_SIZE squared, even if the sine and cosine values are precomputed. Potentially faster techniques like Hartley transforms and fast Fourier transforms (FFTs) are well-documented—this is just a simple example to show the principle:

```
float in[BLOCK_SIZE], x[BLOCK_SIZE], y[BLOCK_SIZE];
const float angleStep = 2.f * PI / BLOCK_SIZE;

for (unsigned k=0; k<BLOCK_SIZE; ++k)
{
  x[k] = y[k] = 0.f;
  for (unsigned j=0; j<BLOCK_SIZE; ++j)
  {
    x[k] += in[j] * cosf(j * k * angleStep);
    y[k] -= in[j] * sinf(j * k * angleStep);
  }
}
```

Result pairs are needed since the outputs are 2D vectors, commonly termed the "real" and "imaginary" parts of the result. They could equivalently be represented as power and angle pairs:

```
float angle[k] = atan2f(y[k], x[k]);
float power[k] = (x[k] * x[k] + y[k] * y[k]) / BLOCK_SIZE;
```

The resultant power array can be used to make a spectrum graph, with bars bouncing up and down as signal frequency components vary. Bin 0 corresponds to the constant DC offset in the input and subsequent bins to increasing frequencies; for N samples captured at H Hertz, bin 1 represents the component at H/N Hertz, then H/(N*2) and so on. These raw numbers are ideal for conversion back to the time domain after analysis or manipulation but need some massaging to correspond to what we hear because of the way pitch and volume are interpreted logarithmically by our ears.

To convert the data into psychoacoustically weighted form, sum contents of increasing numbers of bins at higher frequencies, so the interval between bars is chromatically consistent. For instance, on a 64-bin output you could create a six-bar graph with one bar per octave, plus an initial one for the varying DC component, by outputting bins 0 and 1 directly then summing 2 and 3, 4..7, 8..15, 16..31. Bins 32 to 63 are redundant in this application, as they correspond to frequencies above the Nyquist limit. For perceptually linear amplitudes, convert the power to dB by taking its logarithm.

Nyquist Limit

The Nyquist limit is a principle of sampling theory.[6] Since periodic waves must go both up and down, and we find their pitch by counting the number of ups (or downs) in given time, we must sample at at least twice the frequency of any component wave in a signal, or we'll miss some of the ups and downs. This causes the apparent pitch to fall below the limit rather than above. The more the pitch exceeds the limit, the more it falls. So a 15-kHz wave downsampled into a 24-kHz stream will be heard at 24–15 = 9 kHz—a most annoying whistle.

This "aliasing distortion" becomes especially apparent in environmental simulations, because Doppler shifts cause the distortion to move non-harmonically in the opposite direction, up or down in pitch, to the lower frequencies of the signal. Sadly, it's commonplace in resamplers and decoders with poor filtering (usually justified on speed grounds). Stay alert for it, and consider downsampling assets or cross-fading between low- and high-octave versions when you hear it happening.

Peak Metering

To make peaks easier to see the animation should have a fast attack, moving immediately to a new peak level and only slowly descending between peaks. This can easily be done once you're working in decibels. You need a "history" value which is discounted each frame by a constant number of dB—how many depends upon the size of your bar graph and the rate at which you update it.

At each update, compare the new peak value with the discounted history value. If the new peak is highest, copy that to the history. Then display a bar corresponding in height to the updated history value. This will jump up smartly and smoothly decline between peaks, like the Peak Programme Meters used in broadcasting.

For cosmetic reasons, you might also progressively boost the display for higher frequencies, so that all the bars bounce up and down about the same amount. If you don't do this the highest bars will (accurately but less excitingly) peak lower, because less energy is present at higher octaves in typical audio.

There are several standards for the display range and rise and fall time of meters. Generally we implement digital meters with delayed decay, and the fastest attack time the display can support—if you're refreshing the screen 100 times per second, the attack time must average at least 5 ms. The exception is the Vu (Volume unit) meter, designed to give an indication of average volume on a telephone line, which takes the same 300 ms to go up or down its -20 dB-to-+3 dB range. Peak Programme Meters suit the fast attack and limited headroom of digital systems, though the German DIN specification allows 80 ms rise-time, so it may under-read on brief spikes. Table 16.1 shows some standard meter ranges (between calibrated marks—physical meter needles may go further) and representative rise and fall times.

Table 16.1: Ranges and rates of standard audio meters

Name	Standard	dB range	ms rise	ms fall
Vu Meter	ANSI-C16.5-1942	23	300	300
BBC PPM	BS-4297	28	12	2,850
DIN PPM	DIN 45 406	55	80	1,000
Nordic PPM	N9	45	5	1,750

Constant-Q Transforms

Fourier transforms can also be used to extract pitch and key information and simulate graphic equalisers, albeit with bands evenly spaced in Hertz rather than in musical chromatic intervals such as fractions of an octave. Conventional Fourier transforms have poor resolution at low frequencies and gratuitously high resolution at the top—for instance, if BLOCK_SIZE is 512 and the sample rate 48 kHz, the bands will be at intervals of 93.75 Hertz (48000/512).

The second band will be an octave higher, at 187.5 Hertz; but the last two useful bands before the Nyquist limit at 24 kHz will be at 23,812.5 and 23,906.25 Hertz, an imperceptibly small interval. The intervals are even more arbitrary for 44.1-kHz CD audio. I had special crystals made for Silicon Studio Ltd audio cards so they could sample at 51.2 kHz and spit out bins at nice round intervals of 50, 100 or 200 Hertz. But don't be misled by the decimals—there's always some leakage to other bins when measuring real-world audio signals this way.

If you need to perform pitch analysis, a "constant-Q" transform may suit your purposes better than an FFT, even though it's harder to optimise. Constant-Q implies chromatically spaced bands at intervals of some fraction of a semitone rather than every so many Hertz. Judith Brown did breakthrough research into this tech at MIT,[7] then worked out a quick way to adapt FFT output to suit with Miller Puckette of IRCAM.[8] See the references for details and a "sliding" DFT implementation particularly well suited to parallel computation.[9]

Faster Transforms

Fast Fourier Transforms work best when BLOCK_SIZE is a power of two. They run at a speed proportional to BLOCK_SIZE * log(BLOCK_SIZE) rather than $BLOCK\_SIZE^2$, which makes a big difference as BLOCK_SIZE increases. The speed-up stems from a binary approach to the data, akin to Quicksort or Binary search algorithms. The first pass combines values offset by BLOCK_SIZE/2 samples, the next uses a BLOCK_SIZE/4 offset and so on until the last pass works on adjacent pairs. Each operation on a pair of vectors is called a butterfly. This approach scrambles the order of the results, but the reordering can be countered by reversing the significance of bits in the index, either when the input is presented or by shuffling the output.

A complementary iFFT inverse process converts back from the frequency domain to a series of samples. FFT and iFFT implementations abound; chipmakers Intel and ARM have free binary implementations for their old and new processor architectures.[10,11] KISS (keep it simple, stupid) FFT is a free library, portable but lacking platform-specific optimisations.[12] FFTS has optimisations for 32-bit ARM and 64 bit x86 processors.[13] FFTW is faster on many platforms, but commercial use requires a license from MIT.[14] Apple offers VDSP FFT.[15]

The ideal choice of FFT depends upon the required frequency resolution, channel count and platform architecture, so I provide representative links. Before porting one of these, let alone writing your own, check to see if there's one already available in the system libraries for each system you're targeting. This is likely to be highly optimised for that platform, and as FFT behaviour is well documented, there's no reason to use the same implementation on all platforms. Some "fast" Fourier transforms are a lot quicker than others—experiment!

Windowing

Before we move on from FFTs it's important to note that inaccuracies, known as spectral leakage, occur for any real-world signal that doesn't pack an exact number of wave cycles into the block chosen for analysis.[16] Typical signals are not that tidy, so we smooth the incomplete waves and abrupt start and end of the sampled data by fading the input to the FFT up and down in a process called windowing. Similarly we overlap output blocks after conversion back to the time domain via the complementary iFFT process.

The shape of the fade up and down is a trade-off. The Hann window is a simple cosine bell.

The Hamming window adds an offset, trading narrower frequency peaks for greater spillage to distant bands.

```
for (unsigned k = 0; k < BLOCK_SIZE; ++k)
{
  out[k] = in[k] * (c1 - (c2 * cosf(2.f *
    PI * k / (BLOCK_SIZE-1) )));
}
```

For a Hann (sometimes called Hanning) window, c1 and c2 are both 0.5. For the Hamming window, c1 = 0.54 and c2 = 0.46. As for the Fourier transform, slow trigonometric calls can be eliminated by using a look-up table for the BLOCK_SIZE required angles. Other windows use more trigonometry to compute their curve shape but are no slower once the scaling table is computed and readily converted to SIMD. Fredric Harris's IEEE paper compares the performance of many windowing functions.[17] Try a few and see what yields best results—the only difference between them is the table.

Overlapping Windows

Since windowing suppresses data at the start and end of the block, it's necessary to overlap blocks when converting them from the time domain to frequency and back again. For most purposes, a 50% overlap—where each window of data starts mid-way through the previous one—is adequate. But *Dance Factory* used a 75% overlap, advancing by only a quarter of the block size, to get more accurate timing for changes in frequency distribution. This is a vital aspect of a game that creates dances from the user's music collection.

Windowing is crucial to avoid very audible high-frequency distortion in FFT-processed signals. Reverb makes this even more obvious, leading to a sound like rivets bouncing in a tray if you skimp on the windowing. Windowing is also used to trim theoretically infinite impulse responses for use with short FIR filters, like our Sinc interpolator.

Vectorising Fast Fourier Transforms

Vector instructions can also be used to speed-up Fourier transforms. The obvious way is by making some of the arithmetic operations in the butterfly step work in parallel, but this generally yields a speed-up factor of less than two and complicates the code.

When several same-sized transforms are to be performed, as in the usual case of processing several channels or overlapping windows, much more benefit can be achieved, relatively simply and effectively, by interleaving four sets of input data consecutively, then applying the scalar FFT code to that block but using a vector type like float32x4_t in place of the scalar float.

The transform code is just like that for the scalar version, only with a vector datatype and step size between elements. Source for a templated version can be identical but much faster. This approach does four (or eight, with AVX, data permitting) FFTs in parallel, yielding output interleaved just like the input. As fast Fourier transform techniques involve shuffling the input or output data, so called bit reversal, the extra interleaving comes at low marginal cost since both re-ordering operations can be performed in the same step.

This is one of the techniques used to optimise frequency analysis in *Dance Factory*. Since this uses a 75% window overlap for accuracy of timing, the input data overlaps too, and the data for four consecutive windows occupies only 75% more space than that for one. This was crucial in allowing a long FFT, chosen for good low-frequency resolution, to keep all its data in the small PlayStation cache. It meant *Dance Factory* could (I later discovered) use twice the transform size of Sony's voice-tracing *SingStar* technology and still keep up with data coming as fast as the CD drive could deliver it, greatly reducing the time to analyse full tracks and generate corresponding dances.

References

[1] *Ambisonics in Multichannel Broadcasting and Video; Michael A. Gerzon, JAES, ISSN 1549-4950, Volume 33 Number 11, November 1985, pages 859–871*

[2] *Interpolator Comparisons*: http://yehar.com/blog/wp-content/uploads/2009/08/deip.pdf

[3] *Musical Applications of Microprocessors; Hal Chamberlin, Hayden 1980, ISBN 0810457539*

[4] *Nigel Redmon*: www.earlevel.com/main/2017/08/16/sampling-theory-the-best-explanation-youve-ever-heard-part-1/

[5] *GCC/Clang-LLVM Vector Extensions*: https://gcc.gnu.org/onlinedocs/gcc/Vector-Extensions.html

[6] *The Mathematical Theory of Communication; Claude E Shannon and Warren Weaver, University of Illinois Press, 1949, ISBN 9780252725487, page 86*

[7] *Calculation of a Constant Q Spectral Transform; Judith C Brown, Journal of the Acoustical Society of America, ISSN 0001-4966, Volume 89 Number 1, January 1991*
[8] *An Efficient Algorithm for the Calculation of a Constant Q Transform; Judith C Brown and Miller S Puckette, Journal of the Acoustical Society of America, ISSN 0001-4966, Volume 92 Number 5, November 1992*
[9] *Sliding with a Constant Q; Russell Bradford, John Ffitch and Richard Dobson, DAFx-08 Conference Proceedings, ISSN 2413-6700, Finland 2008*
[10] *ARM NEON FFT*: https://projectne10.github.io/Ne10/
[11] *Intel MKL FFT*: https://software.intel.com/en-us/mkl-developer-reference-c-fourier-transform-functions
[12] *KISS FFT*: https://sourceforge.net/projects/kissfft/
[13] *FFTS*: http://github.com/anthonix/ffts
[14] *FFTW*: www.fftw.org
[15] *Apple FFT*: https://developer.apple.com/library/content/documentation/Performance/Conceptual/vDSP_Programming_Guide/UsingFourierTransforms/UsingFourierTransforms.html
[16] *FFT Misconceptions*: www.nicholson.com/rhn/dsp.html#6
[17] *On the Use of Windows for Harmonic Analysis with the Discrete Fourier Transform; Fredric J Harris, Proceedings of the IEEE, ISSN 0018-9219, Volume 66, 1978, pages 51–83*

CHAPTER 17

Interactive Audio Codecs

Codec stands for "COder/DECoder." It refers to a matched pair of systems which either compress (enCOde) or expand (DECode) data. Their compression ratio is the size of the input, in bytes, divided by that of the encoded output. Codecs can be either lossless or lossy. A lossy codec gains a greater compression by throwing away the least obvious part of the input. Codecs are usually asymmetric, decoding much faster than they can encode, as most samples are encoded once to be decoded many times.

Another way to squeeze audio is by downsampling, which removes high frequencies. Speech recorded at 48 kHz remains intelligible, though a little more muffled, when down-sampled to 8000 samples per second, giving a 6:1 compression ratio without changing the format. Games use a mixture of both techniques—downsampling and transcoding—to maximise the amount of variety they can deliver to the player. The 150,000 lines of dialogue used in a sport commentary system might otherwise exceed the capacity of an entire DVD-ROM.

Codecs trade memory space, processing time and audio quality. They allow more audio to fit into available space on disc or in memory and speed up loading. There's no "right codec" for all circumstances. Consoles and embedded systems tend to come with some codecs licensed and built in by the manufacturer. Such optimised decoder implementations often use otherwise-inaccessible custom hardware, so it makes sense to use them.

Codec writing is an art, so resist the temptation to roll your own unless there's really no alternative or a simple cross-platform approach delivers all the benefit you need. There's one case where you may save processing time as well as RAM, though not without some cost in quality. That's the first codec we'll discuss.

The Mu-Law Codec

Mu-law is a simple and fast audio compression scheme. It offers only 2:1 compression and you can hear the difference it makes to a sound, but the changes are not objectionable, in the opinion of sound designers I've consulted. This holds true even when it's used for Multichannel Ambisonics like B-Format, which rely heavily upon consistent inter-channel deltas to derive spatial information. The *Rapture3D* runtime system, for example, supports B-Format Mu-law as an Ambisonic encoding for 2D and 3D soundfields. A full-3D implementation neatly compresses four B-Format channels into the exact space an unpacked stereo stream would need. Mu-law is a Latinisation of μLaw, sometimes transliterated as uLaw.

Like all lossy compression schemes, Mu-law exploits the non-linear response of human ears. A CD-quality 16-bit linear (LPCM) sample can encode amplitudes from -32,768 to +32,767. The difference we hear between two adjacent values depends upon their relative magnitude. The step from 1 to 2 is a 6.0206 dB increase in level, very obvious if your amp is turned up high enough. But that from 32,766 to 32,767 is the same size to the computer but much less than a thousandth of a decibel, inaudible even to the most golden eared. Linear encoding gives much greater precision for large values than it does for small ones.

Mu-law redistributes the output values corresponding to encoded samples logarithmically to spread the ratio of differences between nearby values more evenly across the output range. The levels of the decoded output are quantised to fewer steps than they were in the input, but the steps are largest for loud signals where they won't introduce proportionately more distortion than the smaller steps for quieter parts of the wave.

Since each sample is encoded and decoded independently of those before or after, loop points can be sample accurate, and there's no risk of low-frequency distortion introduced by block-boundary adjustments. Codecs that work on blocks of several samples at a time can eliminate more redundancy in the signal, giving a higher compression ratio. But they also generate audible lower-frequency distortion as they recalibrate for consecutive blocks. This aspect does not trouble Mu-law, because its block size is a single sample.

Mu-law was invented for 8-kHz telephony, where it is commonly known as G-711 u-law, and was also used for many years to distribute BBC World Service radio programmes to outposts of empire. A-law is a similar but incompatible scheme prevalent in the United States, which gives slightly less dynamic range but marginally lower small-signal distortion.[1] Unix's /dev/audio character device neatly uses Mu-law to tunnel audio through an interface designed for 8-bit text.

Mu-law works equally well at any sample rate. It provides automatic byte-order conversion between big-endian and little-endian processor architectures, so the same data can be used for either ARMs or PowerPCs, for example. Only the 512-byte decoder translation table needs to change between them.

Decoding is so fast that it's typically quicker to read Mu-law, decode it, and output 16-bit PCM than it would be to copy uncompressed PCM unaltered from the input to the output! This is because the translation table is small enough to be entirely held in fast Level 1 cache once decoding starts, while the amount of relatively slow main-memory bus bandwidth used to read the input is halved for Mu-law compared to PCM.

My implementation takes two pointers, pIn and pOut, addressing the input and output respectively, and a count of the number of samples to be decoded.

```
inline void MuDecode(unsigned char* pIn, short* pOut, unsigned count)
{
  unsigned char* pPastLast = pIn + count;
  while (pIn < pPastLast)
  {
    *pOut++ = muTable[*pIn++];
  }
}
```

Notice the function computes the address of pPastLast, the address one after the last input sample we wish to decode, *outside* the loop. This means there's no need for a third variable, the remaining count, to be updated at each iteration. The input and output pointers change, necessarily, so the test for completion could check if either has traversed the entire buffer.

If circumstances make it easier for you to work one sample at a time, perhaps because you're doing other work in parallel with the decoding, this pre-processor macro lets you use Mu-law anywhere you'd otherwise expect a short word:

```
#define MU2PCM(s) (short) muTable[s]
```

I've assumed that the "short" datatype refers to a 16-bit value in memory, as that's true for all the 32- and 64-bit architectures I've encountered, but unless you're sure that's true for all yours too, you should use an explicit-sized type known to your compiler (often either s16/u16, S16/U16 or sint16_t/uint16_t if you previously #include stdint.h). If your compiler conforms to C++ 11 or later standards, the following incantation assures this at compile time:

```
static_assert(sizeof(short) == 2, "short type is not 16 bit");
```

The magic's in the table, which contains 256 16-bit values corresponding to each of the possible input bytes. The data is easier to mix if considered to be signed, but fussy compilers prefer it to be marked unsigned and still let us copy it to a signed short, so who cares?

```
const unsigned short muTable[256] =
{
  0x8284, 0x8684, 0x8a84, 0x8e84, 0x9284, 0x9684, 0x9a84, 0x9e84,
  0xa284, 0xa684, 0xaa84, 0xae84, 0xb284, 0xb684, 0xba84, 0xbe84,
  0xc184, 0xc384, 0xc584, 0xc784, 0xc984, 0xcb84, 0xcd84, 0xcf84,
  0xd184, 0xd384, 0xd584, 0xd784, 0xd984, 0xdb84, 0xdd84, 0xdf84,
  0xe104, 0xe204, 0xe304, 0xe404, 0xe504, 0xe604, 0xe704, 0xe804,
  0xe904, 0xea04, 0xeb04, 0xec04, 0xed04, 0xee04, 0xef04, 0xf004,
  0xf0c4, 0xf144, 0xf1c4, 0xf244, 0xf2c4, 0xf344, 0xf3c4, 0xf444,
  0xf4c4, 0xf544, 0xf5c4, 0xf644, 0xf6c4, 0xf744, 0xf7c4, 0xf844,
  0xf8a4, 0xf8e4, 0xf924, 0xf964, 0xf9a4, 0xf9e4, 0xfa24, 0xfa64,
  0xfaa4, 0xfae4, 0xfb24, 0xfb64, 0xfba4, 0xfbe4, 0xfc24, 0xfc64,
  0xfc94, 0xfcb4, 0xfcd4, 0xfcf4, 0xfd14, 0xfd34, 0xfd54, 0xfd74,
  0xfd94, 0xfdb4, 0xfdd4, 0xfdf4, 0xfe14, 0xfe34, 0xfe54, 0xfe74,
  0xfe8c, 0xfe9c, 0xfeac, 0xfebc, 0xfecc, 0xfedc, 0xfeec, 0xfefc,
  0xff0c, 0xff1c, 0xff2c, 0xff3c, 0xff4c, 0xff5c, 0xff6c, 0xff7c,
  0xff88, 0xff90, 0xff98, 0xffa0, 0xffa8, 0xffb0, 0xffb8, 0xffc0,
  0xffc8, 0xffd0, 0xffd8, 0xffe0, 0xffe8, 0xfff0, 0xfff8, 0x0000,
  0x7d7c, 0x797c, 0x757c, 0x717c, 0x6d7c, 0x697c, 0x657c, 0x617c,
  0x5d7c, 0x597c, 0x557c, 0x517c, 0x4d7c, 0x497c, 0x457c, 0x417c,
  0x3e7c, 0x3c7c, 0x3a7c, 0x387c, 0x367c, 0x347c, 0x327c, 0x307c,
  0x2e7c, 0x2c7c, 0x2a7c, 0x287c, 0x267c, 0x247c, 0x227c, 0x207c,
  0x1efc, 0x1dfc, 0x1cfc, 0x1bfc, 0x1afc, 0x19fc, 0x18fc, 0x17fc,
  0x16fc, 0x15fc, 0x14fc, 0x13fc, 0x12fc, 0x11fc, 0x10fc, 0x0ffc,
  0x0f3c, 0x0ebc, 0x0e3c, 0x0dbc, 0x0d3c, 0x0cbc, 0x0c3c, 0x0bbc,
  0x0b3c, 0x0abc, 0x0a3c, 0x09bc, 0x093c, 0x08bc, 0x083c, 0x07bc,
  0x075c, 0x071c, 0x06dc, 0x069c, 0x065c, 0x061c, 0x05dc, 0x059c,
  0x055c, 0x051c, 0x04dc, 0x049c, 0x045c, 0x041c, 0x03dc, 0x039c,
  0x036c, 0x034c, 0x032c, 0x030c, 0x02ec, 0x02cc, 0x02ac, 0x028c,
  0x026c, 0x024c, 0x022c, 0x020c, 0x01ec, 0x01cc, 0x01ac, 0x018c,
  0x0174, 0x0164, 0x0154, 0x0144, 0x0134, 0x0124, 0x0114, 0x0104,
  0x00f4, 0x00e4, 0x00d4, 0x00c4, 0x00b4, 0x00a4, 0x0094, 0x0084,
  0x0078, 0x0070, 0x0068, 0x0060, 0x0058, 0x0050, 0x0048, 0x0040,
  0x0038, 0x0030, 0x0028, 0x0020, 0x0018, 0x0010, 0x0008, 0x0000
};
```

If your mixer works with float rather than integer values, as makes sense on most modern systems, the table can be re-expressed as 1024 bytes of floats, giving direct conversion from 8-bit Mu-law to 32-bit float format without involving a hardware type-conversion and even more efficient use of main-memory bandwidth. The only complication is that the values in the above table with the most significant bit set need to be re-expressed as negative floats, e.g. 0xfff8 becomes -8.f not -65528.f (the unsigned interpretation of that integer constant). You could do this by hand or write a trivial program to read and rewrite the source.

Encoding is the reverse process. You could look up the best approximation to a 16-bit value by searching the decoding table and pick the nearest, then record its index as the encoded value, or build a big table with a byte for each possible input value (converting from float to short if necessary, as a 4-GB table would be excessive) and use the 16-bit input as a direct index to the corresponding Mu-law byte value.

Most of you will do it in a digital audio workstation or using a command-line tool such as *SoX*,[2] and large data tables do not make for readable books or enjoyable re-typing, so I won't spell out the entire 64 kilobyte table, over 1 MB of source code, except to observe that only 16K is needed if you exploit the fact that Mu-law decodes to 16-bit values but the two most low-order output bits are always zero, so you can shift those out with >> 2 before indexing and condense the table to a mere 16,384 entries.

Unless you're encoding a lot of values at once, the cost of loading and keeping the big table in memory might make the approach of searching the smaller table more attractive, especially if you strip out the sign of the sample value so you can do a binary chop through the relevant half of the table. The complication comes because Mu-law uses one's complement signed values, rather than two's complement, which is a standard in computing but slightly more complicated to implement in hardware.

Games like *DiRT3* use Mu-law to quickly encode sound for a replay video for upload to YouTube, but unless you're encoding on the fly you'll probably only need to **de**code at runtime, as all the encoding will be done offline as part of "baking" the game data for distribution. The following command, good in Windows CMD or bash on Linux or macOS, uses SoX to compress input.wav to output.raw without changing the sample rate:

```
sox input.wav -e u-law output.raw
```

Note that the output is a "raw" headerless file, so you need to stash the sample rate somewhere else. Telephony and Unix systems tend to assume 8000 samples per second. SoX can also do resampling, channel mixing and many other tricks at the same time as encoding a file—see the manual page for all the details.

ADPCM Codecs

The term ADPCM was introduced in a 1973 Bell Systems Technical Journal, initially for speech compression.[3] ADPCM stands for "Adaptive Differential Pulse Code Modulation," where adaptive refers to varied scaling for each block of data and differential to the storage of deltas, rather than absolute values, within the block. ADPCM is a class of codecs rather than a single format. Nintendo and Sony use proprietary variants, with short blocks amenable to hardware decoding and efficient looping.

Most of the free source for ADPCM codecs describes the IMA4 variant used by Apple (AIFC), Intel (NVADPCM) and Microsoft (XADPCM). IMA stands for the Interactive Multimedia Association. Tim Kientzle's article contrasts Apple and Microsoft variants including the way they handle stereo, block boundaries and variable block sizes. It includes C++ source to encode and decode both versions.[4]

IMA-ADPCM was added to the Xbox late in development and renamed XADPCM though it's compatible with IMA encoders. It's most commonly used with 34-byte blocks. Two bytes setup the block, followed by 64 4-bit compressed sample values. The open-source *vgmstream* project can decode more than 50 types of video game audio stream.[5] It relies on third-party libraries for some of these and assumes a little-endian host processor.

David Bryant's ADPCM-XQ project, also on GitHub, pushes ADPCM with noise-shaping, long blocks and look-ahead techniques.[6] It's slow, particularly when encoding, but gains quality as a result. It also includes code to extract raw sample data from the RIFF chunks in a Microsoft WAV file; these are based on the IFF format used in Apple AIFF and compressed AIFC files. The key difference is that IFF is big-endian whereas RIFF stores bytes of a data word in ascending order of magnitude, Intel little-endian style. Bryant's code suits systems expecting either byte order.

Sony's VAG ADPCM

The PlayStation ADPCM format is similar in principle but different in detail from the IMA ADPCM favoured by Microsoft.[7] It's derived from the Super Nintendo ADPCM—which used an eight-voice Sony SPC700 sound chip—but works with blocks of 16 bytes which decode 28 ADPCM samples from 4-bit deltas. That was also used in the Green Book interactive CD standard, again based on Sony hardware. The 16-byte size is a convenient power of two, which makes loading and copying of data more efficient.

The earlier SPC700 unpacks 16 samples from nine bytes—the odd byte (or two on PlayStation) is used to flag the end of the sample, whether or not it should loop automatically from the end back to the start, and to provide a "scale factor" for all the sample deltas inside that block. Electronic Arts games have used a variant of this ADPCM internally, stripping the encoded block down to 15 bytes by tidying up the flag encoding, giving a 3.7:1 compression ratio rather than the usual 3.5:1.

Either way we can find samples in this format on a game disc by looking for a characteristic "wwwwww" ASCII byte pattern it uses to pad the ends of one-shot samples onto 15- or 16-byte boundaries. The close resemblance between the pairs of 4-bit compressed samples and 8-bit PCM mean that a determined investigator can locate samples by listening to the raw bytes, at 192 kHz to save time, getting on with something else till recognisable audio emerges from the random screeching.

This only helps if the audio data has not been recompressed with some general-purpose packer like ZIP, but that's not usually a problem, as the audio-specific compression wrings most of the redundancy (and some of the quality!) out of PCM data, and it is normal practice to leave audio sample data in its raw format on the game disc.

Sony's ADPCM generally uses the .VAG file extension. Sony's PS3 library handles multi-channel interleaved samples with up to eight channels, suitable for 5.1, 7.1 or Ambisonic surround file formats. Older systems are limited to mono VAG samples, which can lead to intermittent unpleasant phasing artefacts when stereo files are played. Chapter 7 explains how to avoid those along with ADPCM looping requirements.

Glitchfinder General

Deliberate distortion can add gritty realism to sounds, but it tests the limits of codecs designed for well-behaved music and speech. Loud high-frequency samples may end up scratchy or clicking after encoding, especially between compression block boundaries, when there's only four bits to encode the difference between samples. There is a quick way to locate and quantify codec distortion and several ways to fix it.

> A quick way to locate and quantify codec distortion.

Intercom-style radio voice processing upsets the XADPCM codec, causing clicks and glitches on some samples where the audio dynamics and spectrum did not fit the prejudice of the codec at the required sample rate. To quantify this, encode the original PCM sample to ADPCM and decode it back with flipped phase so that peaks are troughs and vice versa. The transcoded result is then additively mixed with the original so that everything undistorted cancels out, laying bare the glitches.

Most digital audio workstations can do this as a batch process, checking an entire set of recordings semi-automatically, reporting just the maximum deviation for each sample. We found the XADPCM issue went away when we denormalised the entire set, preserving a consistent level between the samples but reducing the magnitude of deltas between them so that the codec could cope better. This required a compensating boost of a few dB in the speech group volumes on that platform—ugly but effective.

Alternatively, if only a few samples are affected, you might prefer to filter the problematic assets. You could also consider reducing the severity of the processed voice effect. Providing there's enough to make its origin clear, clarity might trump gritty authenticity.

Frequency Domain Codecs

To get better than 4:1 compression at decent quality demands a more sophisticated approach, working in the frequency domain to exploit psychoacoustic masking, squashing the data by identifying and approximating or removing frequencies which the ear won't miss, in context. This means larger blocks of samples and far more processing effort.[8]

The main contenders are AAC, ATRAC, MP3, Ogg Vorbis,[9,10] Opus,[11] WMA and XMA. As for ADPCM, there are variants to choose within as well as between the formats. Opus and Ogg Vorbis have well-supported portable implementations, with ARM NEON optimisations in Opus and SSE in Vorbis; MP3 implementations are plentiful but old and were until recently patent encumbered, as encoders for AAC remain; cross-platform support for the rest is patchy. As proprietary codecs are hand optimised for the host, they're often several times faster than off-the-shelf software; platform-specific custom codecs can deliver great quality and performance benefits.

Platform-specific custom codecs can deliver great quality and performance benefits.

AAC decoding is supported by phones, macOS and Web Audio. It's a more flexible and better-quality update of MP3, handling up to 48 interleaved 96-kHz channels. ATRAC was invented for the Sony MiniDisc and persists on Sony consoles. Ogg Vorbis is the non-proprietary favourite of free software advocates; it delivers good quality, but decoding tends to be slow and spiky. WMA is the old Windows psychoacoustic codec, and XMA is a variant supported by Xbox console hardware. Both take shortcuts which are sometimes conspicuous, especially at high frequencies.

The table compares the compression ratios of eight codecs for 16-bit audio. For obvious reasons, these are all fixed-ratio compressors, which are convenient for efficient streaming and reliable worst-case performance. Variable bit rate compressors like Ogg Vorbis and MP3 are more complicated but can yield higher ratios for a given sound quality—depending upon the input data. AAC and Opus support both fixed and variable bit rates in more combinations than can readily be tabulated.

Table 17.1: Constant bit rate codec properties

Codec	Samples in	Bytes out	Compression ratio
LPCM	1	2	1:1
Mu-law	1	1	2:1
VAG_ADPCM	28	16	3.5:1
IMA_ADPCM	64	36	3.56:1
IMA_STEREO	128	72	3.56:1
ATRAC3_HIGH	1024	192	10.7:1
ATRAC3_MED	1024	152	13.5:1
ATRAC3_LOW	2048	192	21.4:1

The compression ratios in the table presume that input samples are in 16-bit linear pulse code modulated (LPCM) format, like CD audio. ATRAC3_LOW and IMA_STEREO are intended for stereo music and so encode twice as many samples as their mono counterparts, as left/right pairs. ATRAC's rare combination of constant bit rate and high compression stems from its initial implementation for Sony's MiniDisc system; those recorders were popular with radio journalists because they had good built-in facilities for speech editing, another consequence of the constant bit rate.

Flexible Rates

Modern codecs intended for speech and music tend to support just a small set of sample rates, because the rate interacts with the frequency-sensitive aspects of psychoacoustics. But optimal asset compression and streaming require arbitrary rates, and once listeners and sources get moving, Doppler shift means few game assets play at quite their original rate, anyway.

> Deliberately lying to the encoder can squeeze a quart into a pint pot.

ATRAC3 officially supports only 48- and 44.1-kHz samples, but we squeezed much greater compression ratios out of it, at acceptable quality, by deliberately lying to the encoder. Contrary to received wisdom, ATRAC3 does a fine job of compressing audio originally at 32 or 24 kHz, providing you pretend the data is at 48 kHz and play it at two-thirds or half-pitch later. You can double the duration of audio you can fit into memory this way, and the same approach works for other limited-rate codecs like XWMA. Pick the closest rate supported, and let your ears decide.

Preserving Punch

Whether constant or variable bit rate, psychoacoustic codecs have a tendency to dull the initial transient at the start of a sample. They analyse overlapping sections or "windows" which are cross-faded for output because of the way Fourier transforms are adapted to process continuous signals. The first section of a one-shot sample has no previous data to overlap, so it fades up from nothing; this is hardly noticeable for music but takes the edge off shots and impact sounds.

> A momentary initial silence restores the punch of one-shot samples.

To restore the "punch" of one-shot samples compressed this way, add a few milliseconds of silence at the beginning of the uncompressed wave before you crunch it. This means the compressed sample is played slightly later but with its original attack. Don't try this on loops!

MP3? Nein, Danke

Fraunhofer's aging MP3 music format is well known to consumers and less-technically-savvy sound designers, but it's a very poor choice for interactive media. It's (arguably, still) patent encumbered, limited to one or two channels and uses eccentric block sizes which do not fit disc loading or audio output well. Typical MP3 (strictly MPEG Layer III, version 1) files have 1152 samples per frame, good for 26.1 ms of output at the CD-standard rate.

Layer III decoders can be memory hungry, requiring five frames of "history" in memory to fully decode the next block. One game which made extensive use of MP3 files was *Operation Flashpoint—Dragon Rising*, built upon FMOD middleware. Initial attempts to store all the game data in MP3 format foundered because about 30K of memory was needed to decode each MP3 channel, more than the duration of most of the one-shot samples in the game and a large chunk of the audio memory budget. The decoders for 100 voices tied up as much memory as between five and ten minutes of compressed sample data, depending upon the sample rate; worse still, the processor load to decode even a few dozen samples periodically exceeded the runtime audio budget, causing the whole game update to stutter as it consumed 25% of that time.

This was not helped by a poor match between the block duration and the main game update period, introducing spikes of demand in the profile as many samples starved, requiring decompression, in some frames but far fewer in others. Codecs with shorter blocks and less history make it much easier to balance the processing load when scores of samples are playing, at varying rates, simultaneously.

The game ended up shipping with a mixture of ADPCM and MP3 assets and decoders, as ADPCM was more efficient in CPU time and even memory than MP3 for short samples, but MP3 still had the edge for long samples providing there were not many playing at once. This compromise brought its own complications, and if FMOD had supported the native codecs we'd have squeezed a lot more sound into less space. Compared with AAC or Opus, at comparable quality MP3 delivers poor compression ratios and conspicuously audible artefacts, especially in the "joint stereo" mode, and it's relatively hard to make samples loop seamlessly.

This is not to say that MP3 was not quite an achievement when it was designed in the 1990s. It still does a reasonable job of delivering moderately compressed stereo music, which is all it was designed for. Constrained rates, audio quality, memory footprint, CPU load, poor looping and multi-channel support make it far less well suited for complex and layered interactive audio.

Console Codecs

Built-in hardware-accelerated codecs are one of the great advantages of working on console hardware. These reduce CPU load and allow more voices at higher quality and compression ratios than would be practical with software alone.

The Xbox One has a dedicated audio sub-system called SHAPE, which almost stands for "Scalable Audio Processing Engine." It's capable of 15.4 GFLOPS, with two scalar and two vector DSP cores, plus hardware rate converters, mixing and XMA decoders.[12] This is entirely distinct from the AMD x64 architecture of the main processors.

A lot of DSP power is dedicated to speech input processing rather than reverb and DSP for more general audio purposes. But fixed-function blocks mean it's also capable of mixing 512 output voices, bringing it up to the performance of a single PS3 SPE. Incremental improvements like Xbox One S and Scorpio are aimed at support for 4K (8-megapixel) displays, so they boost the Graphics Processing Unit (GPU), main processors and memory speed without significant impact on sound.

XMA is the main Xbox codec, but XAudio2's newer XWMA, for Windows as well as consoles, provides substantially higher compression yet better quality. XWMA takes more effort to decode and won't loop seamlessly, but it's still useful for speech and non-looping music. Much of the audio in Bethesda's games *Skyrim* and *Fallout 4* is encoded in XWMA at just 48,000 bits per second.

Detailed specs of current consoles are controlled by non-disclosure agreements, so you need to be signed up to get details. It's well known that Microsoft has shifted a lot of the audio work which was expensive on the general-purpose Xbox 360 PowerPC cores to dedicated hardware, catching up with PS3 and reducing the overhead of Kinect motion tracking, while Sony are treading water, in relative terms; the PS3 SPE was ideal for audio.

The PS4 has a dedicated audio processor, based on AMD's TrueAudio DSP, but this can only be programmed directly by the manufacturers. The PS4 Pro/Neo update is again intended to boost graphics for 4K screens rather than audio.

Graphics and memory have been greatly boosted in the later consoles, but there's no comparable speed-up in audio capability. If you're writing a semi-custom cross-platform game engine for these consoles it makes good sense to take advantage of the hardware audio decompressors, but you'll still find effects like reverb expensive in main CPU time, whether you roll your own or use the system firmware.

Nintendo's Switch has hardware audio decompression, using the modern Opus codec, but this is limited to 24 voices rather than the hundreds the Microsoft and Sony consoles can juggle. The N64-vintage Nintendo ADPCM format is also supported and useful for short memory-resident samples but problematic for streamed samples because of strict limits on the bandwidth available for reading the memory card. So Opus is the best choice for streamed music on that platform.

In general, try to use whatever optimised codecs your platform provides. If they're inadequate or you really need a cross-platform solution, Opus[11] is the best-suited codec for modern games, increasingly supported by default and straightforward to port otherwise.

References

[1] *ITU-T Recommendation G.711*: www.itu.int/rec/dologin_pub.asp?lang=e&id=T-REC-G.711-198811-I!!PDF-E
[2] *Sample Format Conversion*: https://sourceforge.net/projects/sox/
[3] *Adaptive Quantization in Differential PCM Coding of Speech; P. Cummiskey, N. S. Jayant, J. L. Flanagan; The Bell System Technical Journal, ISSN 0005-8580, Volume 52 Number 7, September 1973*
[4] *Inside IMA ADPCM; Tim Kientzle; Dr. Dobb's Journal, ISSN 1044-789X, November 1997*
[5] *vgmstream*: https://github.com/kode54/vgmstream
[6] *ADPCM-XQ*: https://github.com/dbry/adpcm-xq
[7] *MS-ADPCM*: https://msdn.microsoft.com/en-us/library/windows/desktop/ee415711%28v=vs.85%29.aspx
[8] *The Mathematical Theory of Communication; Claude E Shannon and Warren Weaver, University of Illinois Press, 1949, ISBN 9780252725487, page 62*
[9] *Vorbis Audio Compression*: https://xiph.org/vorbis
[10] *An Introduction to Compressed Audio with Ogg Vorbis*: https://grahammitchell.com/writings/vorbis_intro.html
[11] *Opus Codec RFC*: https://tools.ietf.org/html/rfc6716
[12] *Xbox One Silicon, John Sell and Pat O'Connor, Hot Chips 25 Conference, Palo Alto, 2013*: www.hotchips.org/wp-content/uploads/hc_archives/hc25/HC25.10-SoC1-epub/HC25.26.121-fixed-%20XB1%2020130826gnn.pdf

take advantage of the hardware audio decompression [illegible] whether you [illegible] or use the system [illegible].

Windows [illegible] Wave audio decompression [illegible] [illegible] rather than [illegible] format is also supported and used for short memory [illegible] because of [illegible] on the [illegible] [illegible] format.

In general, [illegible] [illegible] [illegible] the best [illegible] [illegible] otherwise.

References

[illegible]

CHAPTER 18

Panning Sounds for Speakers and Headphones

This chapter and the next deal with "panning," from mono to fully 3D speaker arrays, addressing the special needs of interactive audio, for gaming and VR. It explains how we direct appropriate proportions of any mix to stereo speakers, headphones, or surround-sound rigs.

The original audio replay systems used a single speaker—except in rare cases when extra "horns" were used to increase the replay volume of a mono mix. This changed in 1931, when EMI's founding engineer, Alan Blumlein, invented stereo.[1,2]

In fact his patent was not confined to stereo; page 2 refers to "*a plurality of loud speakers or suitable sound sources in suitable spaced relationship to the listener*" and page 5 to "*a complete directional 'sound picture,' i.e. both horizontal and vertical directional effects*" with modifying networks "*to accommodate any difference between the 'lay-outs.'*" The 3D "sound picture" is nowadays referred to as a soundfield, following further work by Michael Gerzon and colleagues.[3] Speaker layouts are informally known as rigs.

Blumlein was not the first to propose the idea of using multiple speakers, but his patent rigorously explains how to capture, record and replay stereo, the advantages to be gained by doing so, and the principles of binaural sound localisation. Oxford mathematician and Hi-Fi enthusiast Michael Gerzon combined Blumlein's insights with the theory of spherical harmonics while experimenting with 1970s "quadraphonic" systems, which derived four signals for speakers in a square around the listener. He realised that only three channels were needed to encode horizontal surround sound and that the fourth quadraphonic channel might more usefully be employed to encode height.

Cinematic Surround

Blumlein and Gerzon's patents have long lapsed, though they're still a useful resource.[1,4] Quadraphonics failed to catch on for commercial and practical reasons; there was no way to deliver three- and four-channel mixes without severe compromises on media designed for stereo and little trade appetite for a new niche format. After problematic experiments matrixing extra channels into stereo and lossily sifting them out later, the 70mm film format offered six magnetic audio tracks, initially intended for five front speakers and one "surround" speaker. It was not till 1979 that Dolby Labs created a discrete six-channel cinema surround-sound system for the film *Apocalypse Now*, settling on an arrangement of three front speakers, one mono low-frequency woofer and two high-pass-filtered "surround" channels.

This 5.1 surround arrangement—counting the low-bandwidth sub-woofer as the 0.1—was implemented for standard 35mm cinema film releases, first by Dolby and later by DTS and Sony, each using proprietary codecs—Sony's was based on ATRAC, while DTS's approach spun out research from Queen's University Belfast, using less compression (and more bits) than the rivals.

Dolby Digital 5.1 channel technology has been supported on game consoles since the original Xbox in 2001, soon followed by a DTS option on later consoles from Sony and Microsoft. These systems gave greater channel separation than earlier "matrixed stereo" surround systems like Pro Logic, though the consoles continue to support stereo as well as surround output.

DTS and Dolby Digital compression schemes were made for cinema and introduce a delay of a fraction of a second, causing audio to lag behind the display. Movies can compensate by delaying the picture, but games have no such freedom. While HDMI still supports those legacy codecs, uncompressed PCM is much preferred for interactive purposes.

Constant Power Stereo Panning

We're going to illustrate six types of panning, from 0 to 3 dimensions, mono to Higher-Order Ambisonic 3D. The PAN_MODE enumeration keeps track of the types, and the static variable **panMode** indicates what output configuration the runtime system currently expects.

The order of these enumerations is important, as it represents an increasing number of active outputs; later examples will use this to decide when they can stop setting output levels. PAN_STEREO through PAN_HRTF drive two, the PAN_2D51 cases drive six and the last two are for eight-output 7.1-channel configurations. There's no theoretical reason not to continue beyond this; additional channels are processed Ambisonically just like the earlier ones but with custom coefficients reflecting their position in 3D space.

```
enum PAN_MODE
{
  PAN_UNKNOWN = 0, // Initial state, before host is queried
  PAN_MONO = 1,    // One speaker - suits 7% of console users
  PAN_STEREO = 2,  // Two speakers, a safe default mix
  PAN_HEADPHONES,  // 30% of PC users mainly use headphones
  PAN_HRTF,        // Improved Head Related Transfer Function
  PAN_2D51ITU,     // 5.1 Home cinema or PC surround (30..40%)
  PAN_2D51AMBI,    // 5.1 Ambisonic variant, better sides/rear
  PAN_2D71,        // 7.1 2D 60-degree surround fills 5.1 gaps
  PAN_3D71,        // 7.1 channel Ambisonic 3D surround sound
  PAN_COUNT
};

static PAN_MODE panMode = PAN_UNKNOWN;
```

Before initialisation **panMode** is set to PAN_UNKNOWN so the startup code knows to query the host configuration. If this fails—so we can't tell how the listener is listening—we default to PAN_STEREO and hope the user will manually select headphones or one of the surround modes in the front end menus, if they notice and care enough. Currently this is the only way to switch between 2D (hexagonal) and 3D (octahedral) 7.1 configurations. The product should remember the selection between sessions, so it doesn't need to be manually set again.

While it's always best to pick the right setting without bothering the user, sometimes this choice depends upon their environment, or requires extra information (e.g. when you can't tell stereo speakers from headphones or work out the arrangement of a 7.1 rig). Ideally this would be a system setting. . . .

Player-friendly games allow the speaker configuration to be changed in game, usually from the pause menu, though sometimes only from the front end, between events.[5] Perhaps you'll want to switch from speakers to headphones to avoid disturbing others or from stereo to mono for a party game with lots of people taking part all over the room. The optimised update must check for this every frame and force recalculation of the panner coefficients for all sources—even if their listener and they've not moved—if **panMode** changes.

Stereo Panning

It might seem trivial to implement stereo panning. If the source is right in front of the listener, send the same signal to both speakers, and otherwise just fade down the signal to the speaker on the opposite side. Figure 18.1 shows a naïve implementation of these rules. The horizontal axis represents positions between left and right, the black line represents the left-speaker gain, a constant 1 for sounds left of centre falling steadily as the sound moves progressively to the right, and the dotted line shows the equivalent gain for the right speaker, tapering off as sounds move left.

The dashed cone might come as a surprise. It shows the total power output from this scheme as sources move from one side to the other. It's four times greater in the middle! Both speakers are working flat out at that point, whereas at either extreme one speaker is muted and the opposite one does all the work. It should be no surprise that it's much louder in the middle—the factor of four comes because power is the square of amplitude. You may be surprised to hear that many games shipped early this century made this mistake, because one of the most commonly used game middleware packages, RenderWare Audio, panned full power into both speakers for a central source. This is a reasonable approach for stereo balance controls but unrealistic and confusing for diegetic sounds.

The panning rule was apparently eventually fixed, but millions of games had been affected by then. RenderWare Audio is long gone—at the peak of its success its developers were purchased by Electronic Arts, who continued to develop it for internal use but shut off their rivals, including many smaller PS2 developers, from later updates. This is a cautionary tale for small firms dependent upon middleware—even if the product is maintained and the supplier stays in business, you might find your larger rivals pull the rug from under you.

Cross-platform audio middleware like GameCoda and ISACT, and the PC-specific Aureal3D, vanished similarly, while Microsoft stripped out 3D audio hardware acceleration from DirectSound3D with the release of the Vista Windows "upgrade," cutting off Creative Labs' trademark EAX (Environmental Audio eXtensions). Even if middleware meets your needs at a viable price, you'd better know how it works, in case takeovers or new platforms leave it stranded.

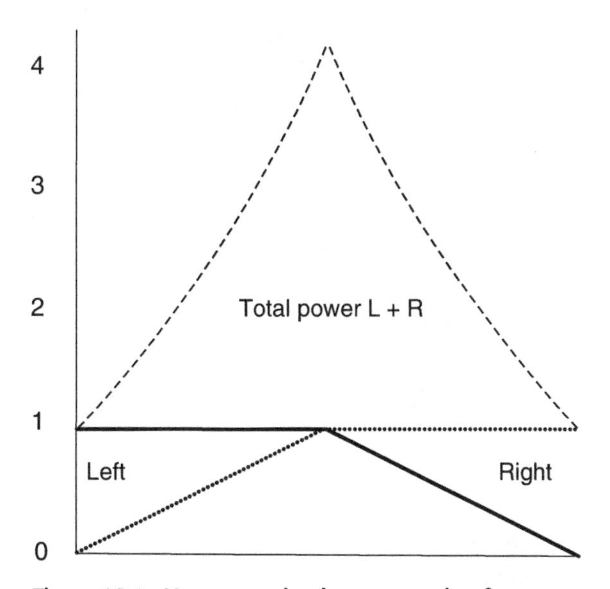

Figure 18.1: How not to implement panning for games

Smooth Panning

Two corrections are needed before the stereo panner sounds smooth to the listener. To preserve constant power, the centre signal should be -3 dB down in both speakers, compared with when it's panned sideways to one alone. This works because power is the square of amplitude and 3 dB attenuation implies an amplitude gain of 0.707. If the un-panned amplitude is 3, the power of a signal using only one speaker will be 3^2, which is 9.

When the mono signal is evenly split between two speakers the total power must be the same, so the perceived volume and source distance does not change with angle. Hence each speaker must contribute half the power (9/2). This occurs when each receives a signal with an amplitude of 2.121, since $2.121^2 = 4.5$ (for each speaker) and $4.5 * 2 = 9$, the total required power. The value 2.121 is the original amplitude, 3, times 0.707, the gain associated with -3 dB attenuation.

The other vital correction may be obvious from the discussion of early sound chip volume controls. Rather than a linear gain taper, the amplitude fading should follow a curve, with left and right curves symmetrical and crossing at a middle gain of 0.707. This value is conveniently also sine(45°) and cosine(45°)—in fact sine and cosine values for a quarter-circle give us just the curves we need, whatever the actual angle between a pair of speakers. This is a constant-power cross-fade and underlies the way a stereo system can seem to position and move sounds between speakers.

Another naïve approach, not shown but sometimes heard, runs the linear fades diagonally between each side rather than down from the centre. This has the opposite flaw—the lines cross at a gain of 0.5, so power from each side is 0.5^2 in the middle. This approach gives a big dip in the middle, where the power is halved compared with either side, since $2 * 0.5^2 = 2 * 0.25 = 0.5$. Oops!

Constant-Power Stereo Panning

Let's see how to do it right. This mono/stereo panner for speakers and headphones takes four inputs: the gain, computed from all the group, listener and voice attenuations; **position.x** of the source—other axes are ignored, as there are not enough outputs for discrete surround; the corresponding distance, in the plane if it's a 3D title; and the panMode. It outputs leftGain and rightGain multipliers for a stereo output.

```
if ((panMode == PAN_MONO) || (distance < 0.1f))
{
  leftGain = rightGain = gain * 0.707f;
}
else if ((panMode == PAN_STEREO) || (panMode == PAN_HEADPHONES))
{
  float pan = position.x / distance;

  if (panMode == PAN_HEADPHONES)
  {
    pan *= 0.67f;
  }
  float angle = (1.f + pan) * PI * 0.25f;
  leftGain = gain * cosf(angle);
  rightGain = gain * sinf(angle);
}
```

Notice how a very short distance is treated as equivalent to mono. In this case both channels are attenuated by 3 dB (0.707 gain) to split the power equally across the speakers, as if it was at POS_CENTRE. Arguably in the mono configuration both channels could use the unattenuated gain directly for maximum power and full dynamic range if only one output is connected, but that approach would cause a doubling of power when switching on the fly to mono from a stereo configuration. That would be more obviously "wrong," though arguably ideal for those with just one speaker, as opposed to users preferring the same sound everywhere in their room. The original Xbox had a mono audio dashboard setting, but the Xbox 360 manual advises listeners to connect *either* of the phono outputs if listening in mono while leaving it to the title to make sure they didn't miss all or part of the mix!

The panner for stereo speakers or basic headphones starts by using **position.x** and **distance** to find a normalised pan value between -1 for full left and +1 at extreme right. These values match the POS_LEFT, POS_CENTRE and POS_RIGHT symbolic names for the cardinal directions. Here's a full 3D set, for future use:

```
const float POS_LEFT   = -1.f;
const float POS_CENTRE =  0.f;
const float POS_CENTER =  0.f;
const float POS_RIGHT  =  1.f;
const float POS_FRONT  = -1.f;
const float POS_BACK   =  1.f;
const float POS_TOP    =  1.f;
const float POS_BOTTOM = -1.f;
```

POS_CENTRE (or POS_CENTER for Americans) is the only one that can be used on all axes—POS_LEFT and POS_RIGHT apply only to X coordinates and so on. The values of POS_FRONT and POS_BACK depend upon the coordinate system. These are for the right-handed OpenGL scheme and should be interchanged—so that positive values denote forwards—when using Microsoft's left-handed Direct3D.

Cross-Fading

Here's a graph of the curves the constant-power panner generates when moving a source from POS_LEFT to POS_RIGHT.

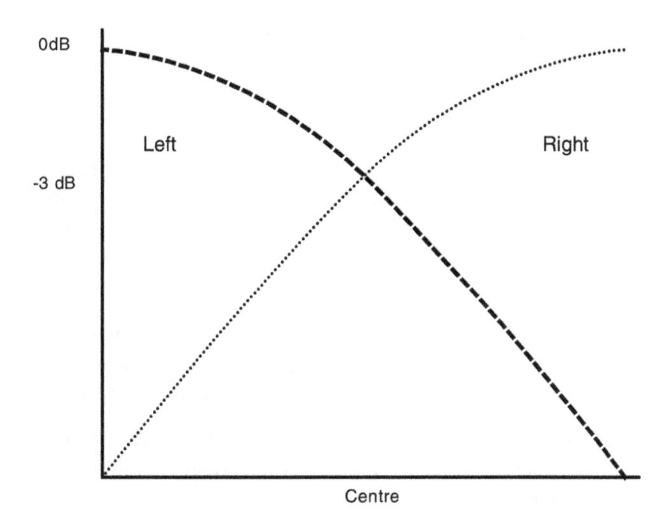

Figure 18.2: Constant-power panning and cross-fading

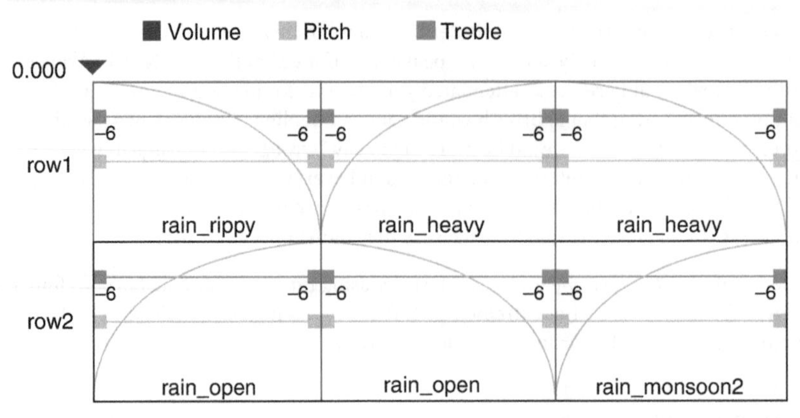

Figure 18.3: Cross-fading weather loops to vary rain sounds from drippy to monsoon

In these minimal unfiltered configurations, the only difference between PAN_STEREO and PAN_HEADPHONES is that the pan range is narrowed/from +/−90° to +/−60° to fix the "hundred-foot-wide car" syndrome, one of the audible improvements between *TOCA RaceDriver 2* and *TOCA 3*. A full binaural implementation, with vertical and front/rear encoding, is explained below. This requires PAN_HRTF and a lot more processing, but any of these three basic output mode settings could still embody directional listener properties in the gain factor; even if there are no speakers behind you, a combination of source, camera or head movements tells an attentive player which sounds are coming from the rear.

Use the same formulae when cross-fading between two mono sounds, assuming they're normalised to the same loudness when fully faded up, to prevent transitional power fluctuations. Figure 18.3 shows a two-voice cross-fade. The Volume, Pitch and Treble controls allow additional variation for the combined mix, independent of the cross-fade, as the control parameter scans across the grid.

Head-Related Transfer Functions

The panning tweak fixes the stereo image width for headphone listeners, but it doesn't improve spatialisation. In theory a headphone mix delivered directly to the listener's ears can include all the directional cues we experience in the real world, including height and front/rear discrimination, but this requires a lot more work.

Research into the head-related transfer function (HRTF) measures the changes in timing and frequency response associated with the direction of sounds from all around the listener. Such differences stem from the shape of the listener's head and ears, so they vary substantially between individuals, but some general principles hold: sounds from behind the ears are attenuated compared with those in front, high frequency sounds on one side are quieter in the opposite ear, and so on.

It's still hard to be sure of the direction of a sound when first you hear it. Confusion between front and rear is particularly common. Providing the source—or, equivalently, the listener—can move, experiment can resolve the ambiguity; we trained our brains to do this in the real world long before games and VR.

Hardware HRTF implementations were pioneered by Aureal in the late 1990s. Their Vortex cards positioned up to 16 sounds in 3D by applying short filters to the mono input, selected according to the source direction to produce appropriate and distinct left and right outputs.

The original Xbox, unlike later models, supported 64 voices of binaural sound, using Sensaura's database of over 1000 pairs of filter characteristics, corresponding to unique source positions in 3D space. These could be simply delivered directly to the players' ears via headphones, or approximately via loudspeakers, providing they sat very still. . . .

Cross-Talk Cancellation

The obvious flaw of delivering HRTFs through the air over loudspeakers is that the listener will hear a mixture of the sounds intended for each ear rather than the clean binaural channels. The challenge is then to cancel-out the lateral interference or "cross-talk." Providing the single listener is motionless at a known distance from both speakers, this can be done by mixing an anti-phase (inverted) copy of the waveform for the opposite ear into each output channel.

Southampton University researchers report that the area over which a soundfield can be controlled is larger when the loudspeakers are close together than when they are spread by 60° in a conventional stereo setup.[6] They recommend a "stereo dipole" of two matched and closely spaced-speakers, spread by 10° in front of the listener. This arrangement delivers clear binaural cues over a range of up to a metre back and forth and half that up and down. The head can turn around 30° to either side without spoiling the effect; greater angles partially obstruct the path to one or the other ear. However, the listener must stay very close to a line right between the speakers—lateral movement more than about 5 cm off this axis prevents the cancellation working properly.

Cross-talk cancellation works best at low frequencies, as air movement and small positional fluctuations inhibit cancellation of short wavelengths. It's more likely to suit an arcade game with fixed seating and speakers, or in-car, than at home. If the listener is not right in the "sweet spot" the tone of the entire mix is compromised.

Anywhere else, mismatched phase cancelation, delays and filter characteristics strongly colour the sound, so it seems unnatural as well as spatially imprecise. For this reason, loudspeaker surround, with real speakers above and behind the listener, still has the edge if you've got room for the extra equipment. Cross-talk cancellation has its place for people limited to stereo speakers at predictable distances—it can even work with twin-speaker phones and tablets.

Head-Related Impulse Response

An HRTF is a pair of time delays and frequency responses associated with a source position, one for each ear. The corresponding time-domain data is called the head-related impulse response, HRIR, conveniently represented as a short convolution filter. To place it binaurally, each incoming sample is multiplied by all the filter coefficients for its source position, separately for each side.

Table 18.1 describes half a dozen sets of HRTF and HRIR measurements, not all of which may be freely used without permission. The second column is the number of different "heads" or ear shapes tested. Azimuths are the number of horizontal directions measured at the equator, and elevations are the number of vertical angles.

Fewer azimuths are measured at extreme elevations. Only one is needed at either pole. Measurements for −90°, directly below the listener, are hard to capture, though handy for zero-gravity games. You may need to interpolate there. The commercial RIEC data only includes elevations from −30 to +90°, in 10° steps, whereas IRCAM goes down to −75°. Fewer elevations are needed, for standing or sitting listeners, because human ears are then horizontally opposed, so their acuity is greater in that plane. This is the rationale for the mixed-order Ambisonic panning used in this book, with one vertical component but six (third-order) horizontal ones, and Facebook's similar second-order approach.

Table 18.1: Composition of HRTF and HRIR datasets

Source	Heads	Azimuths	Elevations
CHIR[7]	111	360	1
CIPIC[8]	45	25	50
FIU[9]	15	12	6
IRCAM-AKG[10]	50	360	12
MIT-KEMAR[11]	2	72	14
RIEC[12,13]	105	72	13

Each set has pros and cons. Not all are public domain. The MIT data was created in 1994 to test the KEMAR dummy head, assuming symmetry and using two different dummy-ear (pinna) shapes. It's supplied as pairs of 24-bit impulse responses for each direction, each 512 samples long. The FIU data is presented as blocks of 256 samples at 96 kHz, from relatively few directions. The CIPIC set has the most elevations but some inconsistencies of measurement.

The CHIR data is horizontal only, measured at 1° intervals for distant sources all around 111 head shapes, but it includes separate measurements for sources close to the listener—at distances of 20 and 30 cm and intervals of 30°. The head response to such close sources is more inflected that at a distance, so if your product puts important sounds that close to the listener you should seriously consider using these measurements close up, with a (tricky!) cross-fade in the transition zone.

Most of the available HRIR data is at 44.1 kHz and in power-of-two block-sizes, typically 256 or 512 samples, though Aureal3D made do with less. You could use it directly at 48 kHz, accepting the errors, or resample it and window away or accommodate the longer impulses. Either way, the silent pre-delay is removed from the start of the data to eliminate lots of needless multiplication by nothing, and the response is smoothly truncated to ease processing. Remember to restore the pre-delay by delaying the point at which you insert the processed output.

To use such data, choose the nearest direction for each source—perhaps with pairwise or triangular interpolation for 2D and 3D, depending upon the granularity of the dataset, and proximity corrections—and convolve the source sample twice, with the left and right binaural data for that position. The cost of this is proportional to the number of voices, but a fast convolver can spatialise hundreds on a modern processor before your colleagues wonder where all the CPU time is going.

Simple Convolution

The following function takes (**outCount** + **impulseLength**) input samples in the array **in**[] and convolves them against the impulse into the array **out**[]. It assumes that there's room for at least **outCount** float results there, and that all the arguments are non-zero.

```
void convolve(float in[], size_t outCount, float out[],
              float impulse[], size_t impulseLength)
{
  for (size_t i=0; i<outCount; ++i)
  {
    float sum = 0.f;
    for (size_t j=0; j<impulseLength; ++j)
    {
```

```
            sum += in[i+j] * impulse[j];
        }
        out[i] = sum;
    }
}
```

The type **size_t** is declared in stddef.h. It's more portable and sometimes faster than **unsigned** for array indexation.

There's a complication when converting this to work on short blocks of input and output. To generate **n** samples it needs (**n**+**impulseLength**-1) input values, since each output is created by multiplying **impulseLength** consecutive input samples with the corresponding **impulse**[] values and storing the total. Assume **n** is a typical mixing block size, say 256 or 512; it's clear that we can't completely convolve the first block till we have **impulseLength** more samples of data to process, and we need to keep that many samples overlap between blocks.

Vector Convolution

Convolution is a potentially slow process, especially for a long impulse, since it involves **impulseLength** multiplication and addition operations for each single output sample. However, none of the multiplications depend upon previous results, and the additions can be done in whatever order is most convenient. This means that we can do the job in parallel if we've got a suitable vector unit: ideally one that can multiply two vectors and add the result to a third in a single step.

The input array and the impulse response must be an exact multiple of the number of floats in each vector. We'll assume VECTOR_SIZE = 4 here, matching AltiVec, ARM NEON and SSE2, the most common Super-Scalar Extension (SSE) sets, but similar approaches suit Nintendo's paired singles (2 FMAs per step) or AVX2 (8) on some x64 processors.

More problematically, vector data is only handled efficiently if each vector is loaded from an aligned boundary—an address which is an exact multiple of the vector size in bytes. We can scream through the convolution for the first output sample, but when we come to do the second—and third and fourth—the input is not aligned with the impulse. Fetching each input vector and rotating it, merging with the next, is complicated and processor-dependent and defeats the speed-up.

The way round this, memory permitting, is to leave the input alone and keep several copies of the impulse. The original data is used first, then a copy shifted one float later in memory—with an extra zero at the start and three at the end to maintain alignment. The third dataset is the original but with two zeroes either end and so on. We increase all the lengths by one vector to eliminate special-case checks at either end. Multiplication by zero is an easy way to ignore an element.

The code is like the scalar version, but the inner loop goes in VECTOR_SIZE steps and accumulates four sums in parallel, storing the total of those at the end. Before each pass, select one of the copies of the impulse cyclically, appropriate to the output alignment, and round the outer index up to a vector-sized multiple. The data shifting takes care of the odd 1, 2 or 3 floats at each end. This brute-force approach is neat for short HRIR convolutions. Longer ones, such as those for environmental reverb, should be done in the frequency domain to save time as well as memory.

Fake Speakers

We need at least two sets of impulse data for each direction, for left and right ears, and hundreds of such pairs for smooth spatialisation in any direction. Each independently positioned voice requires two convolutions to place it. This means a lot of data and a lot of processing, especially for a typical 100+ voice interactive product.

There is a much cheaper way to implement binaural audio. It's nowhere near as accurate for interactive use but adequate for respatialising mixes previously created for small speaker rigs. The trick is to use only two impulse responses, left and right, for each loudspeaker location. Thus YouTube "3D sound" uses eight virtual speakers, arranged in a cube, to spatialise first-order B-Format Ambisonics. Creative Labs use a ring of 12 virtual speakers in their surround-sound renderer, deriving signals for 7.1 or 5.1 real speakers (at various angles in a plane) by mixing those virtual outputs.

Such simplifications reduce the number of measurement points needed to spatialise a single source, from hundreds to a handful. They work fine for sounds intended to come from one of the pre-selected loudspeaker directions but less so for all the positions in between, making them ideal for a passive cinema listener but not for games or XR. Yet again, gamers are foisted off with simple stuff intended for passive consumers and must wait—and pay—for systems that take proper account of interactivity.

Another compromise is described by Bruce Wiggins, who has investigated YouTube and Facebook Ambisonics for VR.[14] His paper points out that Ambisonic decoding and HRTF simulation can be combined, reducing the number of convolutions to one per ear for each Ambisonic channel rather than two per voice.

Symmetry and Personalisation

There's another reason off-the-shelf HRTF data can't suit everyone—it generally assumes that the listener's head is symmetrical. I questioned this when an intern sound designer started working in the next room. One of his pinnae (ear flaps) was flush against his head, as usual, while the other stuck out at right angles like a solitary wingnut. He could hear fine—his brain had become used to this over decades—but would not be a good match for any "generic" HRTF data.

Even if pinnae differences are incorporated by using large ear-enclosing headphones, the associated ERTF (ear-related transfer function) varies between individuals. There have been experiments with multi-speaker headphones and no HRTF processing. Virtuality's pioneering set had two drivers each side. Later Zalman introduced three-element headphones, driven from the left and right front, rear and side signals intended for 7.1 output, with a phantom in-head centre channel. These were unconvincing, perhaps because of naïve panning or lack of attention to recreating the lost local reverb. Combinations of multiple-driver and ERTF technologies seem worth trying in the quest for more consistent headphone 3D.

One of the most unexpectedly clear demonstrations of spatial audio I've heard bypasses both conventional speakers and headphones.[15,16] Demos at the 2015 Derby University Sounds in Space showcase used an array of bone-conduction transducers spread across the skull, from the temples to the sides and mid-rear, above the neck. An additional device on the chin or back of the neck added non-directional bass, without which the audio would still be spatialised—contrary to expectations—but thin sounding.

The soldered-wire prototype was not comfortable but allowed 3D audio cues to be determined, including elevation and front-rear diagonals, without either customisation or occluding the listener's pinnae! Thus it was possible to hear the environment as normal, and an independent soundfield, rendered Ambisonically via the headset, at the same time. This has obvious applications for AR.

The Audio Engineering Society has published a standard, AES69, for personalised HRTF data and 3D room responses.[17] The latter are useful to allow reintroduction of local reverb into a "dry" headphone mix but less relevant to simulations which bring their own room models. AES69 is now supported in dozens of PC games via Blue Ripple's *Rapture3D* OpenAL driver and in applications like IRCAM's *Panoramix* and NoiseMaker's *Ambi Head*. The underlying format is called SOFA (Spatially Oriented Format for Acoustics), and until recently only academics with lots of equipment have been able to create such personalised HRTF files. It's early days, but there are lots of projects underway now to make this process cheaper and quicker, using photography as well as moving sources or listeners. The SOFA website is a great source of fresh HRTFs.[18]

Before long it should be possible to load your own HRTF into extended reality systems and get results which are tailored to suit your own head and ears. They'll still need compensation for the characteristic response of your choice of headphones and maybe their position on the head too. While AES69 won't fix fundamental issues like front/rear confusion, which can occur in the real world, it will be a big step towards binaural rendering systems that suit each individual—even the asymmetrical ones!

> Many are chasing the impossible dream of a "universal" HRTF.

In the short term many teams are chasing the impossible dream of a "universal" HRTF. While this is unachievable, early experience of VR audio demonstrates that players can learn an unfamiliar HRTF, improving in their interpretation of its quirks with practice. While it's possible that this could decrease the effectiveness of their personal HRTF interpretations, learning experiments suggest that people can switch quickly between interpretative models once trained and when the context is apparent. For safety reasons, that's just as well. Personalised HRTFs are still expected to be more accurate, more immediate—and safer.

Speaker Angles

There's no standard for 7.1 speaker layouts, though various recommendations exist. In 1993 the telecommunications and cinema industries settled upon ITU-R BS.775, a layout for "3–2 format reproduction" more commonly known as 5.1 (where the .1 denotes an additional sub-woofer for non-directional low-frequency effects).[19] ITU stands for the International Telecommunication Union, a UN agency.

Soon after I joined Codemasters I found my peers putting gaffer-tape crosses on the floor and measuring up angles with a protractor to set up a new set of five Dynaudios in the recommended layout. The illustration shows the recommended angles for those speakers. The location of the sub-woofer is not critical to spatialisation: put it anywhere convenient that minimises standing waves, so avoid corners or midway between walls.

This is what Francis Rumsey says about that layout, in his book *Spatial Audio*:

> *The left and right loudspeakers are located at ±30° for compatibility with two-channel stereo reproduction. In many ways this need for compatibility with 2/0 is a pity, because the centre channel unavoidably narrows the front sound stage in many applications.*

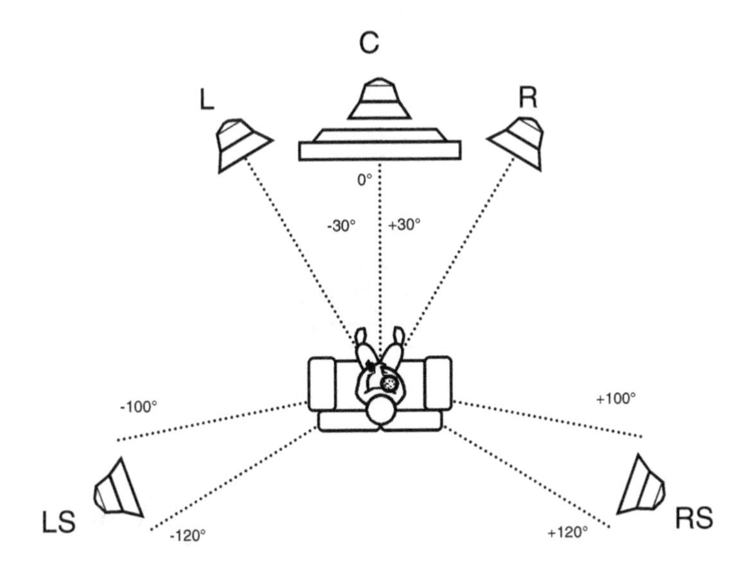

Figure 18.4: ITU 5.1 speaker layout recommendation for cinema surround

> *The surround loudspeaker locations, at approximately ±110°, are placed so as to provide a compromise between the need for effects panning behind the listener and the lateral energy required for good envelopment. In this respect they are more like "side" loudspeakers than rear loudspeakers, and in many installations this is an inconvenient location causing people to mount them nearer the rear than the standard suggests.* [20]

We soon found out that games seldom pan like that. It's rare to find anything that Microsoft, Apple, Creative Labs and Sony can agree upon, but investigations of the panning systems in the main consoles, computer platforms and device drivers reveal that all their gaming systems quietly assume a regular layout of speakers around the listener rather than the asymmetrical cinema surround scheme. This makes sense, as it's easy to program, especially for 5.1 speakers in a square. Just work out the source quadrant from the signs of the coordinates and do a pairwise mix between the nearest speakers in that direction. This preserves constant power in all directions while giving equal importance to them all. It's just not what's generally assumed.

> It's rare to find anything that Microsoft, Apple, Creative Labs and Sony can agree upon.

Once I met a Sennheiser engineer at an industry event. He was working on a system to spatialise existing 5.1 channel signals for binaural headphone listening. It was working as expected for DVD input but unevenly on game audio, where correct positioning of sources was especially important. Of course, he was going by the book and assuming ITU speaker angles! Learned papers explore the challenges of mapping regular Ambisonics to irregular ITU rigs.[21–23] However, most games assume 5.1 speakers are positioned like this:

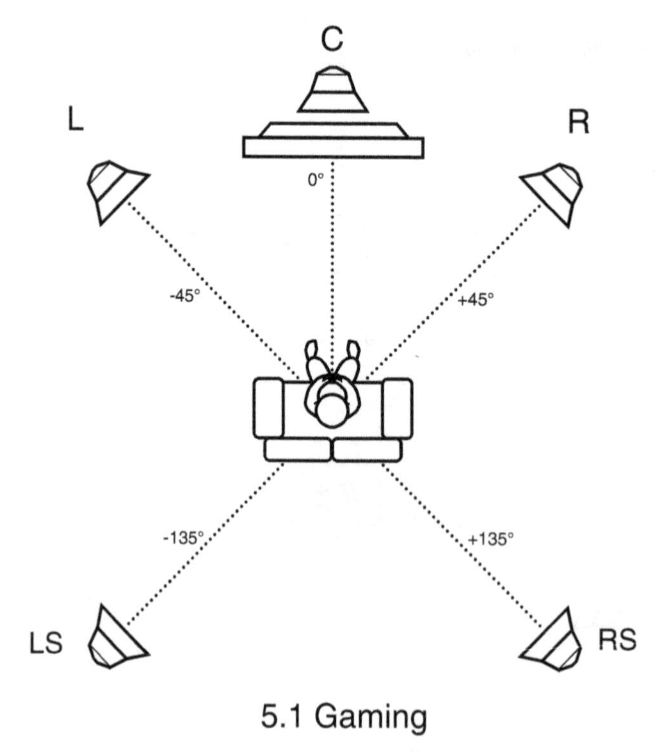

Figure 18.5: Five-speaker horizontal layout preferred for games

Of course, real-world rooms are not like cinemas—some people miss speakers out or put them all in a row or sit way off centre. We can't help that, but we can explain how to get good results for the preferred gaming rigs, which are optimised for a single central listener and give comprehensive directional information even if practical domestic layouts might skew it somewhat. The next chapter explains how to pan sound for any symmetrical array of speakers, in 2D or 3D, using Ambisonics.

References

[1] *Improvements in and Relating to Sound-Transmission, Sound-Recording and Sound-Reproducing Systems; Alan Dower Blumlein, UK Patent 394, 325, 1931*: https://worldwide.espacenet.com/publicationDetails/originalDocument?CC=GB&NR=394325A&KC=A&FT=D&ND=1&date=19330614, *Also in JAES, ISSN 1549-4950, Volume 6 Number 2, April 1958*: www.aes.org/e-lib/browse.cfm?elib=233

[2] *The Inventor of Stereo: The Life and Works of Alan Dower Blumlein; Robert Charles Alexander. Focal Press 1999, ISBN 0240516281*

[3] *General Metatheory of Auditory Localisation; Michael Gerzon, AES Convention Paper, 1992*: www.aes.org/e-lib/browse.cfm?elib=6827

[4] *Shelving filters: "Surround Sound Apparatus"; Michael Gerzon and Geoff Barton, 1993 Patent*: https://worldwide.espacenet.com/publicationDetails/biblio?FT=D&date=19930916

[5] *How Players Listen; Simon N Goodwin, AES 35th International Conference: Audio for Games, 2009*: www.aes.org/e-lib/browse.cfm?elib=15172

[6] *The "Stereo Dipole"—A Virtual Source Imaging System Using Two Closely Spaced Loudspeakers; Ole Kirkeby, Philip Nelson & Hareo Hamada, JAES, ISSN 1549-4950, Volume 46 Number 5, 1998, pages 387–395*

[7] *CHIR*: www.sp.m.is.nagoya-u.ac.jp/HRTF/database.html

[8] *CIPIC*: www.ece.ucdavis.edu/cipic/spatial-sound/hrtf-data

[9] *FIU*: http://dsp.eng.fiu.edu/HRTFDB/main.htm

[10] *IRCAM*: http://recherche.ircam.fr/equipes/salles/listen

[11] *MIT*: http://sound.media.mit.edu/resources/KEMAR.html

[12] *RIEC*: www.riec.tohoku.ac.jp/pub/hrtf/hrtf_data.html

[13] *Dataset of Head-Related Transfer Functions Measured with a Circular Loudspeaker Array; Kanji Watanabe, Yukio Iwaya, Yôiti Suzuki, Shouichi Takane & Sojun Sato, Acoustical Science & Technology, Volume 35 Number 3, 2014, pages 159–165*

[14] *Analysis of Binaural Cue Matching using Ambisonics to Binaural Decoding Techniques, Bruce Wiggins, 4th International Conference on Spatial Audio, Graz, Austria, 2017*

[15] *Inside-Outside: 3-D Music through Tissue Conduction; Ian McKenzie, Peter Lennox and Bruce Wiggins, Proceedings of the International Conference on the Multimodal Experience of Music, Sheffield UK, 2015*

[16] *Spatial Sound via Cranial Tissue Conduction, Peter Lennox and Ian McKenzie, Interactive Audio Systems Symposium, York UK, 2016*

[17] *AES69*: www.aes.org/publications/standards/search.cfm?docID=99

[18] *SOFA*: www.sofaconventions.org/mediawiki/index.php/Files

[19] *ITU-R BS 775-1, cited in "Multichannel surround sound systems and operations," AES Technical Council document AESTD1001.1.01-10, 2001*: www.aes.org/technical/documents/AESTD1001.pdf

[20] *Spatial Audio; Francis Rumsey, Focal Press 2001, ISBN 0240516230, page 89*

[21] *Decoding Second Order Ambisonics to 5.1 Surround Systems; Martin Neukom, AES 121st Convention 2006*: www.aes.org/e-lib/browse.cfm?elib=13814

[22] *The Design and Analysis of First Order Ambisonic Decoders for the ITU Layout; David Moore and Jonathan Wakefield; AES Convention 122, 2007*: www.aes.org/e-lib/browse.cfm?elib=14038

[23] *An Ambisonics Decoder for Irregular 3-D Loudspeaker Arrays; Arteaga, Daniel, AES Convention 134, 2013*: www.aes.org/e-lib/browse.cfm?elib=16818

References

CHAPTER 19

Ambisonic Surround-Sound Principles and Practice

Ambisonics is a generalised way of representing a "soundfield"—the distribution of sounds from all directions around a listener. It can use any number of channels—the more you give it, the greater the spatial accuracy of sources in the soundfield. The following explanation is adapted from one I drafted for Stephan Schütze; his CRC Press book *New Realities in Audio* contains many practical tips for sound designers working in VR and AR.[1]

Before Ambisonics caught on, mixing was a one-step process. Individual recordings—stems, samples, synthesised audio and location recordings—were combined into a mix, usually in stereo but sometimes in mono for radio, or 5.1 or 7.1 channel loudspeaker surround for cinema. This was delivered as finished work to the listener. The accuracy of the result, in terms of relative levels, timbre and source directions, depended upon the closeness of the correspondence between the rig used to mix the "master" copy and the consumer system replaying it.

Ambisonics recognises that there's no longer a standard mix configuration which can be created and then archived as part of the production process. Movies are more often consumed at home via whatever speaker arrangement suits the furniture, or while on the move—in a plane or on a phone or tablet, via headphones or small battery-powered speakers—than they are in a THX-certified cinema. This is even more true for new media, from 3D games on consoles and set-top-boxes to VR, because:

- New media is often interactive and designed to be consumed in more varied environments than old passive content like films or LPs
- Interactivity implies flexibility and customisation; consumers have grown accustomed to this and expect it
- Interactive media is optimised for a single consumer, whereas films get mixed to suit many listeners at once, in theatres large and small, with optional features like LFE and crude adaptations like fold-down
- The player controls the camera and there is no preferred direction where the content can be conveniently located in interactive media. As the player moves, the sides become the centre and the front the back from moment to moment. We can't dedicate a row of speakers at the front for the "action" and reserve a few more for "immersion" so that the people in the cheap seats get some sense of being there—people seeking to explore a virtual environment need accurate spatialisation, not just a vague sense of envelopment.
- Knowing the location of things you can't see—whether to find them or run away—is typically a matter of (virtual) life and death to an interactive player, whereas immersion in passive media is nice to have but not essential—it enhances the experience, but you can still follow the story even if the dialogue does not track the actors or invisible things are not accurately placed in the listeners' ears

Even before VR, there was pressure on ambitious sound designers to cater for multiple listener configurations. PCs and consoles first tentatively embraced quadraphonics, then cinematic 5.1 surround was adopted to deliver legacy content and new experiences, via DSP-enhanced consumer devices like the first Xbox and SoundBlaster Live! cards. Games and XR systems deploy custom panners to suit each, proving the user configuration is apparent.

7.1 surround arrived in several flavours: the horizontal cinema version then early 3D variants like the Aureal layout with a W-shaped arrangement of front speakers, via sound cards like the Vortex, Audigy and X-Fi. Before long 7.1 channel analogue output on PC motherboards became standard, soon followed by one-wire solutions like HDMI which (from release 1.3 in 2006) supports eight-channel 192-kHz 24-bit digital audio by default. HDMI is now implemented on consoles, computers, tablets and even phones.

It was a mess. Even if you had a 7.1 channel mix to listen to and all the amps and speakers or headphone virtualisation tech to deliver it, you could not be sure that the channels delivered would correspond in positions, or even order, to your local rig. Even working out how close you were was a challenge. Dolby and Creative Labs (among others) differed on the locations of the two extra speakers added in transition from 5.1 to 7.1.

Microsoft muddied the waters by changing the assumed positions of those speakers, between the rear and the sides, between two service pack releases of Windows XP! The advice for audio pedants was curt—if this matters to your customers, consider providing a switch to swap side and rear pairs over. Besides shifting responsibility to already-confused consumers, this introduced new ways of getting the mix wrong, irritating enough for passive consumers but potentially deadly for shooters, drivers and explorers.

Combine that with the lack of any official standard (there's no 7.1 equivalent of the ITU 5.1 recommendations) and a proliferation of 3D variants including non-proprietary 3D7.1 and the trademarked Audyssey DSX, Aureal, Auro-3D, Dolby Atmos and DTS-Neo:X arrangements, which offer 5.1 compatibility but add height in diverse ways.

For interactive audio designers and engineers looking to surpass the achievements of Hollywood, in a market fast outgrowing cinema, it was no longer possible to pick a format and hope that consumers would follow suit. They needed a way to hedge their bets, to pass some of the control to the players paying their wages—to avoid making decisions on behalf of listeners that would compromise their experience, whether on a basic TV, headphones or home cinema.

No wonder the BBC, when faced with the challenge of evergreening thousands of archive shows from stereo masters and raw multi-channel sources, were keen to archive soundfields rather than stereo, 5.1, 7.1, Dolby/DTS, binaural and 360VR mixes. That way they capture all the possible source directions and can pick or derive ones a given broadcast customer needs. Even a decade ago it was desirable for designers at Codemasters to author horizontal B-Format mixes rather than the four main mixes needed for contemporary game platforms.

One B-Format ambience, with three or four channels, can be decoded at negligible cost to 5.1 (regular square or irregular pentagon), 7.1, binaural headphones or stereo speakers. You save loads of disc space, as well as design and QA time. Pre-rendered speaker and headphone stereo, 5.1 and 7.1 channel mixes need 18 channels, versus three or four for the soundfield, or maybe six for a hybrid horizontal mix that gives greater precision for irregular 7.1 and ITU rigs. Even allowing for separate localisation tracks (or soundfields if you want those to turn with the listener) the result is a lot more future-proof than a set of custom mixes. Add cross-talk cancelation and you can even render front/back and side surround from a soundfield via just a front stereo pair.

Sega's *Showdown* arcade racer can work this way, as the motorised seat and speaker layout were designed at the same time as the game. Ambisonics facilitate custom decoding, refined by hardware or software that might not exist when the assets are authored.

One Ambisonic soundfield delivers any number of sources, just as one mono, stereo or 5.1 mix can. Unlike any conventional multichannel mix the soundfield can be smoothly rotated in 3D, varying yaw, pitch and roll to any orientation, preserving all the spatial information. This feature has made Ambisonics especially useful for XR sound design. In VR the player controls the orientation of the camera and the corresponding audio mix, typically through fast head tracking, and it's vital that the mix remains coherent regardless.

To understand the significance of this, consider what happens when a four-speaker surround mix is rotated, using conventional pairwise panning between speakers. When all four channels of the mix correspond to speaker positions, the mix sounds as the designer intended. As the listener turns, all the sources are panned to positions between speakers. Like stereo, this works plausibly for a narrow angle between the front speakers but much worse to the sides and rear. It fails when the listener gets in the way of either signal.

The resultant soundfield goes in and out of focus as the channels move to and between particular speakers. Spatialisation of sources between the side and rear speakers is lost, breaking the sense of immersion. For stadium-set games using pairwise mixing (such as *Brian Lara Cricket, LMA* and *Club Football*) it was soon found better to leave crowd ambience pegged to speakers than to try to rotate it with the listener. That's also consistent with the TV-sport presentation those games aped.

The smooth fading of Ambisonics, which involves all the speakers regardless of the source orientation, makes it practical to rotate any soundfield and decode it appropriately on the fly rather than take a prefabricated speaker-specific mix and cross-fade it between the nearest speakers—which doesn't work anywhere near as well. The Ambisonic approach keeps the spatial components of the mix distinct, whereas speaker mixes cannot. Unmixing and remixing is harder and more error prone than making a custom mix from data which keeps the spatial content as separate as possible for a given channel count.

Fold-Down Flaws

A particular issue, which I demonstrated at the 2016 AES International Audio for Games Conference, relates to fold-down and source positions in between speaker positions presumed by the original mix. Re-panning boosts the level of sounds in such positions, interfering with the distance cues which are so important in games and augmented reality—it sounds as if sources in between speakers are closer than those which happen to fall on a speaker location.

Figure 19.1 shows the way five speakers in an ITU-compliant layout are addressed in a generic commercial OpenAL driver, used by hundreds of PC games. Each curve corresponds to the output level for a given speaker, for sources behind the listener at either side, left and right sides as appropriate and in front in the middle of the graph. It shows one approach to front-centre panning and a simple pairwise approach to panning which uses no more than two speakers at a time, with poor side and rear spatialisation except at the surround speaker locations, +120° right and 240° (-120°) left, where everything else is muted.

So far, so-so, but see what happens when this is "downmixed" in proportions recommended for cinema soundtracks, as the driver automatically does for stereo PCs? The 5.0 surround channels are summed into the left and right stereo ones, and the total power oscillates wildly for source positions around the listener. The only place it's consistent is on the long, lonely back path between the surround speakers, where head occlusion, pairwise mixing and the irregular pentagonal provision for cinema emergency exits scupper the prospects for smooth panning, even if we weren't reflecting it all to two close-spaced speakers in front of the listener.

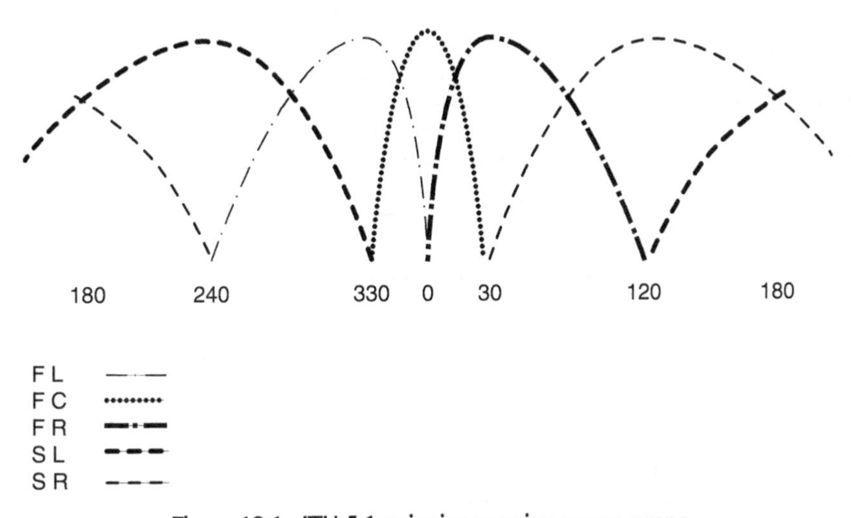

Figure 19.1: ITU 5.1 pairwise-panning power curves

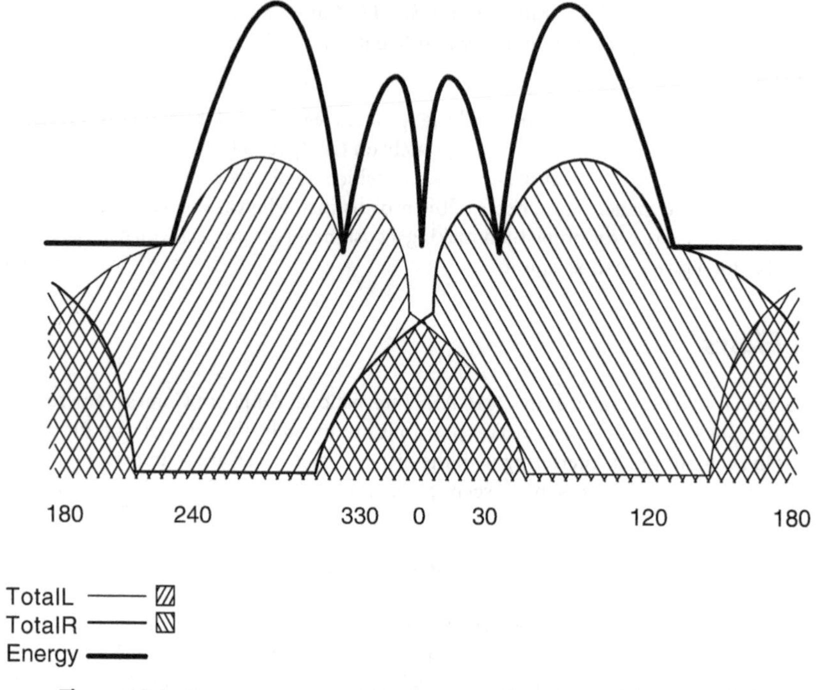

Figure 19.2: Stereo energy and total power after downmixing from 5.1

This is not just a surround-sound issue. When downmixing stereo to mono, the correct proportions depend upon the degree of correlation between the channels. If the stereo image is narrow, with most sounds audible on both sides, a mono mix should first reduce the level of each channel by 6 dB. But if they're uncorrelated, like a pre-pan-pot sixties pop mix, only –3 dB attenuation is needed to preserve constant power. Some BBC engineers compromise on –4 dB; that's never quite right but considered near enough for AM radio.

There are other fold-down formulae, with varying effects on directional balance, but they're all problematic for an interactive listener pointing the camera and microphone. Figure 19.2 shows that as a sound source moves across the front arc around the screen it will wobble up and down in power by a factor of two, slowly across either side and rapidly up and down twice between the speakers. This so-called compatible downmixing works for cinema but confounds interactive players' ability to track the locations of threats which they can hear and not see—most of the scene, inevitably, in a busy world—and even whether they are approaching or going away from the player. All this potential confusion goes away when you avoid making final mix decisions till you know the player's listening setup.

> Avoid making final mix decisions till you know the player's listening setup.

Ambisonic standards in console and PC games allow one mix to be made then custom-decoded for 5.1 and 7.1 mixes with comparable quality, even if the layout and exact speaker angles for the final replay are not known at the time of encoding and unlikely to match that used by the original sound designer. The separation of Ambisonic processing into two stages, encoding and decoding, allows flexible delivery.

Even if you don't know the exact details of the listener's setup when you design a product, a good audio programmer or middleware package like Wwise should be able to customise your design to suit each particular listener—including other designers, testers, producers and fellow team members, during development—without the need or false confidence of standardising listener setups across the dev team. Their experiences will vary, but it'll be far closer than if you attempt a one-size-fits-all mix, with up-mixing or fold-down to plug the gaps, and easier and quicker to build, test and balance than one made with a matrix of Group volume tweaks and asset-swaps to adjust between headphones and all the possible speaker mixes.

Even in a VR experience, optimised for a single listener, there's often a requirement to make a simultaneous conventional speaker mix for players waiting their turn or a stereo or 360 surround (horizontal only) version to upload to YouTube. Since Ambisonics separates encoding and decoding, it's relatively easy to derive these multiple endpoints from the full set of data used to make the 3D mix delivered to the main player. For instance, a stereo mix of arbitrary width can be derived from two of the four B-Format signals—mono and left/right difference.

Ambisonics is *both* a way of mixing *and* a speaker-agnostic intermediate spatial audio format. It fits neatly between sound designers' Foley spot FX and location recordings and final mixes customised for the player's ears and environment.

Rather than some populist one-size-fits-none compromise, designers build an ideal soundfield with no preferential layout or even direction. They monitor and demonstrate it on speakers and headphones as usual, then ship the raw material, in the form of objects and soundfields, with endpoint detection systems needed for the platform to decode it optimally.

If a new custom HRTF or spatial decoder comes along, the consumer may get a better mix than the original designer. That's fine—that's progress! The packaged soundfield is essentially an ideal mix for a perfect replay environment, subject only to your channel and disc budget. It's futureproof in a way that a cinematic 5.1—or 22.2—mix cannot be.

Ambisonic Encoding

Encoding involves combining recordings or positioned sources into a layout-agnostic set of related channels, according to mathematical rules which are optimised to economically preserve spatial information for later use. Decoding takes those and distributes them appropriately for the known properties of the listening environment. The more we know about that environment, the more accurately the mix is delivered from a given set of encoded channels.

Encoding is a very fast process on modern computers and consoles. Profiling of the 5.1 surround panner built into console middleware, which uses standard pairwise mixing techniques similar to those in Apple and Microsoft firmware, showed that games could encode a source to hybrid third-order Ambisonics (eight directional channels) and decode it back for a 7.1 speaker layout, in 2D or 3D and regardless of exact layout, faster than the default 5.1 panner could work out which pair of speakers to address, split the signal and mute the others!

This is partly because Ambisonic encoding and decoding treats each output and input channel equivalently. The code performs the same operations on each audio stream, varying only the intensity and polarity of the signals in each channel of the soundfield. The values of the coefficients computed by the encoding panner, and the decoding renderer, vary the proportions of input signals used and how they reach the player's ears, but the system workload is identical, since it takes the same amount of time for a CPU to multiply by 0.7071068 as it does −0.001.

The coefficients encode the source and output positions; the arithmetic does the rest. As a designer you just provide 3D positions, sizes, velocities for sources and the orientation, speed and position of your listeners, just as you would in any game made in the last 20 years. The encoder uses those and the decoder uses additional information about the replay environment—or environments—to get as much of that information into the player's ears as the input data, intermediate format and output gear can handle.

Ambisonic Efficiency

It's reasonable to expect that this is an expensive process, but actually Ambisonics is almost instantaneous and well suited to the way DSPs work. Output codecs and HRTF processing may add output lag, as they would for any panning system. Modern processors, especially the power-constrained, cost-reduced ones in consoles and mobile devices, are fast at sequential operations but soon get bogged down handling interacting conditions (a consequence of hardware pipelining), making Ambisonics particularly suitable for VR and AR.

It also means that there are no performance spikes which could cause unsteady framerate, breaking the illusion of being there by introducing erratic lag. In game performance, only the worst-case matters, and in Ambisonic games the worst case is both predictable and relatively quick. This allows you to mix more sources for a given CPU load or combine pre-mixed soundfields and foreground object audio sweeteners for a more varied and immersive mix than would otherwise be practical.

Ambisonics also scales well, up or down, which matters on mobile or PC gaming where the difference in capability between a mass-market low-end device and the latest hardware may be a factor of 30 or more. Producers pressure us to maximise sales potential and support millions of low-end devices without sacrificing quality on the best kit which reviewers will have and use to judge you against the top-flight competition. It helps to have quick ways to trade performance and fidelity which do not sacrifice content but only spatial detail in the quest to keep up the frame-rate on older devices.

Ambisonic Decoding

The main differences between Ambisonic systems involve the decoding and playback stages. Filtering, cross-talk processing and other DSP tricks are used in proprietary ways. Since the input—the soundfield representation—is the same for all Ambisonic decoders, this means content producers don't have to worry about the differences in rendering except to note that their work will take advantage of the various proprietary techniques on offer—advanced and customised HRTFs, feature extraction—like Harpex,[2,3] which uses DSP analysis to derive higher-order data from a B-Format mix—or custom adaptation for the listening environment, headphone characteristics, etc. Since you're using standard Ambisonic intermediate channels, you or your customers benefit automatically from standardised encoding now and future advances in decoding yet to come.

Ambisonics is well-established non-proprietary technology, mathematically rigorous yet simple to apply, scaling progressively up and down between mono and any number of channels. As you add channels it becomes mathematically equivalent to wavefield synthesis, an alternative approach to surround sound which relies on a very large number of speakers. One key practical difference is that you can add channels progressively to an Ambisonic representation.

Ambisonics was first used in the era of quadraphonic sound, where it offered forward and backward compatibility—backwards to mono and stereo without the positioning discrepancies associated with fold-down, where the correlation between channels affects balance of mixed sounds (fold-down makes conventionally panned sounds louder if they're between speaker positions)—and forward to larger horizontal speaker arrays, like the hexagon central to modern 7.1 cinema and beyond to true 3D sound with speakers (or sound sources) above and below the listener.

Unlike mono, stereo, quad or transitional arrangements like cinema 5.1 or 7.1, the number of channels in an Ambisonic mix does not correspond to the number of speakers. As soon as you get past stereo, Ambisonics requires fewer channels for a given number of speakers than would be needed for a direct per-speaker mix, and the more channels you have the more speakers—or virtual source positions in an HRTF-based headphone system—it can accurately drive.

B-Format

B-Format is entry-level Ambisonics—a single simple embodiment, a minimal representation to give full-sphere surround sound for a single central listener. This uses four channels with matched format and bandwidth, named W, X, Y and Z. Drop the height component (Z) for horizontal surround. Drop front/rear (X) for stereo. The W channel alone gives mono. FM stereo radio works similarly—the main channel is mono for compatibility and a second optional carrier contains the left/right difference for stereo receivers. The extra hiss in stereo stems from the analogue differencing, which duplicates signal noise. Digital Ambisonics is less prone to such artefacts.

B-Format uses four channels. The assignment of the last three letters to directions follows mathematical rather than gaming conventions, so the four channels are confusingly labelled as follows:

Table 19.1: B-Format channel assignments and order

1	W	Omnidirectional component	Parts of the sound heard equally in any direction
2	X	Front/rear component	Front minus back (separated out in decoding)
3	Y	Horizontal component	Left minus right
4	Z	Vertical component	Up minus down (sometimes empty or absent)

What happens if you get them in the wrong order? It's vital to get the first one right—this typically carries the most power and is the channel against which all the 3D channels push or pull. Swapping the others over is equivalent to re-orienting the listener, leaning forward or back or maybe rolling over—it changes the directions without adding or removing any content. There's no special reason front/back comes second, for example, but for compatibility we must all use the same sequence. It's convenient that height is the last channel in B-Format, so we can extract a horizontal-only soundfield for 360VR or 5.1 surround by just ignoring the last channel, saving a quarter of the bandwidth at the expense of delivering sounds which respond to height.

A-Format

A-Format is an equivalent but different arrangement, more practical for microphone designers. In theory you could make a B-Format microphone with one omnidirectional capsule and three figure-of-eight ribbon mics, but you'd struggle to match those up spatially and acoustically. Most Ambisonic or "soundfield" mics use a tetrahedronal arrangement of four matched directional microphones, as close together as the makers can get them. These also generate four signals, but they're not yet separated out into mono/omnidirectional and orthogonal directions—in particular, the W component comes equally from all four. Electronics or software—ideally customised to each instance of the mic—transcodes such A-Format signals into orthogonal B-Format ones.

G-Formats

G-Formats are output mix arrangements for a specific speaker rig which also contain all the information needed to reconstitute B-Format, and thus derive signals for other speaker positions not explicitly provided. G-Formats are handy for recovering data from old recordings and titles, or customising those to play better in your own listening environment, including HRTF-based 3D headphone systems.

The four channels of a regular quad mix, with speakers at right angles, are a G-Format. So is the regular hexagon implicit in cinema 7.1, stripped of LFE and front centre channels. Likewise the 3D7.1 format games already support can be losslessly decoded back to B-Format, with height, as shown later.

B-Format is the key concept for most purposes, though A-Format is worth understanding if you intend using live captures in your mix. Infinitely many refinements over B-Format are possible by systematically adding channels, corresponding to intermediate directions or internal positions. Such higher-order Ambisonics systems build on B-Format channels as a base, referring back to a common W channel however many extra spatial channels are added.

First-Order Horizontal Ambisonics

The following snippets implement first-order Ambisonic energy-vector panning, suitable for placing a mono sound in the plane of a listener centred in a 5.1 channel speaker setup. It leaves the front centre channel free for speech or similar non-diegetic content. These are the same assumptions made by Microsoft's function X3DAudioCalculate() in a 5.1 channel configuration when called with the flags X3DAUDIO_CALCULATE_MATRIX and X3DAUDIO_CALCULATE_ZEROCENTER set, so you can set up a quick A:B between the default pairwise quad panner built into XAudio2 and this replacement.

For a convincing demonstration of the smoothness of Ambisonic panning to the sides and rear, just send a helicopter or similar sample in a circle around a listener who can switch between this and the default panner at will. *Instant* A:B switching is a valuable capability when evaluating audio solutions. Even short pauses confound comparisons.

Notice how hard it is to locate sounds outside the front quadrant when the default panner is used unless they're coincident with one of the rear speakers. Ambisonic panning addresses this confusion by involving all the speakers, not just the bracketing pair. This makes binaural spatialisation effective even when the listener's head occludes the path from some speakers to either ear.

To pan smoothly within a set of speakers we need to know the order in which they're declared and their angles around the listener. The aim is to use this information and the position and size of the source relative to the centre of the speaker array to generate a gain value for each speaker. These values, which may be positive or negative according to the principles of spherical harmonics, distribute the power of the source between speakers, making the sound appear to come from any direction around the listener.

In common with default panners from Sony, Microsoft, Creative Labs and Apple we shall eschew the cinema-centric ITU 5.1 speaker angles and record our front and rear speakers as equally placed around the listener. This is not just consistent with the behaviour of existing game panners—it also avoids the need to adjust for an asymmetric layout by boosting the level in gaps and attenuating it when speakers are packed close together. In this I'm following Dave Malham's pragmatic advice that it's often better to lie (somewhat) about the positions of the speakers than to distort the path of sources or the symmetry of the decoder.

This equi-spaced assumption is a least-bad fit for unknown layouts and preserves constant power, although the perceived angles will vary in proportion to the asymmetry of an irregular rig. For instance, if the speakers are more widely spaced horizontally than on the front/rear axis, the soundfield will be similarly stretched or compressed spatially. This is not ideal, but it's easy to get used to as the panning remains smooth unless the gaps are very uneven, and in any case the distortion is consistent with the visible geometry in the listening space. The more consistently the speakers are distributed, whatever their number, the better the results. Similarly, the more speakers that are involved, especially at higher orders, the less precisely they need to be placed and the greater the size of the central sweet spot where best results are experienced.

To satisfy Gerzon's diametric decoder theorem,[4] speakers should be added in matched pairs equidistant either side of the listener, and the sum of the signals for any pair should match. This is the omnidirectional W component. The differences between the output of each pair depend upon directional components of the soundfield, appropriately factored for the speaker positions. The more pairs and the more evenly spaced they are the better.

Implementation

We'll start by declaring symbolic names for the speakers. The order of these depends upon the arrangement of output channels on your platform and is not standardised. This example follows the XAudio2 convention. This puts left and right signals as adjacent pairs to exploit the "joint stereo" optimisations in XMA and similar codecs which seek to exploit commonality between such pairs of signals and can be extended from two to six or eight channels by using more of the names, consecutively. There's no redundancy between FC and LFE and indeed front centre signals often have nothing in common with other directions, but this pair still compresses acceptably because the LFE channel is bandwidth-limited leaving plenty of resolution for the front centre signal bundled with it.

```
enum SPEAKER_POSNS
{
  SPEAKER_L = 0,
  SPEAKER_R,
  SPEAKER_C,
  SPEAKER_LFE,
  SPEAKER_LS,
  SPEAKER_RS,
  SPEAKER_LE,
  SPEAKER_BHI = SPEAKER_LE,
  SPEAKER_RE,
  SPEAKER_FLO = SPEAKER_RE,
  SPEAKER_COUNT
};
```

Only SPEAKER_L and SPEAKER_R are considered for stereo. Rival systems may put front centre first or between left and right or order the speakers clockwise or anti-clockwise from any arbitrary direction. Providing you shuffle the names into the same order as the output expects, you can use the subsequent code on any target.

SPEAKER_LE and SPEAKER_RE keep track of the extra pair of speakers added when extending a horizontal rig from 5.1 to 7.1 channels. SPEAKER_BHI and SPEAKER_FLO are aliases used for the back high and front low assignments in 3D7.1, which follows an octahedron rather than a hexagon, for true 3D output while retaining horizontal 5.1 compatibility for legacy passive media.[5]

For first-order Ambisonics, each speaker direction will have four coefficients or weights, associated with the omnidirectional (W) component of the signal, which goes to all speakers and directions in 3D:

```
enum AmbiComponent
{
  AMBI_W = 0,
  AMBI_X,
  AMBI_Y,
  AMBI_Z,
  AMBI_COUNT
};
```

This uses the Classical Ambisonic order of X front/back, Y horizontal and Z for height (in 3D only) rather than the 3D graphics standards of Y for height and Z representing front/back or vice versa depending upon the handedness of the coordinate system.[6] This is necessary for consistency with existing B-Format .AMB files and exporters in the FuMa format, used here and in Wwise. The actual order doesn't matter as long as encoding and decoding use the components consistently. Recent VR systems favour SN3D format files, with channels in the order W, Y, Z, X.[7]

The following array assignments convert angles for six or eight speakers arranged in a plane into first-order Ambisonic coefficients:

```
float ambiCoeffs[SPEAKER_COUNT][AMBI_COUNT];
```

Let's allow angles to be expressed in degrees, clockwise with 0° representing front centre rather than radians where far left must be expressed as -1.5708 rather than -90° or 270°. These trigonometric functions sinDeg() and cosDeg() wrap the C library ones and accept arguments in degrees rather than radians, e.g.:

```
const float degToRad = (PI * 2.f) / 360.f;
float sinDeg(const float d) { return sinf(d * degToRad); }
float cosDeg(const float d) { return cosf(d * degToRad); }
```

Here are preferred angles for a 5.1 speaker gaming system:

```
ambiCoeffs[SPEAKER_L][AMBI_X] = cosDeg(315.f);
ambiCoeffs[SPEAKER_L][AMBI_Y] = sinDeg(315.f);
ambiCoeffs[SPEAKER_C][AMBI_X] = cosDeg (0.f);
ambiCoeffs[SPEAKER_C][AMBI_Y] = sinDeg (0.f);
ambiCoeffs[SPEAKER_R][AMBI_X] = cosDeg (45.f);
ambiCoeffs[SPEAKER_R][AMBI_Y] = sinDeg (45.f);
ambiCoeffs[SPEAKER_LS][AMBI_X] = cosDeg(135.f);
ambiCoeffs[SPEAKER_LS][AMBI_Y] = sinDeg(135.f);
ambiCoeffs[SPEAKER_RS][AMBI_X] = cosDeg(225.f);
ambiCoeffs[SPEAKER_RS][AMBI_Y] = sinDeg(225.f);
```

The front centre channel angle is set for consistency, though the example panner deliberately doesn't use this for positional sounds, leaving it free for localised speech. This is consistent with most games, including my own Ambisonic implementations and *Grand Theft Auto 5*. That blockbuster game uses third-order horizontal Ambisonics internally, with interior panning and support for irregular speaker layouts.

This came about after my hybrid Ambisonics demo at the 2009 Audio for Games conference,[5] which piqued the interest of *GTA5* lead audio programmer Alistair MacDonald from Rockstar North. His game has nine speaker layouts, using Ambisonic coefficients optimised offline by the Tabu search technique explained in Wiggins (2007).[8] There are "narrow," "medium" and "wide" settings for front and rear speakers, where narrow front and wide rear suit the ITU recommendation and medium/medium corresponds to the ideal quad layout for 5.1 speakers.[9]

A 5.1 speaker rig does not have enough speakers for 3D, even with Ambisonics, so the AMBI_Z elevation weights are not used here. A consistent weight, equal to half the square root of two, is assigned to the AMBI_W components for all the speakers, as it's the omnidirectional part.

```
for (unsigned speaker = 0; speaker < SPEAKER_COUNT; ++speaker)
{
  ambiCoeffs[speaker][AMBI_W] = 0.707f;
  ambiCoeffs[speaker][AMBI_Z] = 0.f;
}
```

The LFE channel is not associated with or used in panning, as such deep bass sounds are considered non-directional, but it does no harm to clear the dummy entries in the coefficient table associated with LFE:

```
ambiCoeffs[SPEAKER_LFE][AMBI_X] = ambiCoeffs[SPEAKER_LFE][AMBI_Y] = 0.f;
```

Assuming an eight-channel output to a horizontal 7.1 speaker rig, these angles are recommended:

```
ambiCoeffs[SPEAKER_L][AMBI_X] = cosDeg(330.f);
ambiCoeffs[SPEAKER_L][AMBI_Y] = sinDeg(330.f);
ambiCoeffs[SPEAKER_C][AMBI_X] = cosDeg (0.f);
ambiCoeffs[SPEAKER_C][AMBI_Y] = sinDeg (0.f);
ambiCoeffs[SPEAKER_R][AMBI_X] = cosDeg (30.f);
ambiCoeffs[SPEAKER_R][AMBI_Y] = sinDeg (30.f);
ambiCoeffs[SPEAKER_LS][AMBI_X] = cosDeg(150.f);
ambiCoeffs[SPEAKER_LS][AMBI_Y] = sinDeg(150.f);
ambiCoeffs[SPEAKER_RS][AMBI_X] = cosDeg(210.f);
ambiCoeffs[SPEAKER_RS][AMBI_Y] = sinDeg(210.f);
ambiCoeffs[SPEAKER_LE][AMBI_X] = cosDeg(270.f);
ambiCoeffs[SPEAKER_LE][AMBI_Y] = sinDeg(270.f);
ambiCoeffs[SPEAKER_RE][AMBI_X] = cosDeg (90.f);
ambiCoeffs[SPEAKER_RE][AMBI_Y] = sinDeg (90.f);
```

By convention in 7.1 the extra pair, SPEAKER_LE and SPEAKER_RE, are added either side of the listener, allowing the front FL/FR pair to move forward to match the recommended angles for pairwise stereo panning and the RL/RR to move back. Front centre is not used in panning. The other six speakers form a regular hexagon of opposed pairs, which is an ideal arrangement for Ambisonics. Thanks to lobbying by Mark Tuffy of THX, this arrangement falls within the range of angles recommended for cinema 7.1 surround.

Distance Compensation

Like stereo or any surround panning system, irregular distances between the speakers and the listener will impair panning; the Haas precedence effect will draw sounds towards the nearest speaker.[10] This matters most for large rooms. If the distances are known or can be worked out by centralised measurement of signals directed to each speaker in turn, a compensating delay can and should be applied so that a mono signal reaches the central listener simultaneously from each speaker. Allow 1 ms of delay per foot of reduced distance for each speaker other than the farthest. The Haas effect is also relevant to reverb, as Chapter 21 explains.

Interior Panning

Unlike the most basic Ambisonic panner implementations, this one supports interior panning, so that a sound very close to the listener spreads out among all the speakers rather than clicking from front to rear or one side to the other as it passes through.

It also supports volumetric sounds, using the radius of the source to spread "big" sounds appropriately around the listener as they get close. The larger the radius and the bigger your rig, the sooner this happens. The factor **earSpacing** introduces interior panning even for a point source, when it's closer to the listeners than the distance between their ears—here assumed to be 15 cm—and **balance** normalises the internal and external power to 0.75 for a quad rig and 1.125 for a hexagon, with 6/4 more speakers. We're not panning through centre, but to preserve

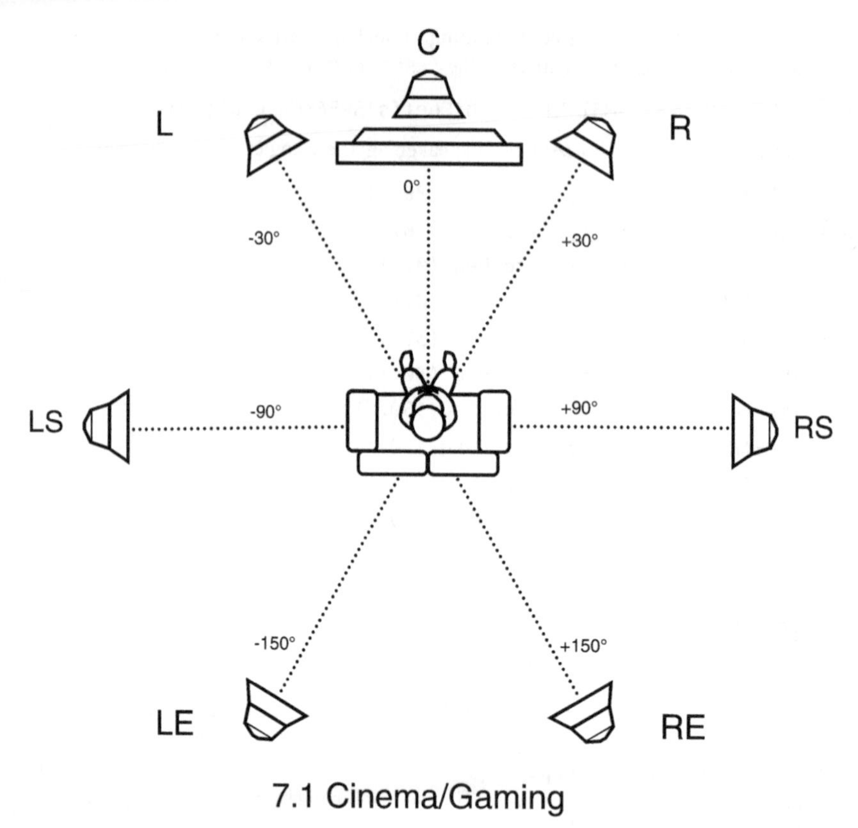

Figure 19.3: The preferred horizontal seven-speaker layout for cinema and gaming

balance with non-3D sounds it's necessary to scale their output gain correspondingly so that a front-centre sound plays at the same volume as a panned phantom centre.

The example uses four input values from the physical voice configuration: **positionX**, **positionZ**, **radius** and **distance**. The output is an array of gains for each speaker:

```
float gains[SPEAKER_COUNT];
```

In XAudio2 this array is already set up for you in pDSPSettings->pMatrixCoefficients; other panners use a similar output matrix, with one gain setting for each speaker output and potentially extra rows or columns for multi-channel input. In this case the matrix of gains is a one-dimensional array as we're positioning a single mono channel, but the same approach could be used to position elements of multi-channel input, such as the *RaceDriver GRID* music stems, by computing sets of gains corresponding to the desired spatial location of each interleaved channel and plugging them into a 2D matrix for each channel. One dimension represents output speakers, the other represents input channels; which index comes first is an arbitrary choice.

The anti-clockwise angle to the source from the central listener is azimuth, expressed in radians, so 0 means front, PI/2 is due left etc. The local variables **fW**, **fX** and **fY** are the Ambisonic encoding weights. There's a slight discontinuity at the inner-panning boundary to match distance-panning conventions. Ideally W should be 0.528f and the other weights should be 0.667f on the cusp, but the transitional glitch is no more than 0.5 dB so inaudible.

These are only used locally, as this implementation includes both encoding and decoding steps to make it easy to plug into a system which does not use an intermediate Ambisonic bus. You might save a little runtime by creating an intermediate three-channel W, X, Y bus and mixing all the positional sounds into that after encoding. Thus the decoding from Ambisonic format to speaker feeds need only be performed once for each output configuration. In practice it's so fast that the overhead of using an intermediate bus is comparable to that for duplicating the decoding steps unless you're using a plethora of sources or multiple Ambisonically derived outputs.

```
const float earSpacing = 0.15f;
const float balance = sqrtf(2.f/3.f);
float internalRadius = radius + earSpacing;

float azimuth = PI - atan2f(-positionX, positionZ);

float fW, fX, fY;

if (internalRadius < distance)
{
  fW = 0.501f;
  fX = cosf(azimuth) * 0.707f;
  fY = sinf(azimuth) * 0.707f;
}
else
{
  float profile = (distance * distance) /
    (distance * distance + internalRadius * internalRadius);
  float directionalWeight;
  if (profile < 0.001f)
  {
    directionalWeight = 0;
  }
  else
  {
    directionalWeight = balance / sqrtf(1.f + 1.f / profile);
  }
  fW = (1.5f * balance) / sqrtf(2.f + 2.f * profile);
  fX = cosf(azimuth) * directionalWeight;
  fY = sinf(azimuth) * directionalWeight;
}
```

Now we're ready to encode the input into three Ambisonic channels by scaling the input samples by **fW**, **fX** and **fY**, respectively, and mixing them into an Ambisonic output bus. This example is intended to drop in place of an old-fashioned panner which goes directly to speaker feeds, so we'll carry straight on and decode in place from Ambisonic components to specific speaker levels.

If you use an Ambisonic bus the following process only really needs to be performed once after encoding all the channels and for however many speakers the outputs require. In practice it's so simple that duplicating the effort for a hundred or so voices barely impacts the overall performance profile. Notice that trigonometrical functions were needed for the encoder, because the source could be at any angle relative to the listener, but not for decoding, as the speaker positions don't change:

```
gains[SPEAKER_L] =  0.707f * ((fW * ambiCoeffs[SPEAKER_L][AMBI_W])
                            + (fX * ambiCoeffs[SPEAKER_L][AMBI_X])
                            + (fY * ambiCoeffs[SPEAKER_L][AMBI_Y]));
gains[SPEAKER_R] =  0.707f * ((fW * ambiCoeffs[SPEAKER_R][AMBI_W])
                            + (fX * ambiCoeffs[SPEAKER_R][AMBI_X])
                            + (fY * ambiCoeffs[SPEAKER_R][AMBI_Y]));
gains[SPEAKER_LS] = 0.707f * ((fW * ambiCoeffs[SPEAKER_LS][AMBI_W])
                            + (fX * ambiCoeffs[SPEAKER_LS][AMBI_X])
                            + (fY * ambiCoeffs[SPEAKER_LS][AMBI_Y]));
gains[SPEAKER_RS] = 0.707f * ((fW * ambiCoeffs[SPEAKER_RS][AMBI_W])
                            + (fX * ambiCoeffs[SPEAKER_RS][AMBI_X])
                            + (fY * ambiCoeffs[SPEAKER_RS][AMBI_Y]));
gains[SPEAKER_C] =  0.f;
```

Figure 19.4 shows the gain for each channel for all directions around the listener, clockwise starting at the rear and with front centre in the middle. It's clear that, unlike a pairwise panner, all the speakers are contributing to the soundfield, not just the nearest two. Perhaps more surprisingly, at some points a small negative gain is applied to the opposite speakers, so that those "suck" while the others "blow" and vice versa. This follows from the theory of spherical harmonics and contributes to the constant power and smooth panning in all directions, providing you don't get too close to the speakers. Gordon Monro's ICMA paper[11] discusses "corrections" for this, but if your panning matrix supports signed gains they're not needed.

Eric Benjamin's 2010 AES paper demonstrates that this anti-phase component actually improves spatialisation by ensuring that the *correct* polarity of signals arrives at both the listener's ears when the frequency is low enough for relative phase to be a key directional cue. The paper compares Ambisonic panning at all frequencies with pairwise and virtual binaural positioning and illustrates how Ambisonics matches natural hearing more closely than either, especially for sources to the side which are otherwise prone to pop between front and rear positions.[12]

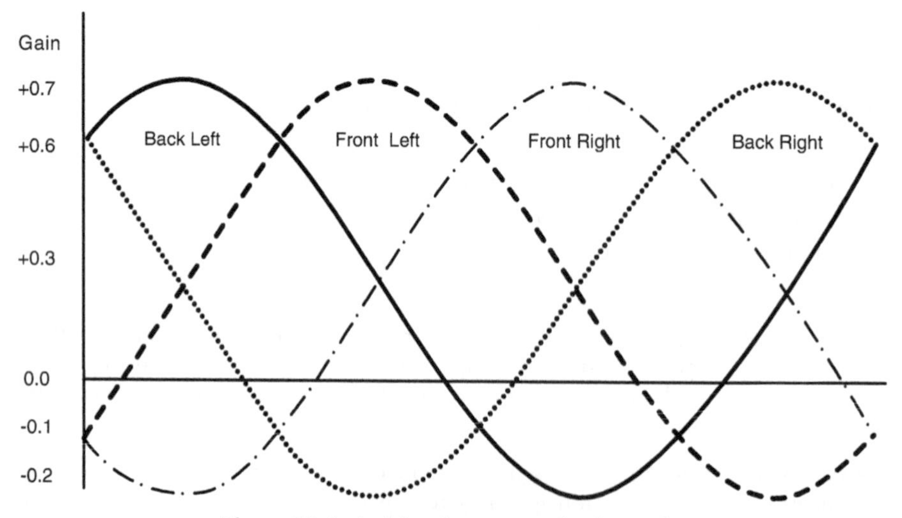

Figure 19.4: Ambisonic output gains by angle

Non-Diegetic Channels

The reserved front centre output is muted by clearing **gains[SPEAKER_C]**; this will be directly addressed by non-3D mono voices played without a 3D Listener, such as speech, event notifications and UI confirmations.

Set the sub-woofer send **gains[SPEAKER_LFE]** to whatever your designers specify in the logical layer, perhaps modified by proximity to give an extra thump for sounds close to the listener. Remember that the sub-woofer is an optional component of a 5.1 or 7.1 system, so not all listeners will experience this effect even if they have a "surround" system. One producer detached his LFE for use as a kitchen seat. Live with it, and without.

The following snippet adjusts optional LFE output in dB according to the physical voice distance and radius properties, adding a cheap and effective butt-kicking effect when a sub-woofer is available:

```
void CalculateLFE()
{
  m_FinalLFE = m_LFE + m_FinalVolume; // relative!
  float proximityBoost = m_pGroup->m_ProximityBoost;
  if (proximityBoost > 0.f)
  {
    float dl2 = distance * distance;
    float profile = 1.f - (dl2 / (dl2 + radius * radius));
    m_FinalLFE += profile * proximityBoost;
  }
  m_FinalLFE = std::min(m_FinalLFE, 0.f);
}
```

The clamp at the end prevents the **m_LFE** and **m_ProximityBoost** overloading the output. A perfect proximity-boost would be infinitely loud at zero distance, but this approach merely soaks up spare headroom. Since LFE is optional, how can we lose?

2D and 3D7.1 Encoding

The next function is a first-order Ambisonic encoder for 7.1 channel output which works in either 2D or 3D. The extra code to support height is skipped unless panMode == PAN_3D71. Extensions to support second- and third-order encoding are discussed later.

It starts by converting the direction vector into polar azimuth and elevation angles; azimuth is the anti-clockwise Angle in radians, where 0 denotes front, PI/2 due left and so on. The optimisation which computes cosElev using one multiplication, a subtraction and a square root operation was suggested by Peter Corlett. This is faster than a call to cosf() on typical modern processors that implement sqrtf() in hardware but need a subroutine for trigonometry. Don't worry about checking for divisions by zero or optimising the reciprocals like 1/zeroBase—your compiler should eliminate the runtime divisions as part of its constant-folding efforts.

If you have a normalised direction vector (magnitude 1) you can skip the trigonometry entirely and get fX, fY, fZ from the corresponding vector fields. But the vector *must* be normalised, ignoring height if working in 2D, when fZ = 0.

```
void AmbisonicEncoder(const Vector &direction, const float distance,
            const float gain, const float radius, float encoding[4])
{
```

```
const float zeroBase = 0.707f;

if (distance < 0.01f)
{
  encoding[AMBI_W] = gain / zeroBase;
  encoding[AMBI_X] = encoding[AMBI_Y] = encoding[AMBI_Z] = 0.f;
}
else
{
  float azimuth = PI - atan2f(-direction.x, direction.z);
  float fX, fY, fZ; // Encoder gains
  if (panMode == PAN_3D71)
  {
    float cosElev = sqrtf( 1.f - (fZ * fZ));
    fX = cosf(azimuth) * cosElev;
    fY = sinf(azimuth) * cosElev;
    fZ = direction.y / distance;
  }
  else
  {
    fX = cosf(azimuth);
    fY = sinf(azimuth);
    fZ = 0.f;
  }

  float centreSize = radius + earSpacing;
  float innerWeight, outerWeight;

  if (distance >= centreSize)
  {
    innerWeight = gain * zeroBase;
    outerWeight = gain;
  }
  else
  {
    float centreFade = distance / centreSize;
    innerWeight = gain * zeroBase * (2.f - centreFade);
    if (innerWeight > (1.f / zeroBase))
    {
      innerWeight = (1.f / zeroBase);
    }
```

```
            outerWeight = gain * centreFade;
        }

        encoding[AMBI_W] = innerWeight;
        encoding[AMBI_X] = fX * outerWeight;
        encoding[AMBI_Y] = fY * outerWeight;
        encoding[AMBI_Z] = fZ * outerWeight;
    }
}
```

First-order 2D and 3D 7.1 channel Ambisonic Encode

Interior panning within **centreSize** is handled by computing two weights, following Clarke and Malham.[13,14] **innerWeight** scales the W component and **outerWeight** fades down the directional components as the centre is approached. Volumetric sources are simply implemented by adding their radius to the internal panning distance. This works, but the approach used in the 2D example is smoother. Adjust the minimum **centreSize** to suit your game, but expect clicks when sources pass through the centre if you set it too low. This encoder has a gain of about 3 when driving six speakers.

Shelving Filters

The ear can determine the directionality of low-frequency sounds from the relative phase with which they reach each ear, but as the wavelength approaches the distance between the ears intensity, rather than phase, becomes the primary spatial cue; sounds above 700 Hertz seem to be sucked towards the nearest speaker rather than appear to be positioned between them. Gerzon recommended that truly Ambisonic decoders should compensate for this by applying a gradual (first-order) shelving filter to each Ambisonic component, boosting the W component at frequencies above 700 Hertz while slightly reducing all the directional components. The associated factors are different for horizontal and 3D rigs, as the 3D mix incorporates more directional channels.[15]

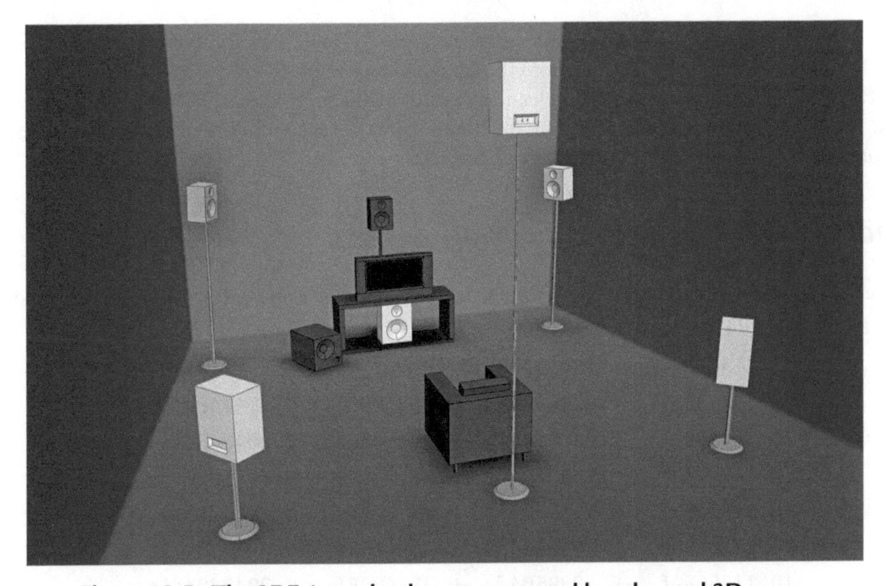

Figure 19.5: The 3D7.1 speaker layout supported by advanced 3D games

For his 2D rig, Gerzon suggested W should be boosted by 1.76 dB and X and Y attenuated by 1.25 dB. This delivers 150% power in W and 75% in X and Y, but only above 700 Hertz. In 3D he boosted W by 3 dB and trimmed the others by 1.76 dB.[15] The power in W doubles at high frequencies, falling to two-thirds in X, Y and Z. This approach can improve image stability but reduces the resilience of the decoder to inaccurate speaker placement, so it's best implemented as an option.

The 3D7.1 layout has one subtle property which makes it particularly suitable as a 3D audio output format. The six directional channels, excluding front centre and LFE, can be losslessly decoded to speaker-agnostic B-Format and then re-encoded for any number of speakers located anywhere on a sphere around the listener—the more the better. This means that any game which has 3D7.1 output, including *Bodycount* and some *F1* and *DiRT* games for PS3, can be re-interpreted for any other 3D speaker rig, correcting for asymmetry if necessary just as we would for first-order output, and potentially using advanced decoding tech like Harpex[3] to extract more spatial information than would readily be available from a first-order signal.

Of course, this only works if you know how the 3D7.1 outputs were encoded.[16] Most of the games already supporting 3D7.1 output presume the following Cardioid (in-phase) decoder matrix:

```
const unsigned BFORMAT_CHANS = 4;
const unsigned OCTAHEDRON_VERTS = 6;

const float m[BFORMAT_CHANS][OCTAHEDRON_VERTS] =
  {{1.414f,  1.414f,  1.414f,  1.414f,  1.414f,  1.414f},
   {0.577f,  0.577f, -0.577f,  0.577f, -0.577f, -0.577f},
   {0.707f, -0.707f,  0.0f,    0.0f,    0.707f, -0.707f},
   {0.408f,  0.408f,  0.816f, -0.816f, -0.408f, -0.408f}};
```

This matrix is derived from the "tilted" 3D7.1 rig specification demonstrated at the 2009 AES Audio for Games conference and Table 2 in the companion paper.[5] The rows correspond to four B-Format channels in W, X, Y, Z order. The columns weight those channels for the six directional outputs, in this order: Front Left, Front Right, Front Centre (low), Rear Centre (high), Rear Left, Rear Right.

The first 3D7.1-compatible games used the slightly different layout of Table 1,[5] with the two speaker triangles matching in height rather than their distance from the front and rear walls. If you decode those speaker feeds with the matrices presented here, you'll need to lean backwards a little for the soundfield to be accurately oriented around you. The Table 2 approach and corresponding layout match the recommended angles for legacy 5.1 cinematic content more closely, so those angles are preferred here.

Transcoding 3D7.1 Back to B-Format

To transcode the G-Format 3D7.1 output to B-Format, decode the six speaker feeds made with the decoder matrix above, through the following pseudo-inverse matrix:

```
const float t[OCTAHEDRON_VERTS][BFORMAT_CHANS] =
          {{0.1178689f,  0.2888504f,  0.3536068f,  0.2042484f},
           {0.1178689f,  0.2888504f, -0.3536068f,  0.2042484f},
           {0.1178689f, -0.2888504f,  0.0f,        0.4084967f},
           {0.1178689f,  0.2888504f,  0.0f,       -0.4084967f},
           {0.1178689f, -0.2888504f,  0.3536068f, -0.2042484f},
           {0.1178689f, -0.2888504f, -0.3536068f, -0.2042484f}};
```

This time, the columns correspond to Ambisonic components W, X, Y and Z and the rows to the speaker feeds. For instance, the W output is the sum of all six feeds multiplied by 0.1178689. This matrix was generated with MAT-LAB and singular value decomposition. Note how the middle two speakers have 0 coefficients in the Y (left/right) column. This is because they're centred on the front/rear axis.

If the original mix used the Cardioid in-phase encoding above, without filters, you get B-Format out. If filters were used the result is filtered B-Format, with extra omnidirectional high-frequency energy. Compensating filters may be able to undo this, but my PS3 products did not use filtering, and it's optional in *Rapture3D*.

The same technique could be used to extract first-order horizontal B-Format from an Ambisonic speaker mix intended for a quad layout—sadly, there's no height information, but the matrix is only a dozen elements in size; with it you can rotate the mix, after it's been encoded, to match a room where it's not practical to place the screen exactly between the two front speakers.

A higher-order Ambisonic mix for a 7.1 speaker system, with the main speakers in a regular hexagon, can be decoded similarly to second-order horizontal Ambisonics with some additional azimuthal accuracy, though not all the higher-order information can be retrieved from just six channels, even in 2D. So let's see how we can encode higher-order Ambisonics and fully extend our output into 3D.

FuMa and SN3D Formats

There are two commonly used Ambisonic file formats, known as FuMa and AmbiX. It's easy to convert between them, by scaling and re-ordering the channel data, but essential to know which you have and what your code expects, so they can be made to match, AmbisonicToolkit[17] and O3A[18] plugins include free lossless converters.

Furse-Malham "FuMa" format uses the last 16 letters of the alphabet to label channels up to third-order. This "Classical Ambisonics" scheme is the one preferred in this book. It's usually supplied in a RIFF container with the extension .AMB or .WAV. It supports 3, 5 and 7 channel 2D and hybrid (6, 8 and 11 channel) 3D as well as full first-, second- and third-order 3D with 4, 9 or 16 channels. The channel count indicates the order and degree of 3D support. Sample data is limited to 2 Gb due to RIFF's 32-bit signed lengths.

VR developers favour the newer alternative AmbiX format, which supports longer files and arbitrarily higher orders (though not hybrid schemes) and has a more logical channel ordering derived from the use of spherical harmonics in geology, known as SN3D, which stands for Schmidt 3D semi-normalisation. AmbiX is SN3D used for sound, in a .CAF (Apple Core Audio Format) container.

The only differences between first-order FuMa and AmbiX soundfields are the channel order and normalisation factor for the omnidirectional component. Where FuMa uses 0.707 for zeroBase, the channel W weight, allowing twice the power in that channel, SN3D uses 1.0. SN3D also shuffles the order, so channels W, X, Y and Z in a B-Format .AMB file would appear in the order W, Y, Z, X in a first-order AmbiX .CAF file.

Horizontal second-order Classical Ambisonic channels U and V are the fourth and fifth channels in AmbiX, scaled by a factor of 0.866 (0.5*sqrtf(3)). Horizontal third-order channels P and Q appear tenth and eleventh in AmbiX attenuated by 0.791 (0.25*sqrtf(10)). Sursound mailing-list members have collated scale factors for higher orders.[19]

Higher Orders

High orders allow more accurate positional encoding. Ambisonics becomes mathematically equivalent to wavefield synthesis[20] as Ambisonic orders and number of wavefield outputs increase.[21] The practical difference is that Ambisonics gains accuracy more quickly as channels are added; each additional component benefits several directions in ever-finer detail. Facebook uses second-order, several of my games have used a variant of third-order,

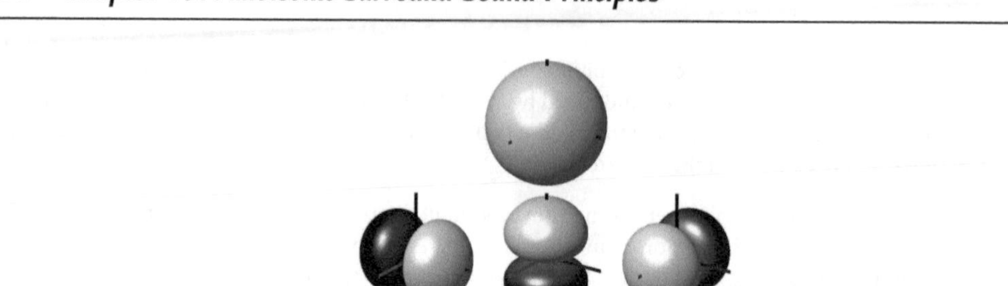

Figure 19.6: How first-, second- and third-order Ambisonic components interlock spatially

and products like O3A,[18] SuperCollider,[22] Rapture3D,[23,24] MATLAB,[25] GNU Octave[25] and B2X[26] plug-ins support up to fifth order (36-channel) Ambisonics.

Figure 19.6 shows how the first three orders interlock to convey increasing amounts of spatial resolution. Once you understand this everything about Ambisonics becomes a lot clearer—the ease of rotation, scalability, limits of resolution, directional decoding, volumetric and proximal interpretations all stem from this spatial representation.

The blobs represent directional lobes or areas of the soundfield affected by that channel. At the top is the zeroth or W component—the part of the sound that can be heard from all directions, typically carrying much of the bass. The next three pairs of blobs, in the second row, represent the first-order Ambisonic components on each 3D axis. Simplified shapes are shown—they actually follow a Cardioid pattern, narrowing in the middle like a hot-air balloon above its basket—but these suffice to reveal the spatial arrangement.

For stereo, use the Omni and the left/right (Ambisonic Y) channels and ignore the up/down and front/back ones. Combinations of all four position a sound in a sphere around the listener. The positioning is no more precise than stereo unless higher orders are used but in full 3D.

Five more channels are needed for second-order Ambisonics. They add detail in the 19 directions shown in the third row. The middle one, with two lobes bisecting a donut, represents the "in- or outside the plane of the listener" property. The others add detail in four directions around an axis. In combination all nine first and second-order channels express 25 directions.

The last row, third-order Ambisonics, adds seven more channels and 44 directional discriminations. Imagine all 16 clusters overlapping, like the surface of a berry. Each channel, each cluster, contributes proportionately to the sound from any position around the berry. Higher orders use more channels for greater resolution. Drop orders and the entire mix remains, losing only spatial acuity.

I'll leave to your imagination the 9 + 11 additional clusters for fourth and fifth order; they follow the same pattern and so on for higher orders indefinitely—until you run out of channels or acuity. It's beautifully elegant and practical too.

Hybrid Third-Order Ambisonic Encoding

The next example is inspired by code shared by Dave Malham at York University, which supports hybrid third-order Ambisonics. The term "hybrid" refers to the way it encodes only first-order elevation but first-, second- and third-order horizontal components. Sometimes these schemes are known as mixed-order. They represent a compromise between speed and spatial acuity optimised for the most likely hardware.

A full third-order encoder would generate all 16 polar patterns shown. Hybrid third-order requires only eight—specifically, the top four—B-Format—and the first and last in each horizontal row below this, which are the second- and third-order *horizontal* components.

Horizontal second-order components, known as U and V by extension of the W, X, Y, Z first-order convention, characterise the two outer polar patterns in the third row of the spherical harmonics diagram. Each has four lobes around the vertical axis. Likewise, the third-order components P and Q characterise the outer pair in the final row, with six lobes each. In combination, these components refine spatialisation in and in between 20 directions around the listener. There's no improvement in height resolution over first order, because the eight inner patterns in those rows are not considered—but you won't hear the difference unless you have a lot of loudspeakers!

This is how the extra horizontal Ambisonic components are computed:

```
float fU = cosf(2.f * azimuth) * cosElev;
float fV = sinf(2.f * azimuth) * cosElev;
float fP = cosf(3.f * azimuth) * cosElev;
float fQ = sinf(3.f * azimuth) * cosElev;
```

The multiplication factors applied to the angles generate the higher-order spherical harmonics which fill in the gaps between the first-order lobes. This improves horizontal spatialisation and the size of the sweet spot as additional speakers are added and smooths panning when irregular rigs such as ITU 5.1 cinema layout cannot be avoided. The first-order vertical weighting **cosElev** is applied to the higher-order horizontal components to preserve balance in all directions.

The encoding factors for the extra four channels (U, V, P and Q in Classical Ambisonic naming) are just the component values **fU..fQ** scaled by the master gain value, interior pan weight and an order-specific base factor. These factors determine the interaction between first and higher-order components in any direction. Higher orders have less influence on the output. The third-order PlayStation implementations, in games like *GRID* and *DiRT*, used empirically chosen fudge-factors suggested by Dave Malham:

```
const float secondBase = 0.58f;
const float thirdBase = 0.36f;
```

These values are rig dependent and have only been tested for symmetrical square, rectangle, hexagon, octagon and 3D7.1 layouts. They're not generally applicable to arbitrary rigs. The AllRAD paper[27] explains how to balance rigs which you know to be asymmetrical.

There's no **firstBase** factor explicitly scaling distance in the earlier examples, as that factor is 1.0 for B-Format components X, Y and Z, and multiplication by 1 happens unnoticeably all the time. **secondBase** is the weight applied to the second-order U and V components, and **thirdBase** applies to P and Q. For a distant source, this snippet scales the four additional higher-order components appropriately for gain, interior panning and the source radius, using **outerWeight**:

```
encoding[AMBI_U] = fU * secondBase * outerWeight;
encoding[AMBI_V] = fV * secondBase * outerWeight;
```

```
encoding[AMBI_P] = fP * thirdBase * outerWeight;
encoding[AMBI_Q] = fQ * thirdBase * outerWeight;
```

Since the same **outerWeight** is applied to all four components and the Base factors are constant, this scaling and the resultant write (to a quad-float-aligned address) can be done more efficiently with four-way vector operations. The higher the order, the more valuable vectorisation becomes, especially if you minimise shuffling between scalar and vector variables.

Higher-order encoding involves quite a lot of trigonometry for each source position update. There are two ways to optimise that code. Consider the pair of sin/cos calculations for each horizontal order: working out both the sine and cosine of an angle is almost twice as fast as working them out independently. This could be a vector function like FastSinCos(), taking several angles and returning four corresponding sines and cosines. This single function call does all the trigonometry needed for hybrid third-order encoding; one sine value, corresponding to elevation, is unused, but even after the bundling and unpacking of inputs and outputs it's likely to be faster than just two calls to the standard library sine and cosine functions.

If you have unit-normalised direction vectors to the sources, with distance separated out, you can avoid trigonometry entirely by using the Cartesian representations tabulated in *Encoding Equations for Third Order SN3D* in Blue Ripple's *HOA Technical Notes*.[7]

Hybrid Third-Order Decoding

Higher-order decoding uses the same formula for any speaker, so I won't write it out for each position in a 7.1 rig. The decoder is identical for 2D7.1 and 3D7.1, or any even number of speaker pairs in any symmetrical spatial arrangement, as speaker position compensation is baked into the decoding coefficients. This computes the gain to apply so that the correct proportion of the input is played through each loudspeaker (**ls**) for its contribution to the soundfield:

```
gain[ls] = fW + (fX * XCoeff[ls]) + (fY *
  YCoeff[ls]) + (fZ * ZCoeff[ls]) + (fU * UCoeff[ls]) +
  (fV * VCoeff[ls]) + (fP * PCoeff[ls]) + (fQ * QCoeff[ls]);
```

For each speaker position, compute the corresponding azimuth and elev(ation) as shown previously and encode those positions into Ambisonic component coefficients just like you did when encoding the source position:

```
float cosElev = cosf(elevation);
XCoeff[ls] = cosf(azimuth) * cosElev;
YCoeff[ls] = sinf(azimuth) * cosElev;
ZCoeff[ls] = sinf(elevation);
UCoeff[ls] = cosf(2.f * azimuth) * cosElev * secondBase;
VCoeff[ls] = sinf(2.f * azimuth) * cosElev * secondBase;
PCoeff[ls] = cosf(3.f * azimuth) * cosElev * thirdBase;
QCoeff[ls] = sinf(3.f * azimuth) * cosElev * thirdBase;
```

There's no point optimising the trigonometry here, as these values are constant unless the speakers move. The **Coeff**[] arrays are calculated once for each speaker layout. Here are the azimuth and elevation angles, in degrees, for the 3D7.1 layout documented in my AES paper,[5] since supported in 40 PC games,[28] in the order of SPEAKER_POSNS. The fourth entries, for SPEAKER_LFE, are dummies, as we don't pan through the subwoofer. The "extra" 7.1 speaker channels SPEAKER_FLO and SPEAKER_BHI are mapped to front low and back high positions to complete the octahedron:

```
azimuth3D71[] = {45, 315, 0, 0, 135, 225, 180, 0};
elevation3D71[] = {24, 24, 24, 0, -24, -24, 55, -55};
```

Be sure to convert these angles to radians before plugging them into the speaker Coefficient calculator. Here are the azimuths for 2D7.1, a regular hexagon plus front centre and LFE, which suits both games and cinematic content:

```
azimuth2D71[] = {30, 330, 0, 0, 150, 210, 90, 270};
```

Elevation values for horizontal 7.1 are all zero; in this case we're ignoring height and working with cylindrical rather than spherical harmonics. As **cosElev** is 1.0 there's no need to multiply by it.

Soundfield Decoding

Sometimes the runtime system is used to decode pre-authored soundfields. These play a role similar to the graphical background "skybox" which envelops a 3D world. *DiRT2* made pioneering use of pre-rendered B-Format soundfields, most notably for paddock ambiences. Sound Designer Andy Grier assembled these from scores of samples mixed across 16 Nuendo tracks using free VST plug-ins from Derby University.[29] *Formula One* games took this idea further; the 2011 edition used Ambisonic soundfields for the paddock and garage ambience, terminal damage and other transitions including loading and scenes after each race and qualifying session—a total of 14 soundfields.

Each pre-rendered B-Format soundfield consists of four channels which can be decoded to represent the sound from any point or area on a sphere around the listener. The loudness and orientation of the soundfield can be controlled by the arguments **gain** and **orientation**, respectively. **Orientation** is a rotation matrix; strictly only a 3 × 3 matrix is required. Games often pass such information around in an even more concise four-float Quaternion format, but this example is simplified if we reuse the matrix structure introduced for listener orientation. We can make this from a Quaternion if desired: start with an identity matrix and update the bottom right 3 × 3 rotation part, by converting it from the Quaternion in the conventional way.[30]

The result is an array of INPUT_CHANS * SPEAKER_COUNT gains—each of the INPUT_CHANS subsets corresponds to a loudspeaker, and each element of that sub-array determines the amount of each Ambisonic component, W, X, Y or Z, which is mixed into the output to that speaker. INPUT_CHANS and hence the total

Figure 19.7: Soundfield authoring with Wigware VST plug-ins

size of the array depends upon the underlying host. Some allocate a fixed-size array regardless of the actual output configuration, typically 8*8 to allow 7.1 input and 7.1 output, but XAudio tailors the array to match each source and output configuration.

```
void SetAmbiGains(unsigned speaker, float gain,
 Matrix orientation, float* output)
{
  Vector coefficients, settings;

  coefficients.w = 1.f;
  coefficients.x = ambiCoeffs[speaker][AMBI_X];
  coefficients.y = ambiCoeffs[speaker][AMBI_Y];
  coefficients.z = ambiCoeffs[speaker][AMBI_Z];
  ApplyMatrix(&coefficients, &settings, &orientation);

  output[AMBI_W * SPEAKER_COUNT + speaker] = settings.w;
  output[AMBI_X * SPEAKER_COUNT + speaker] = settings.x;
  output[AMBI_Y * SPEAKER_COUNT + speaker] = settings.y;
  output[AMBI_Z * SPEAKER_COUNT + speaker] = settings.z;
}
```

In order to decode the soundfield, SetAmbiGains() is called for each active loudspeaker. We start by clearing the entire levels array with memset() from string.h, so there's no output except where we've explicitly requested it. The next steps depend upon the **panMode**—PAN_MONO is simple, as it uses just the omnidirectional W component of the soundfield. Otherwise we know we need at least the front pair of outputs, so we work out levels for those, without special adaptations for headphones in this simplified case. The left stereo channel is derived from W+Y and the right from W–Y, since W = L + R and Y = L – R. Scaling Y down reduces the image width.

We add the rear pair only for surround configurations and the last two only for 2D7.1 (side) or 3D7.1 (high/low) rigs. The scale factor gives the decoder a consistent gain of 2, however many speakers it's driving. We don't need to worry about the various 5.1 and 7.1 geometries, as the Ambisonic coefficients precomputed to suit the rig cater for them automatically. The same code handles flat and 3D 7.1, as their extra speaker enumerations have matching values: they represent the same speaker outputs, but in different positions. Replace SPEAKER_BHI and _FLO with SPEAKER_LE and RE if you prefer; the code is identical either way.

```
void DecodeSoundfield(float* levels, const Matrix& rotation, float gain)
{
  memset(levels, 0, BFORMAT_CHANS * SPEAKER_COUNT * sizeof(float));

  if (panMode == PAN_MONO)
  {
    levels[AMBI_W * SPEAKER_COUNT + SPEAKER_L] = sqrtf(2.f) * gain;
    return;
  }
  float scale = sqrtf(1.f/2.f); // 2 speakers
  if (panMode >= PAN_2D51ITU)
```

```
    {
      scale = sqrtf(1.f/4.f); // 4 speakers
    }
    if (panMode >= PAN_2D71)
    {
      scale = sqrtf(1.f/6.f); // 6 speakers
    }

    SetAmbiGains(SPEAKER_L, gain * scale, rotation, levels);
    SetAmbiGains(SPEAKER_R, gain * scale, rotation, levels);

    if (panMode >= PAN_2D51ITU)
    {
      SetAmbiGains(SPEAKER_LS, gain * scale, rotation, levels);
      SetAmbiGains(SPEAKER_RS, gain * scale, rotation, levels);

      if (panMode >= PAN_2D71)
      {
        SetAmbiGains(SPEAKER_BHI, gain * scale, rotation, levels);
        SetAmbiGains(SPEAKER_FLO, gain * scale, rotation, levels);
      }
    }
  }
```

Directional Soundfield Manipulation

We've seen how to rotate soundfields and adjust their gain in all directions. We can also collapse the soundfield from a sphere to a plane by suppressing the vertical component of the B-Format signal—or any other chosen axis. But what if we want to suppress sounds from a specific direction? This is useful if a large or nearby mobile object partially obscures sound from some directions or when a gap in surrounding geometry means less reverberation in that direction. Such dynamic changes maintain listener interest and make the simulation more realistic and informative about the immediate spatial environment.

The following formulae sculpt the directional characteristics of a first-order ambience or impulse response. fW, fX, fY and fZ are the B-Format components, and dir is a unit vector (length 1) pointing in the direction from which you wish to suppress sounds. These axes must be oriented in the conventional mathematical sequence rather than left-handed or right-handed graphics order. You will need to shuffle X, Y and Z and maybe negate Z depending on whether you're using a left- or right-handed coordinate system: Direct3D and OpenGL differ in this respect,[6] and we've seen that Ambisonic follows mathematical convention with a *vertical* Z axis.

Finally we need a "dimming factor" d, where 0 makes no difference and 1 suppresses as much as possible of the signal in the chosen direction. The original signal in the dimming direction, s, is

```
float s = 0.5f * (1.414f * fW + dir.x * fX
                + dir.y * fY + dir.z * fZ);
```

This can be taken from the original soundfield by adjusting the component weights as follows:

```
float sd = s * d;
fW -= 0.707f * sd;
fX -= dir.x * sd;
fY -= dir.y * sd;
fZ -= dir.z * sd;
```

It's not perfect because of the limited spatial acuity of first-order Ambisonics, but given the way sounds propagate indirectly as well as directly it's a quite realistic volume-only manipulation for varying ambient and reverberant soundfields. It's very cheap to implement: just ten multiplications, three adds and four subtractions per four-channel update frame. This may be faster than conditional code to detect the gap.

These examples show the essential principles of Ambisonics and pan far more smoothly than pairwise mixing for regular surround arrays with a central listener. They're fast and simple enough to explain here. They can be improved by using low-pass filters to reduce the amount of directional signal sent to the speakers at frequencies above 700 Hertz. Tweaks to the coefficients improve support for irregular speaker layouts and off-centre listeners. Ambisonic O-Format, devised by Dylan Menzies, encodes source shapes and directivity in sophisticated ways.[31]

References

[1] *New Realities in Audio—A Practical Guide for VR, AR, MR and 360 Video; Stephan Schütze & Anna Irwin-Schütze, CRC Press 2018, ISBN 9781138740815*

[2] *High Angular Resolution Planewave Expansion; Svein Berge and Natasha Barrett, Proceedings of the 2nd International Symposium on Ambisonics and Spherical Acoustics, May 2010*

[3] *A New Method for B-Format to Binaural Transcoding; Svein Berge and Natasha Barrett, AES 40th International Conference: Spatial Audio, October 2010*: www.aes.org/e-lib/browse.cfm?elib=15527

[4] *Practical Periphony: The Reproduction of Full-Sphere Sound; Michael A. Gerzon, AES 65th Convention, February 1980*: www.aes.org/e-lib/browse.cfm?elib=3794

[5] *3-D Sound for 3-D Games—Beyond 5.1; Simon N Goodwin, AES 35th International Conference 2009*: www.aes.org/e-lib/browse.cfm?elib=15173

[6] *Direct3D and OpenGL Coordinate Systems*: https://msdn.microsoft.com/en-us/library/windows/desktop/bb324490%28v=vs.85%29.aspx

[7] *SN3D Cartesian Encoding Equations*: www.blueripplesound.com/b-format

[8] *The Generation of Panning Laws for Irregular Speaker Arrays Using Heuristic Methods; Bruce Wiggins, AES 31st International Conference 2007*: www.aes.org/e-lib/browse.cfm?elib=13946

[9] *The Sound of Grand Theft Auto; Alistair MacDonald, Game Developers Conference 2014*: www.gdcvault.com/play/1020587/The-Sound-of-Grand-Theft

[10] *The Influence of a Single Echo on the Audibility of Speech; Helmut Haas, JAES, ISSN 1549-4950, Volume 20 Number 2, March 1972*: www.aes.org/e-lib/browse.cfm?elib=2093

[11] *In-Phase Corrections for Ambisonics; Gordon Monro, International Computer Music Conference Proceedings, 2000*

[12] *Why Ambisonics Does Work, Eric Benjamin, Richard Lee and Aaron Heller; AES 129th Convention, November 2010*: www.aes.org/e-lib/browse.cfm?elib=15664

[13] *Control Software for a Programmable Soundfield Controller; John Clarke and Dave Malham, Proceedings of the 8th Institute of Acoustics Autumn Conference on Reproduced Sound. Windermere, UK, 1992*

[14] *Ambisonics in the New Media—"The Technology Comes of Age"; Dave Malham, AES UK 11th Conference 1996*: www.aes.org/e-lib/browse.cfm?elib=7086

[15] *Periphonic Sound, Wireless World, ISSN 0959-8332, May 1980, pages 50, 75*

[16] *3D7.1 Setup*: www.blueripplesound.com/3d7.1

[17] *Ambisonic Toolkit*: www.ambisonictoolkit.net

[18] *O3A Plugins*: www.blueripplesound.com/product-listings/pro-audio
[19] *FuMa and SN3D*: http://sursound.music.vt.narkive.com/b6uKzORY/converting-furse-malham-coeficients-to-sn3d
[20] *Wave Field Synthesis; Karlheinz Brandenburg, Sandra Brix and Thomas Sporer; IEEE Xplore, ISSN 2161-2021, June 2009*
[21] *Further Investigations of High Order Ambisonics and Wavefield Synthesis for Holophonic Sound Imaging; Jérôme Daniel, Rozenn Nico, and Sébastien Moreau, AES 114th Convention, 2003*: www.aes.org/e-lib/browse.cfm?elib=12567
[22] *SuperCollider*: https://musinf.univ-st-etienne.fr/lac2017/pdfs/13_C_D_141345N.pdf
[23] *Rapture3D*: www.blueripplesound.com/gaming
[24] *Building an OpenAL Implementation Using Ambisonics; Richard Furse, AES 35th International Conference 2009*: www.aes.org/e-lib/browse.cfm?elib=15174
[25] *MATLAB/GNU Octave*: http://research.spa.aalto.fi/projects/ambi-lib/ambi.html
[26] *B2X*: www.radio.uqam.ca/ambisonic/5b.html
[27] *All-Round Ambisonic Panning and Decoding; Franz Zotter and Mattias Frank, JAES, ISSN 1549-4950, Volume 60, Number 10, October 2012*: http://www.aes.org/e-lib/download.cfm/16554.pdf?ID=16554
[28] *PC 3D7.1 Titles*: www.blueripplesound.com/compatible-games
[29] *Derby Wigware Plug-ins*: www.brucewiggins.co.uk/?page_id=78
[30] *Quaternions*: www.cprogramming.com/tutorial/3d/quaternions.html
[31] *W-Panning and O-Format, Tools for Object Spatialization; Dylan Menzies; AES 22nd International Conference: Virtual, Synthetic, and Entertainment Audio, 2002*: www.aes.org/e-lib/browse.cfm?elib=11151

CHAPTER 20

Design and Selection of Digital Filters

Filtering is the selective removal of frequencies from a signal. In the simplest case, a filter only affects frequencies above or below a defined rate. These filters work like a simple "tone" control, progressively attenuating high frequencies. Such controls usually work by muffling the signal a fixed amount and then adding back a proportion of the high notes they'd otherwise cut as the control is advanced; a software version can work this way or by allowing adjustment of the point at which frequencies are considered "high," giving more control at little extra cost.

Such a "top-cut" or "low-pass" (beware confusion) filter is just the start. Consumers are familiar with treble and bass tone controls which allow the high or low frequencies to be cut or boosted. In the jargon, the bass knob controls a "low-pass" filter which affects deep notes, low frequencies, while the treble control just affects high frequencies.

This intuitive control pairing was invented by Peter Baxandall in 1952, using thermionic valves to provide boost as well as cut.[1] In this section, I'll explain how to implement a software equivalent of his circuit, as well as how to add additional control for in between frequencies and tweak the response by adjusting the onset of the low- and high-frequency effects.

Before explaining how to program a five-knob "state-variable" filter, it's worth exploring the simplest and fastest type of software filter, the first-order infinite impulse response (IIR) filter, which is simpler than its name might suggest.

The IIR term references the time delay and feedback inside the filter, which sometimes means it's described as "recursive," though in computer-science terms the algorithm is iterative. The "infinite" bit refers to the fact that every previous input sample has some influence over the output, though that decreases over time. A filter that only took account of a finite number of previous samples would be called an FIR—finite impulse response—filter. We'll consider those later.

First-order means that a single delay and mixing step is used, making the code fast but limiting its effect. The transfer function, or slope, of such a low-pass filter is such that the magnitude of the output is halved for each doubling of frequency. More technically, the gradient is -6.03 dB per octave. Empirically, this is a gentle, stable filter with weak pitch discrimination. We'll discuss higher-order filters later.

Low-Pass Filter

All filters work by smoothing out some fluctuations of the incoming signal. High frequencies correspond to fast fluctuations, so they are preferentially suppressed by a low-pass filter of the type we'll explore first. It works by delaying part of the incoming signal in a reservoir—a variable in software, a capacitor in hardware. Incoming signals add to the content of the reservoir, while a proportion of the content is drained away at every update, so the reservoir does not overflow. The rate of leakage determines the frequency response of the filter—the greater the leak, the more low frequencies will be suppressed.

```
float LowPass0(const float input, const float p)
{
  static float r = 0.f;
  r = input + r * p;
  return r;
}
```

Here, input is the incoming sample value, **r** is the reservoir and **p** is the proportion of its content to preserve at each update. This is directly related to the rate of leakage, which we'll call **k**, since it's the proportion leaKed each time, capitals are reserved for constants and **l** would be prone to misreading:

```
k = (1.f - p);
```

Although LowPass0() is the simplest implementation of a digital filter, requiring just one multiplication and one addition per update, it has an annoying property. It has substantial gain, since it adds the entire input sample value every time and only subtracts an amount from the reservoir proportional to **k**.

The smaller the value of **k**, the greater the gain. For a centre frequency of 1 kHz, the output level would be more than eight times greater than the input, about sixteen times for 500 Hz, and so on. Since there's likely to be a gain control downstream of the filter, you could use that to correct for the gain; you need to scale by the reciprocal of **k**, and the implied division only needs to be done when the filter frequency is changed, so it's not expensive if you don't mind the complication.

Chamberlin offers an alternative version that keeps the potentially large value in the reservoir from overflowing the output.[2] This version needs one extra operation—a subtraction as well as the multiplication and one addition—and takes **k** rather than **p** as an argument:

```
float LowPass(const float input, const float k)
{
  static float r = 0.f;
  float output = r * k;
  r += input - output;
  return output;
}
```

The values of **p** and **k** depend upon the desired cut-off frequency—the pitch at which the filter starts to bite—and the sample rate, which is also the rate at which the filter code is called, to accept one input sample and generate one filtered output sample each time. We'll call the cutoff frequency **fc** and the sample rate **fs**, making the formula:

```
float p = 2.f * sinf(PI * fc / fs);
```

With one division, two multiplications and a library call, this code is much slower than the LowPass() function, but that doesn't significantly impact performance because the value only needs to be re-computed when the filter setting needs to be changed, not as each sample is processed.

If the sample rate is 48 kHz and we're updating the controls at most 50 times per second, 960 samples are processed between each control tweak. In fact, if the output rate is constant—as is typical for specific consoles and mobile devices—the entire subexpression (2 * PI/fs) only needs to be worked out once, substituting the constant 48000 for **fs**. If we store that in **p48k** (about -0.0001309 in this case) we're left with two multiplications and a function call to perform when **fc** changes:

```
float p = 2.f * sinf(fc * p48k);
```

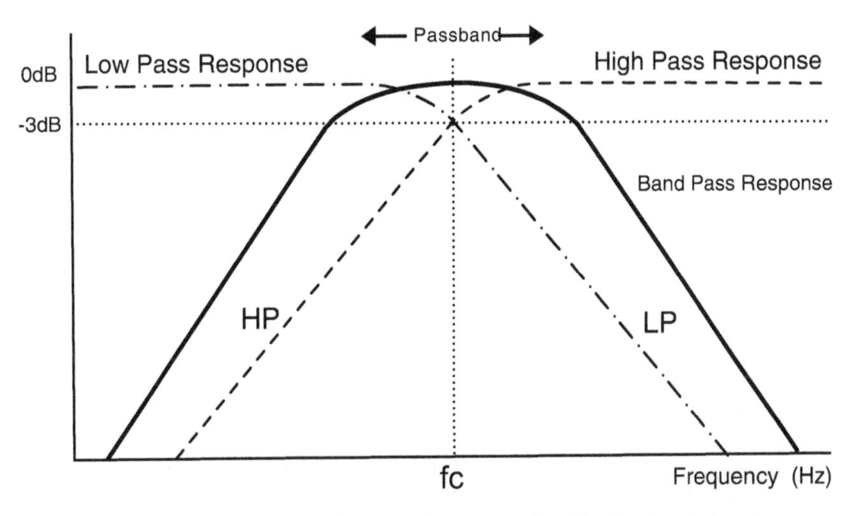

Figure 20.1: State-variable filters implement Paula chip filtering in hardware

I've shown the reservoir as a static member of the filter function, but you'll need to move that out, e.g. by making it a class instance member, if you want to use this code on several sample streams. Each needs its own reservoir. It's also vital to clear the contents of the reservoir back to zero before reusing the filter, or your new output will be corrupted by data accumulated from the previous use.

This is the simplest sort of digital filter, suitable for muffling sound with distance but limited in effect. It has a gentle slope of –6 dB per octave—in other words, the output at **fc** * 2 will be half the magnitude of that at **fc**, three-quarters of a signal at four times **fc** will be suppressed, and so on.

For creative purposes, it would be useful to have a steeper slope and be able to extract more than one frequency band from the input. In fact, we can trivially derive a high-pass output from the LowPass() function, since subtracting the low-pass output from the unfiltered input leaves that part of the signal the filter discarded—the high-pass component. So one additional subtraction yields separate low- and high-frequency output signals, either side of the control frequency **fc**.

This is a minimal two-band tone control, though not a particularly user-friendly one. More useful treble and bass controls leave a middle band unaffected, typically in the range 250..2500 Hertz, rather than pick an arbitrary middle of 500 or 1000 Hertz. Baxandall's controls shift the upper and lower shelf frequencies dynamically.

To get more bands, steeper slopes and control of the frequency interval between the low and high ones, we need to step up to the five-knob state-variable filter. This divides an incoming signal into up to three frequency-dependent components, treble (**hp** or high-pass), middle (**bp**, band-pass) and bass (**lp**, low-pass), with two additional controls which set the centre frequency and width of the middle "passband," thereby also determining the interval between bass and treble. These two controls are **fc** and **fw**, respectively, both expressed in Hertz.

State-Variable Filter

You don't need to expose all five controls and will be able to save a few cycles if you only use a subset. The code to implement them all is surprisingly simple, so you may end up using it as a general-purpose building block—indeed, I've used it for decades, in almost every audio system I've programmed, since first encountering it in Hal Chamberlin's encyclopaedic 1980 tome *Musical Applications of Microcomputers.*[2] This book is still highly recommended for those more comfortable with electronics than mathematics. Here's the kernel of the filter function, configured to return a high-pass filtered output for each input sample **s**:

```
float SVF0(const float s, const float p, const float qp)
{
  static float r1 = 0.f;
  static float r2 = 0.f;

  r2 += p * r1;
  float hp = s - r2 - qp * r1;
  r1 += p * hp;
  return hp;
}
```

Two arguments, **p** and **qp**, control the centre frequency and pass-band width. The variable **hp** contains the high-pass output. The band-pass output is in reservoir **r1** after each update, and **r2** contains the low-pass output. To use this code on more than one sample, make **r1** and **r2** object instance members and clear them before you start filtering a fresh sample. Adjust the return argument to suit yourself.

Robert Bantin alerted me to a subtlety in this conventional kernel which introduces distortion at high frequencies when **r2** is updated midway. The difference is slight, but swapping two lines should fix this:

```
float hp = s – r2 – qp * r1;
r2 += p * r1;
r1 += p * hp;
```

Sometimes it's useful to have a notch filter, which excludes tones around a given frequency. This is the opposite of the band-pass output and can be computed by summing the high- and low-pass components:

```
float notch = hp + r2;
```

It's one thing to know the names of the variables and controls and their qualitative effects and quite another to be able to accurately predict the effect of each setting. The **p** coefficient is calculated as before. The centre bandwidth, resonance or "Q factor" 1/**qp** can be calculated several ways. Some prefer the conventional expression of a "Q factor" as a value from 0.5, with the centre band extending from 0 Hertz up to 2 * **fc**, to some high value; the higher, the narrower the passband and the more "resonant" the filter is said to be. Criterion's action shooter *Black* models the player's loss of health by passing almost the entire game mix through sweeping resonant filters. This gives it an extra unique dimension to inform the player with.

An infinite Q factor, **qp**=0, makes a perfectly resonant filter—whatever you put in, the output is a sine wave that loud, at **fc**. In practice, a Q factor of 17 gives a pass-band less than a semitone wide. Commonly used Q values are in the range 0.5 to 10, yielding **qp** values from 2 to 0.1:

```
float qp = 1.f / qFactor;
```

A more direct way to express the width of the passband is as an interval in Hertz, which we'll call **fw**. From this and **fc** we can find the value of **qp**, after checking that the centre frequency is at least half **fw** (or the passband would extend below 0 Hertz, implying an equally nonsensical Q less than 0.5) and that **fw** is not zero, which would cause a division-by-zero error.

```
assert(fc >= fw * 0.5f);
assert(fw > 0.f);
float qp = fc / fw;
```

If you're looking for conventional treble and bass controls with a middle gap, for instance working below 250 and above 2500 Hertz, respectively, it's more convenient to pass the low-pass and high-pass reference frequencies, **fl** and

fh, and use those in conjunction with **fs**, the output rate, to work out **p** and **qp**. Putting it all together and making sure both band frequencies are positive and the upper one is above the lower, we get:

```
assert(fl > 0);
assert(fh > fl);
float fw = fh - fl;
float fc = (fl + fh) * 0.5f;
float qp = fc / fw;
float p = 2.f * sinf(PI * fc / fs);
```

Making an Oscillator

If all you want is consecutive sine or cosine values at progressive intervals, you don't need to keep calling sinf() or cosf()—a filter with 100% feedback will do the job much more quickly and is an ideal source of smooth LFO (low-frequency oscillator) modulations like vibrato, tremolo, reverb decorrelation and spatial effects. Since **qp** and **s** are both zero (infinite Q, no input) the code's a simplification of the state-variable filter:

```
r2 += p * r1;
r1 -= p * r2;
```

Either **r1** or **r2** must be non-zero for oscillation to start. Set either to the absolute magnitude of the wave you require, zeroing the other, or respectively to sinf() and cosf() of the initial angle (0..PI*2) if you want the waves to start at a configurable phase. Phase control makes LFOs easier to use in combination.

LFO rates are conveniently represented as time periods rather than frequencies, and designers are more comfortable with angles in degrees. This sets the range of oscillation to ±r; phase is the angle, in degrees this time:

```
static float x = r * sinDeg(phase);
static float y = r * cosDeg(phase);
```

Those initialisation lines won't be needed in subsequent updates. For an appropriate new pair of values every **u** milliseconds for a total period or wavelength (1/frequency) of **w** milliseconds, set this frequency factor **p** in radians:

```
p = 2.f * sinf(PI * u/w);
```

For instance p = 0.0524 for a 0.5 Hertz LFO, rising and falling every 2 seconds, updated at 60 Hertz, as 0.0524 = 2 * sinf(PI * 16.67 / 2000), where 16.67 is 1000/60, the number of ms per 60-Hertz frame, and 2000 is the cycle duration, also in milliseconds, 1000/0.5. The formula matches that based on frequency—only the units have changed, consistently.

> You can never have too many Low-Frequency Oscillators.

At each update, for any frequency, update the sine and cosine components **x** and **y** like this:

```
x += p * y;
y -= p * x;
```

This needs just two persistent variables, two multiplications, one addition and a subtraction per update. The staged feedback between **x** and **y** skews the geometry for high frequencies (values of **p** above about 0.15), stretching and shortening opposite quadrants of the sin/cos circle, but it's very smooth at lower rates. The asymmetry is rarely sonically obvious, even for "fast" LFOs—and you can never have too many LFOs!

Frame-Rate Compensation

Each voice in a mix might use several LFOs for pitch and volume variation—often cascaded at different rates, for more complex variation in time, which maintains a consistent range determined by the addition or multiplication of their outputs. As long as the update is called at a regular rate, say 100 Hertz, everything goes smoothly. But what if the update is uneven due to extra CPU load and "drops a frame" because of a bottleneck somewhere?

While a perfectly monotonic frame-rate is the ideal, games are complex, and it's much easier to make one that runs smoothly 99% of the time than it is to eliminate all rare but possible timing fluctuations. The video system can hide the hiccup by holding the current frame, but uneven wobbles in game sounds are more obvious because of the different way the ear interprets time.

To hide this, initialise **p** to reflect the expected framerate, and increase it temporarily only when the last frame took significantly longer than expected. The system update reads the real-time clock hardware each time it runs and uses the difference between successive time stamps to work out the actual frame rate. The audio update uses this time difference; if it's no more than a few per cent above expectation, the oscillators run as above.

When the previous frame was slow, we update **u** and hence **p** to catch up. This preserves a consistent period for the whole wave, at the expense of a rare, momentarily and unavoidable uneven step. This approach works nicely for the odd dropped frame and small values of **p** but still introduces distortion. Rather like the TX-0 oscillator algorithm, this is a trade-off between consistency and averages.

Lissajous Fly Torture

LFOs are handy in 3D, for instance to whirl a swarm of wasps round a player's head. Run three LFOs at relatively prime rates to obtain tumbling but predictably bounded X, Y and Z offsets. Centre sources on the listener and vary **r** depending how much you want to freak the player. This is the audio equivalent of a graphics "particle effect" and similarly amenable to optimisation.

Current audio mixing systems can render hundreds of flies, with varied samples, Doppler and distance filtering effects. If that's not enough, consider spatialising outputs from a multi-tap delay. To make the onset more progressive and spread the overhead of triggering many new sounds in a single frame, limit the number of flies spawned per update. For more variety, slow flies ascending and speed them up on the way down. This is an informative test for VR proximity HRTFs.

Alternative Scales

Filter mathematics uses angular factors easily converted from frequencies expressed in Hertz. These come naturally to many engineers but are needlessly precise and not psychoacoustically weighted—the 25 Hertz interval from 50 to 75 is very distinct, a major fifth, while that from 10,000 to 10,025 Hertz is imperceptible to the most golden of ears. Depending upon their training, some sound designers will be happier to express pitches by note names, e.g. A4 for "concert pitch," 440 Hertz.

Coders beware: scales are numbered from C to C rather than A to G, so C0 is a much *lower* pitch than B0, but D0 is a tone *higher* than C0. Still, if that's what your customers are most comfortable with, that's what you should expose. It's fairer to let the system programmer and processor do the translation, and less error-prone too. If not, you may need to swap jobs!

That conventional scale extends from C0, 16.35 Hertz, up to B8 at 7,902 Hertz, the highest note on a piano keyboard; higher frequencies are sometimes useful, so note names up to B9, 15,804 Hertz, could be useful on systems with output at 44.1 kHz or higher. That'd be too close to the Nyquist limit for stability, or much useful effect, on a 32-kHz mixing system like a GameCube.

Pitch Representations

MIDI note numbers use the same semi-tone interval and C0 start point; some sound designers with a sequencer and tracker background consider those the natural way to express pitch; if so, they're easily catered for. This formula uses the C standard library powf() function to convert an integer MIDI note number **m** to a frequency **f** in Hertz, assuming standard concert pitch, when MIDI note 69, A4, is 440 Hertz:

```
float f = powf(2.f, (m-69) / 12.f) * 440.f;
```

MIDI notes are numbered 0..127, from C0 to G10; the *Guinness Book of World Records*[3] says Georgia Brown can sing that ultrasonic note, though at 25,088 Hertz no one can hear it without a bat detector or similar.

Audio interfaces like EAX expect filters to be controlled by setting an attenuation at a reference high frequency of 5 kHz. To keep us on our toes and allow the value to be passed as an integer on vintage systems, the value is expressed in millibels, 1% of a decibel.

This helper finds **fc** from **mB**, the attenuation in millibels, and **fr**, the reference frequency, for a first-order filter:

```
assert(mB <= 0);
const float fr = 5000.f;
const float factor = 0.0005f;
float fc = fr / powf(10.f, mB * factor);
```

Replace **factor** with 0.00025 to set a second-order filter. You may need this formula for compatibility with I3DL2 filters and reverbs, which come for free on some platforms, but it's not an ideal convention, as it implies that the filter is always partly closed even if the attenuation is zero, since **fc** cannot exceed 5000.

Sensible Values

If you arrange the signal to bypass the filter entirely if no attenuation is requested you'll save some processing time and preserve the higher frequencies, but then the effect of the filter will be immediate rather than gradual as soon as even a tiny attenuation is required; the audibility of the "step" depends upon the proportion of frequencies above 5 kHz in the input. The problem is inherent in the specification.

It's problematic to determine a safe upper limit for the centre frequency, as it depends upon content, and game designers are not above exploiting glitches creatively—except to observe that **fc** should never be allowed to exceed **fs/2**, the Nyquist limit, and weird whistles are increasingly likely above fs/4. Code to catch and reject such extreme values with ERANGE should be included in the appropriate parameter Set() methods.

The All-Pass Filter

There's one more type of filter worthy of discussion, even though it typically has a flat frequency response. It makes up for that by scrambling the phase of the input in complicated ways, making it ideal for use in synthetic reverberation and the decorrelation of sounds to make them sound larger.

In this example, **r** is a persistent-delay register like **r1** and **r2** above, **s** is the sample input presumed to be in the range -1..+1, and **c** determines the proportion of input and delayed output mixed and hence the cross-over frequency. The value 0.5 is commonly used, as this makes hardware implementation especially easy.

```
static float r = 0.f;
float c = 0.5f;

float output = r + c * s;
r = s - c * output;
```

This is a first-order filter with the low-pass and high-pass outputs mixed, so that only the frequency-dependent phase change remains. The scalar **r** gives only one sample of delay; in reverberation systems an array is used for longer delays, and the decorrelation achieved varies with the delay length. At last we're ready to discuss reverberation and FIR filters.

References

[1] *Negative Feedback Tone Control—Independent Variation of Bass and Treble Without Switches; Peter Baxandall, Wireless World, ISSN 0959-8332, October 1952, pages 402–405*: www.learnabout-electronics.org/Downloads/NegativeFeedbackTone.pdf

[2] *Musical Applications of Microprocessors; Hal Chamberlin, Hayden 1980, ISBN 0810457539, pages 441–444*

[3] *Top Notes*: www.guinnessworldrecords.com/world-records/greatest-vocal-range-female

CHAPTER 21

Interactive Reverberation

Reverberation, or reverb for short, is the pattern of echoes heard when sounds play in a reflective environment. Stuart Ross, sound designer on the BAFTA audio award–winning *Crackdown* and *Grand Theft Auto: Vice City* games, likes to say that reverb is to sound as lighting is to graphics. It's thus very subjective, combinatorial and influenced by many aspects of the scene.

Done well, reverb adds greatly to the listener's ability to find their way around a complex environment and their sense of immersion or being there. Done the more usual way, it makes it harder to work out where sounds are, conceals important details and may even drive the listener into the "uncanny valley," where experience and the simulated world are so far at variance that the player's ability to suspend disbelief and enter fully into the simulation are jarringly at odds.

> Reverb is to sound as lighting is to graphics.

In games we must cope with environments large and small, moving listeners, many mostly moving sound sources—and sometimes even moving walls (usually downward).

Reverberation is a complex topic. This chapter describes four ways to implement it, with special reference to real-time constraints, dynamic changes and the parameters that homebrew or off-the-shelf reverb units expose. It tells you what to expect of a good reverb, where to get code to implement one and the pros and cons of doing so. It also leverages techniques introduced earlier, including digital filtering and interpolation.

This chapter pays special attention to directional and 3D reverb and the real-time requirement for glitchless parameter changes. Such aspects are rarely addressed in the literature but are of paramount importance for enhanced reality.

Reverb Usage

Table 21.1 shows the number of high-quality multichannel reverbs used in eight of the games previously cited. The drop between *DiRT2* and *DiRT3* reflects the addition of split-screen multi-player to the latter game. The PC figures depend upon the underlying operating-system driver—at least one reverb was always available, but Rapture3D software mixing was needed for three or four. Rapture3D dynamically adjusts the quality of reverbs, spatialisation and voice mixing at startup to make sure it will need no more than 10% of the available CPU time to process all the requested voices and effects. Otherwise, updates to the "optional" reverb units were ignored when Creative Labs drivers were being used.

These are not the only reverberant effects in these games—some samples have baked-in reverb, and a "reflection system" tacked on to the output of the engine and exhaust sound synthesis spatialised distant echoes, particularly adding to the excitement in circuit-racers like *GRID* and Codemasters' *F1* series. A long delay line with eight taps allowed copies of the player- and AI-controlled car sounds to be delayed according to the path to the nearest reflector and then replayed from that position. Slight periodic wandering of the tap points, independent of the reflection distance, modelled air movement along the path and curtailed resonances that might otherwise be caused by unnaturally precise comb filtering.

Table 21.1: Reverb usage by game and platform

Game	PlayStation	Xbox	PC
DiRT	2	2	1
DiRT2	4	4	1..4
DiRT3	2	1	1
Showdown	2	1	1
GRID	1	1	1
F1 2010	3	3	1..3
F1 2011	3	3	1..3
Bodycount	1	1	N/A

Rockstar North's *GTA5* uses three internal reverb buses for spatialisation, each four channels wide, pre-set for small, medium and large reverberant spaces. Each of the 100 physical voices can be routed to any of those, in varying proportion to suit the environment and distance from the listener.

Reverb on the Cheap

In my career I've used a mixture of simple delay effects and platform-specific reverbs, finding that cross-platform ones tend to be both slow and unrealistic. This is a trade-off between speed, compatibility, tweaking time and playing to the strengths of each platform. In recent years I've made increasing use of impulse-response reverbs, which were once considered too big, slow and inflexible for real-time use but which make good use of modern vector mathematics hardware. If you've got 25,600 MFLOPS to burn—and just one co-processor of the decade-old PlayStation 3 delivers that—convolution is a good way to soak those up.

Reverbs are the most expensive effect, typically shared between several groups. Unless you'll accept harsh and hollow mono feedback delays, each reverb may take a few per cent of the available processing time on a typical CPU core, or a fixed slot on a DSP. You're lucky to have five of these at once and may need to shuffle them between roles depending upon the game mode—menus, main game, replays—indoor and outdoor environments, and often when switching between first- and third-person camera views too.

Mobile games might use no dynamic reverb at all but can still sound reverberant by baking reflections into the sample data or adjusting the mix balance—the ultra-cheap "tunnel effect" in *Colin McRae Rally* for Android just tweaks the balance to emphasise player-car sounds which would be reflected from the bridge walls and roof. On a low-end mobile phone this may be all you can afford.

Mid-period *McRae* games shipped with paired banks of "kickup" samples, played in varying intensity to simulate the sound of stones and gravel being thrown against the bodywork of the player's car. These were treated offline to sound appropriate for external or in-car listeners and swapped from disc while the player switched camera views. Thus only one memory slot was needed for each category of kickup or for stage-specific variants for different types of terrain and weather. The sound designer or build system had to identify all the competing kickup banks and make sure that this slot was big enough for any of the banks.

Typically, this swapping would take only a fraction of a second, even with the slow speed of optical media, but if the game was in the middle of streaming music or ambience, loading speech or other game data, there might be a second or so without any kickup sound. This hole could be masked by playing a transition sound from another bank, but generally it is unobjectionable because the change is synchronous to a user action—the camera switching—so it does not come as a surprise.

Grand Theft Auto 5 uses dynamic bank swapping similarly, to fetch granular synthesis data for freshly hijacked cars. There are too many cars to keep all their data in memory at once. This loads during the animation of the player getting in, so it's ready for use by the time the player is in the driving seat. Switching between the layered music, streamed as eight stereo stems, to the in-car radio channels is masked by speech, usually by waiting for pre-programmed dialogue to momentarily mute the music, allowing the assets to be swapped, or by spinning in some DJ banter if that's not convenient. The aim is to give the impression that everything could play at any time while managing transitions to eliminate worst-case spikes.

Kickup sounds, like all "particle effects," are nice to have but not essential, so the Virtual Voice system naturally reduces their number when there's a lot else going on. Over and above the priority system, the game should limit the number of kickup sounds freshly triggered in a single update; this helps to spread the processing cost, limiting update time spikes, and reduces the incidence of unrealistic pulsing, as lots of samples are triggered together when the player gains speed, then comes a lull as they tail off before the voices are released for reuse and another wave of clattering is enabled. This is an interactive equivalent of the "wait an hour for a bus, then three come at once" issue.

Early and Late Reverberation

Academics would have us believe that reverb consists of two main parts—distinct early reflections (echoes) and a sort of warm mush (reverberation) when echoes of echoes rapidly build up as the original sound and early reflections bounce around and interact. In practice the dividing line between these parts is illusory, but it's still helpful to consider early reflections specially as they are especially relevant to our perception of 3D positions, and sometimes we can get away with just an early-reflection or late-reverb simulator, depending upon the context.

The timing of the first reflection tells the ear how far away the source is from the nearest reflector. In a racing game this is vital information, telling you how close to a collision you are getting in your tussle for the tightest racing line. Both fixed and mobile geometry should be interrogated to work this out unless you're in a single-vehicle game like *Rally*. The fluctuating sound of your own car reflecting off another alongside is quite a spur to manoeuvre.

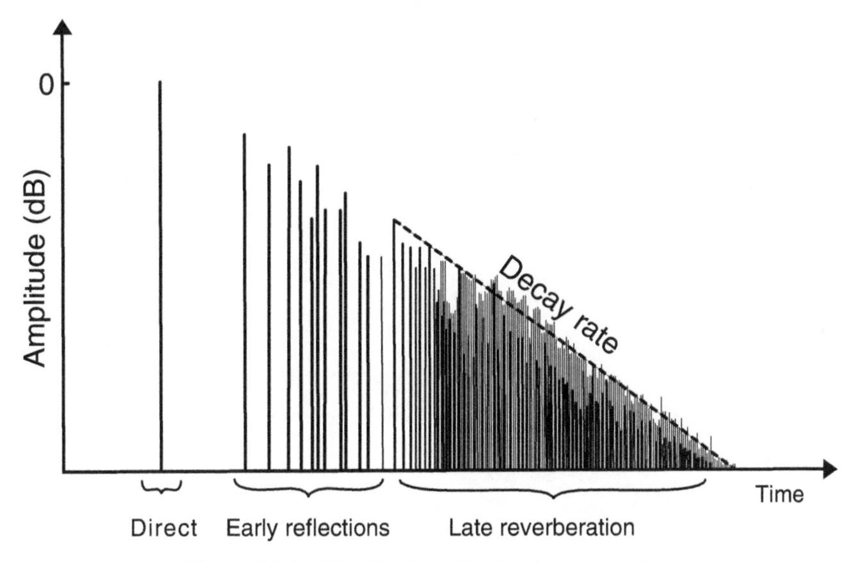

Figure 21.1: Distribution of reflections over time

One of the innovations introduced to *Colin McRae Rally 4*, since followed in many games, was to model reflections from the left and right side of the track. This was laboriously implemented at first by defining I3DL2[1] reverb pre-sets for thousands of close-spaced positions either side. As the player drives along these pre-sets are cross-faded, with adjustments for proximity to the racing line, so that the reflected sounds either side change as the player passes walls, buildings and even individual trees.

This sounded great but made for six months of very dull and painstaking work for the designer, Ben McCullough, who had to create and test all the mark-up at "cats' eye" intervals for 88 tracks, each several miles long. These days we prefer to create such data via automated processes, as explained in the next chapter.

At distances less than a metre the separate quality of the reflection will be lost as it interacts with the direct sound more quickly than the ear can distinguish them. This is another cue that you're running out of road.

There are four main ways to implement reverb. Unless you have access to specialist audio processors with dedicated RAM, all of them are quite processor intensive, mostly because of their scattered memory access patterns, which we'll explore later. The underlying technologies are delay lines, all-pass filters and convolution.

Delay Lines and Early Reflections

Delays are implemented by buffers, which store samples for later use. The maximum delay in seconds is determined by the sample rate and the buffer size in samples, which we call MAX_DELAY. The size directly relates to the longest path between source and listener, so:

```
maxDistance = MAX_DELAY / SAMPLE_RATE * speedOfSound;
```

The simplest delay line has a single output or "tap" from which delayed samples are fetched. Once you've gone to the trouble of creating the delay line, it's easy to add additional taps. Providing the tap offsets are continuously varying or relatively prime, to minimise their interaction, this lack of common factors conceals the comb filtering caused by mixing in a delayed signal.

This example supports TAP_COUNT taps. Reflections are inevitably less intense than the direct sound, since they travel further, so an attenuation—a "gain" factor less than one—is applied to the output from each tap before they're mixed together. So each tap needs two values, the gain and a temporal offset in samples multiplying the required delay in seconds by the sample rate.

```
const unsigned TAP_COUNT = 1;
const unsigned MAX_DELAY = 1024;
static_assert(!((MAX_DELAY-1) & MAX_DELAY));

struct delayTap
{
  float gain;
  unsigned offset;
};
```

Given those definitions, here are the instance variables for a single mono delay line:

```
delayTap delays[TAP_COUNT];
float buffer[MAX_DELAY];
unsigned offsetIn;
```

These three variables should be object members, so that they persist between update calls and are not shared between instances. The static assert checks that MAX_DELAY—the number of samples the delay line can hold—is a power of two (e.g. 4096 or 8192). This requirement makes subsequent updates faster because the wrapping of each index from the end of the array back to the start can be done with a bitwise and assignment operator, &=, rather than the modulo assignment %=, which requires a relatively slow division operation to compute each remainder.

Each element of the **delayTap** array contains an offset in the buffer from which an output sample will be read and a gain by which it will be scaled. Realistic tap delays should be exponentially spaced, at increasing intervals. **offsetIn** keeps track of where the next input sample should be slotted into the buffer. Initialisation of the object involves clearing the buffer and copying TAP_COUNT offsets and gains into the array of delays.

```
void SetupDelayLine(delayTap taps[])
{
  for (unsigned i=0; i<MAX_DELAY; ++i)
  {
    buffer[i] = 0.f;
  }
  offsetIn = 0;
  for (unsigned i=0; i<TAP_COUNT; ++i)
  {
    delays[i].gain = taps[i].gain;
    assert(taps[i].offset<MAX_DELAY);
    delays[i].offset = taps[i].offset;
  }
}
```

Call UpdateDelayLine() for each sample to be output to add a new sample for the current time and return the mix of delays:

```
float UpdateDelayLine(float in)
{
  buffer[offsetIn++] = in;
  offsetIn &= (MAX_DELAY-1);
  float mix = 0.f;
  for (unsigned i=0; i<TAP_COUNT; ++i)
  {
    mix += delays[i].gain * buffer[delays[i].offset++];
    delays[i].offset &= (MAX_DELAY-1);
  }
  return mix;
}
```

Tom Zuddock's article "Virtual Audio Through Ray Tracing" describes a simple approach to scanning a room's geometry to work out reflection sources and delays.[2]

Spatial Reflections

The example weights delayed samples according to their gain and combines them into a single mix value. This could them be spatialised using a panner, or individual taps could be panned separately for a spatially diffuse sound. Work out where the earliest reflection is coming from by scanning the geometry for the closest reflective surface and applying the law of reflection (angle of incidence equals angle of reflection relative to a surface normal) to trace it onward to the listener, preferably with a filter consistent with the properties of the reflective surface; thus you can position it independently of the direct sound.

Figure 21.2 shows the pop of a balloon echoing around a room. A vast number of other paths are possible. We'll model the most psychoacoustically important and fake the rest plausibly.

This is potentially a big step forward in spatial realism for medium-sized acoustic spaces, where the first delay is beyond the range of the Haas precedence effect. The cost is manageable, as later reflections fall within the Haas window, so the brain won't care where they come from. Dave Malham's *Reflector* project explores the potential of this technique.[3] James Moorer's 1980 paper *About this Reverberation Business* discusses spatialised early reflections as well as critiquing Schroeder's earlier work.[4,5] Processes which were once unacceptably costly for real-time processing are now practical on modern VR and gaming systems, though you still need to pick which parts of the mix deserve the most elaborate treatment.

Buffer Considerations

As noted earlier, this implementation assumes that MAX_DELAY is a power of two. If not, you must remove the static assert which checks that and replace both instances of:

```
&= (MAX_DELAY-1);
```

with

```
%= MAX_DELAY;
```

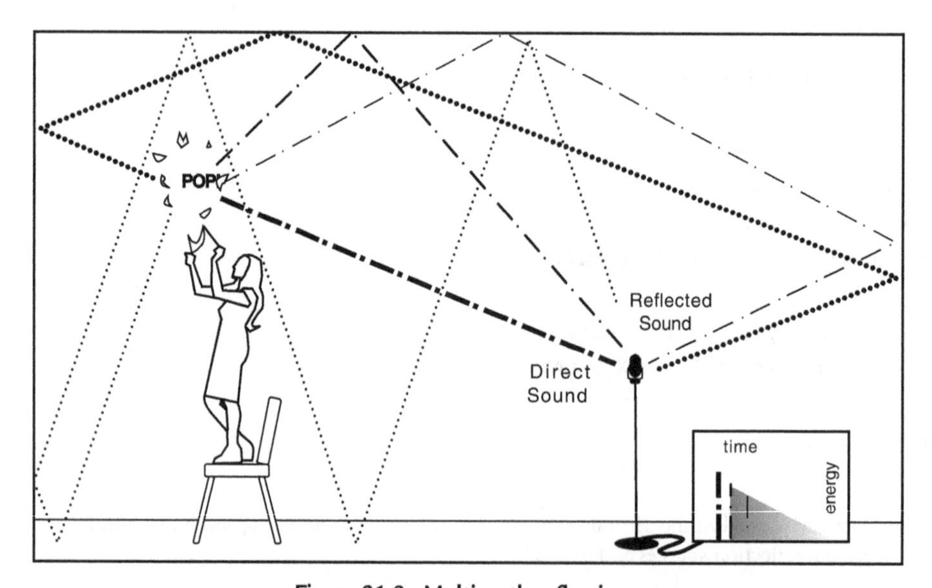

Figure 21.2: Multi-path reflections

The implied divisions will quite likely take longer than the rest of the function, so unless memory is extremely tight the extra memory to round it up to the next power of two will be a good investment. Since it's a circular buffer, all entries are eventually used, and the only requirement is that the distance in samples between the input and the latest output tap is less than MAX_DELAY.

Extending the buffer decreases the cache efficiency of the code, since the values in the delay line will be distributed over more memory, but caches are little help when echoes and reverberation are being computed since the values used at each update are necessarily scattered through memory, reflecting their offset in time. Reverb is a test of the worst-case random-access performance of a memory sub-system and exposes the shortcomings of cost-reduced general-purpose processors which are optimised to read bursts of consecutive words quickly rather than pick up individual dispersed floats.

The PS2's 128-bit Emotion Engine actually used 16-bit memory in fast bursts to read eight consecutive short words. Access to main memory was 60 times slower than reading from the cache and fetched 16 bytes even if you only asked for one. As processor clock speeds climbed, things got worse.

The Xbox 360 introduced a 128-bit bus, shared by the GPU and six CPU threads, and was likewise optimised (in general terms, not ours) to transfer and cache "lines" of 128 **bytes**—1024 bits—by eight back-to-back bus transactions. This process, assuming no contention, took over 500 processor cycles and was liable to throw out cached data which other threads would like to reuse in favour of audio samples most of which would be ignored and which would not be accessed again for milliseconds.

So the Xbox 360 gained a special machine instruction to read memory by bypassing the Level 2 cache. Manually adding **__xdcbt** intrinsics reduced the impact of random access on other threads, but this late addition was susceptible to the Sceptre processor architecture bug[6] a decade before that was widely known, and even when it worked it didn't directly benefit code with a substantial random-read requirement, like the built-in Eventide reverb. So much for GHz, or raw bandwidth measurements.

The Xbox One and PS4 have a 256-bit internal data bus, so each read or write transfers at least 32 bytes, usually in pairs, as the processor caches 64-byte lines. If the data you want is in Level 1 cache it's available three clock cycles—around 2 nanoseconds—after you ask for it. But if this is not the case the processor must look to main memory, which takes at least 30 times longer—potentially hundreds of clock cycles if the GPU or any of the other seven processor cores competing for main-memory access are already waiting, as is often the case in complex games with a large active dataset.

Passby Reflections

Urban and circuit-racing environments bounce echoes of vehicle sounds around, so that it may sound that many more cars are milling around than actually exist. Game programmer Rob Pattenden and sound designers on the *F1* and *GRID* games came up with a system of virtual "reflection lines" drawn around the map, recording the changing locations from which these echoes might be heard. As loud sound sources move, the games compute the nearest point on these lines and place a virtual emitter for the reflected sound there.

Having just implemented a granular engine synthesiser, Rob's first idea was to run multiple copies of that, re-using the same source sample data for multiple positions. Once we'd refined the granular prototype to match the quality of earlier simpler schemes, this became expensive, and I proposed adding multi-tap delay lines to the output of the granular synths. The offset within the delay line determines the time delay, and each output could be spatialised independently at relatively low cost. To add depth and smooth out the tone these tap positions were made to wander back and forth over a small range between updates.

Natural Reverberation

Panned delays are an effective way to improve spatialisation, but even with randomisation they lack the complex tone and immersive quality of late reverberation. This aspect demands more complicated techniques, which we'll review now, with dataflow diagrams and example timing. Writing a reverb is non-trivial and tuning one to sound good by whatever definition is harder still—let alone optimising it for several processor architectures and memory systems—so there's a strong case for using whatever's already packaged for your platforms. But you still need to understand how these ready-made reverbs work and are controlled.

There are at least four ways to implement natural-sounding reverberation. The most widely known was invented by Manfred Schroeder in 1962, later refined by James Moorer. In a 1972 *Studio Sound* magazine article Michael Gerzon proposed the use of all-pass filters in a loop, which influenced David Greisinger's design of the classic Lexicon 224 and later products from Ensoniq and Alesis.[7]

In 1992 Jean-Marc Jot devised Feedback Delay Networks,[8] which were used by Creative Labs EMU-based sound cards and underlie the I3DL2 abstract reverb specification.[1] Recent increases in processor power have made a fourth technique, impulse response reverb, viable—even on embedded systems. We'll explore all these in the rest of this chapter.

All-Pass Loop Reverbs

Alesis engineer Keith Barr sadly died in 2010 but left a lucid explanation of the all-pass loop reverb concept on his Spin Semiconductors web site.[9] The basic idea is to build up a loop of delays and all-pass filters, injecting the input into the loop and extracting decorrelated reverberant outputs from taps on the delay lines. Each loop contains several filter elements, most of them with the structure shown in Figure 21.3.

The diagram shows two all-pass filters in series with a third delay. A full implementation would use two or more of these blocks in a ring, connected between the feedback loop points. The total length of all the delays in the loop should be at least 200 ms (requiring at least 37 K of float buffer memory for 48-kHz mono processing), or the sound will be tinny or prone to subsonic "flutter," which is intrusive at rates between 4 and 8 Hertz. Longer loops slow this down so the tone is heard to evolve rather than flutter.

The principle is that all-pass filters with long delays let through all frequencies but vary their phase in complex ways when combined, so the tone of the original sound is retained without the comb-filtering artefacts of Schroeder's approach. The factor **g** in each all-pass filter should match, with opposite sign, in each path around the associated delay.

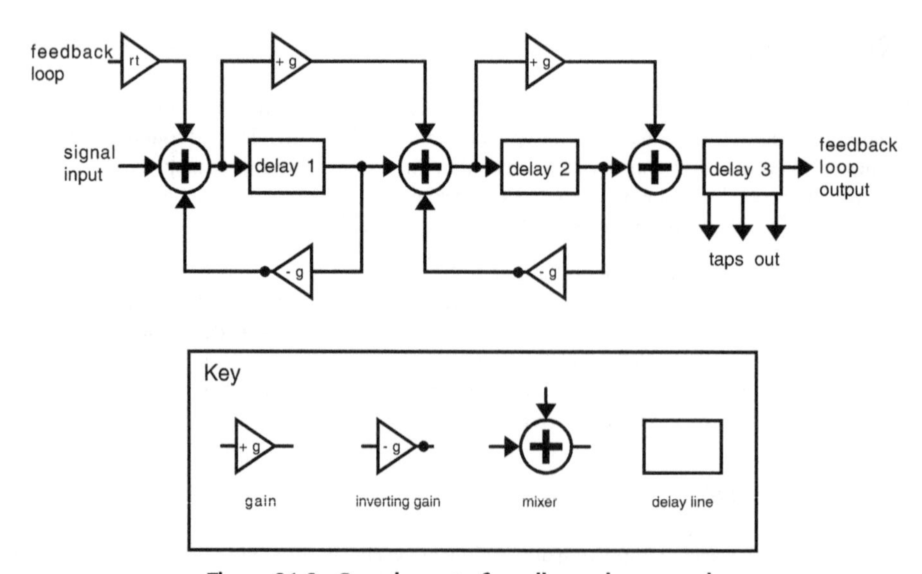

Figure 21.3: One element of an all-pass loop reverb

Hardware implementations are fastest when **g** is +/− 0.5, as that's a trivial multiplier in binary arithmetic, but this causes the output to build over time, which may sound unnatural. A tapped delay line can add early reflections to mask this, at the expense of the flatness of the frequency response. Alternatively, try increasing the value of **g**—Keith Barr found 0.6 accelerated the build-up and made the early output sound "fatter." Values above 0.7 give more immediate output but unbalance the frequency response, adding a "ringing" tone.

Tuning the Reverb Loop

There are many ways to make the sound more natural, but their choice is a matter of trial and error and depends upon the desired sound. There's no perfect general-purpose arrangement—feedback reverb design is as much art as science.

Sinusoidal variation of one or two of the delay lengths, or the positions of the taps used to extract output, can smooth the sound but may be audible in the output unless the rate and range are sympathetically chosen. Use the LFO Sine/Cosine outputs and interpolation techniques of Chapter 20 to implement this. Shelving high- and low-pass filters in the loop make the decay over time more realistic.

The attenuation factor **rt** determines the Reverb Time by controlling the feedback. Inputs and outputs can be applied and extracted at any stage, perhaps in more than one place. Multiple output taps provide decorrelated signals for surround speakers, giving a sense of space and envelopment without accurate spatialisation, which is best achieved by panning early reflections. Providing downmixing is avoided and you have time to experiment, this reverb topology is both fast and flexible, though it lacks the fine control over parts of the sound possible with Schroeder and in particular Feedback Delay reverberators.

I3DL2 Standard Reverb Parameters

We have Jean-Marc Jot to thank for more flexibility and independent control over aspects of the sound with human-friendly names. He devised the I3DL2 reverb specification, first used in OpenAL and Creative Labs EAX audio extensions for Microsoft's Direct Sound, subsequently reimplemented on many platforms including Xbox and PlayStation, and middleware like FMOD and Wwise.

Table 21.2 shows the properties of a reverb built to the I3DL2 specification.[1] This is by far the most commonly used cross-platform interface for the control of interactive reverb, first supported in Creative Labs and NVIDIA versions of OpenAL, then Android's *audiofx* class, CryEngine, OpenSL ES 1.0.1+, Rapture3D, Unity and XAudio, among others.

Table 21.2: I3DL2 reverberation controls

Name	Units	Minimum	Maximum	Description
Room	mB	−10,000	0	Room effect attenuation
Room rolloff factor	ratio	0	10	Distance level adjustment (1=none)
Room HF	mB	−10,000	0	High frequency attenuation
HF reference	Hz	20	20,000	Onset of "high frequencies"
Diffusion	%	0	100	Late reverb echo density
Density	%	0	100	Late reverb frequency density
Decay time	s	0.1	20	Time to attenuate to -60 dB
Decay HF ratio	ratio	0.1	2	High to low frequency fade ratio
Reverb	mB	−10,000	2,000	Late reverb output level
Reverb delay	s	0	0.1	1st reflection to late reverb time
Reflections	mB	−10,000	1,000	Early reflections output level
Reflections delay	s	0	0.3	Pre-delay before first reflection

Some implementations add extra controls. FMOD has a dry level for the direct path. EAX adds a "direction" vector which neatly controls the spatial distribution of reverberation around the listener. If the vector has a magnitude of zero, the default, reverb is omnidirectional. The maximum magnitude is 1, which directs all reverb to come from the specified direction. Jean-Marc Jot gave me this elegant formula:

```
directionalMagnitude = (2.f * sinf(0.5f*a)) / a;
```

The angle **a** is in radians. Use (-0.637, 0, 0) to place the reflections over a 180° semicircle to the left of the listener (+0.637, 0, 0) for the opposite side, 0.90 (PI/2 radians) for a 90° spread in whatever direction you wish.

Before you get too excited, be warned that not all implementations of I3DL2 follow the full spec. HF Reference is often hard-wired to 5 kHz. On investigation, later confirmed by Microsoft, I found that XAudio's Eventide reverb ignored the Room Rolloff Factor completely. So does FMOD. The *CryEngine 3* manual explicitly warns that some I3DL2 parameters are ignored in some reverb implementations.[10]

Given this and the difficulty of making any two reverbs sound alike, you need to keep your ears tuned and accept the trade-off between having one standard reverb that performs equally poorly everywhere and platform-specific optimised versions that may need tweaking for each configuration. If you're aiming for the best sound on each platform and wish to take advantage of optimised code for each, this is one case where target-specific configuration is justifiable, even if this leads to some surprises when you get twiddling the knobs. It's up to you how many and how you name them—my client sound designers agreed that *damping* was a better description than "decay HF ratio," but this depends upon what you're used to.

Platform-specific parameter sets, and sometimes transfer curve adjustments, may sound more consistent—and, crucially, sweeter—than throwing the same numbers at rival black-box implementations. Sometimes a choice of reverb models is needed to support a range of target devices, especially PCs. Platform reverbs tend to be faster and sound better than middleware ones but are less consistent.

Reverb pervades the sound of your product, and only a small number of high-quality reverb units run at any time—I've not used more than four, along with many simpler delay and reflection systems, at any one time in my work, though sometimes this has involved dynamic reconfiguration with changes of camera view.

Low-spec platforms might limit this further—the "tunnel effect" in the mobile remix of *Colin McRae Rally* had to work on an iPhone 4, so it was implemented at trivial cost by varying the engine and exhaust volumes, using headroom otherwise reserved for bodywork damage effects, on approach and exit to each bridge or tunnel. PlayStation versions used hardware reverb, since it was available almost for free. You make the best of what you can get.

The I3DL2 specification is freely available from the MIDI Manufacturer's Association. Its first appendix includes detailed reverb descriptions, defaults and some pre-set values—padded cell, quarry, cave, bathroom etc.—which are useful starting points.

The time-related parameters often cause audible glitches if they're suddenly adjusted. This is a consequence of most reverbs being designed for static rather than interactive use. When I raised this with Sony and Blue Ripple tech partners, they added update smoothing, limiting the rate of change, concealing this artefact. Microsoft laterally introduced a "new reverb" with no time-related controls! Since pre-delay is a strong cue to environment size, it's worth writing your own simple slew-limited delay and using that in front of impulse response reverbs, all-pass loops and other types lacking in smooth time control.

Reverb Costs and Mitigations

High-quality reverb comes at high cost—each XAudio stereo reverb burned 14% of a 3.2 GHz Xbox 360 hardware thread after months of optimisation work and 808,464 bytes of RAM. Remember you may need several at once, and other game systems will be competing for these resources.

This is basically a stereo reverb, so early reflections are only spatialised at the front, and the surround channels are derived by matrixing the late reflections across the rear. My measurements revealed that the rear left channel was a mix of the front left reverb plus half the front right signal, added in anti-phase (mix factor -0.5), with symmetrical shuffling the other side.

That's cheap to implement and delivers cinematic immersion on a 5.1 speaker system but caused great problems for a prototype discrete 5.1 to HRTF headphone encoding system. The phase reversals and left/right and front/rear matrixing made it very hard to decide where any sound not in front of the listener was coming from.

Of course, anything intended to shuffle discrete speaker feeds for headphones will struggle to create a plausible soundfield for moving listeners. Too much information has been thrown away in the 2D panning process. That's why it's essential to create a custom mix for each listening configuration, taking account of each source direction, listener position and motion. This is practical for games and VR, where the listening configuration is known and we can focus on each listener rather than a cinema crowd.

After profiling and discussions with Microsoft's Scott Selfon and Tom Mathews, the corresponding XAudio2 reverb was optimised, without limiting its best ability, by "cost-reducing" options, such as mono input and decorrelated (fake) stereo output, decoupling of the late and early reflection units and reduced-rate reverb processing, so the reverb could run at 24 kHz even if output was mixed at 48. Thus we could use four stereo units at once—it's unfortunate that we had to run two separate reverbs for left and right reflections, but it was certainly worth it in terms of the dynamic sound. Such overhead can be eliminated if you have access to the "black box" of the stereo reverb or populate it yourself.

8-In, 18-Out HDMI Reverb

My 7.1 channel PlayStation HDMI reverb implementation, first used in *DiRT2*, uses a different approach. Eight distinct reverb inputs deliver three soundfields via 18 output channels. The bulk of the processing effort goes into six mono directional reverberators, each with a separate input derived from sounds directed to either side and four corners around the player, and correspondingly spatialised outputs. This extends the idea of left-right reflections to fit all six major directions in a 7.1 system, greatly benefitting spatialisation as well as immersion.

Distant enveloping environmental reverb uses a mono-in, six-channel-out approach, with separate taps to decorrelate the output for each speaker direction. A similar arrangement suits cabin reverb, with a hollower timbre and faster decay. Some games save resources by switching between environmental and interior reverb depending upon the camera position. Allow a short pause for things to settle down before you turn the Late Reverb level up again.

DiRT Rally favours impulse response reverb for the in-car sound, which indirectly found its way into major manufacturers' automotive audio simulators. Even impulse responses from a lecturer's old Subaru had a transformative effect on the sound of supercar simulators, not that we told the metal benders where they came from. Since then we've captured 3D impulse responses for real rally and race cars, using the car engine as a source (beware of overheating!) and a soundfield microphone inside to capture the 3D reverb and the way it changes as windows and doors are opened (or broken).[11] In conjunction with tweaks to the external sound balance and "sense of speed" wind noises, this is an important part of realistic "damage" modelling.

Freeverb

The most commonly used implementation of synthetic reverb—the choice of major electronics manufacturers, academics and middleware vendors alike—is known as Freeverb.[12] This is a conventional Schroeder-Moorer reverb, except that months of work have been put into choosing "magic numbers"—mix factors and delay times—which sound good to the human ear. It probably helps that it's free, well documented and efficiently written in C++, but the sound is the secret of its success.

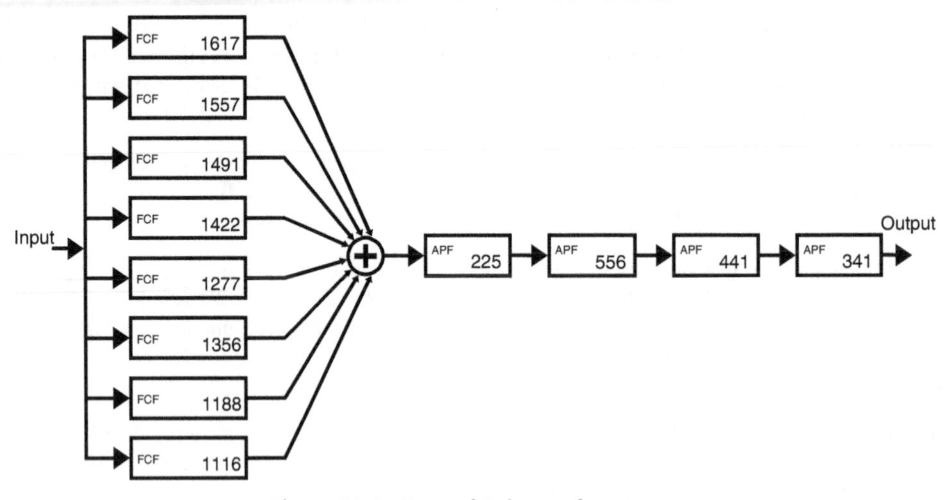

Figure 21.4: Freeverb's internal structure

Earlier examples and academic papers explained the concepts well but sounded disappointing, uneven in pitch and amplitude, with a tendency to unrealistic resonances on real-world input and problematic interactions between supposedly independent parameters of control.

Freeverb uses a conventional Schroeder-Moorer architecture, starting with a parallel array of eight feedback comb filters (FCF) which spread the same input sample into delays of varying length then extract it, low-pass-filtered, through an all-pass filter chain. Julius O. Smith III explains this neatly in his book *Physical Audio Signal Processing.*[13]

Each of the parallel filters incorporates a separate delay line, between 1000 and 2000 samples long, as shown in the FCF blocks. The output from all eight delays is mixed into four modified all-pass filters (APFs) with shorter delays, a few hundred samples long. Lengths are adjusted for the perceived size of the reverberant space. The interaction between all these delays and associated comb and IIR filters determines the sound and linearity of a reverb.

What Freeverb's author, Jeremy "Jezar" Wakefield, did exceptionally well was choose those, by experiment, to give an open and realistic sound for most control values and input signals. And then release clear code to the public domain—he's the Phil Zimmermann of reverberation.

Freeverb mixes stereo input to mono before processing but can still diffuse the output by using two sets of delay lines, left and right, with slightly longer ones on the right channel. The **stereoSpread** parameter defaults to add an additional 23 samples of delay in each line. Reducing it and combining some of the left/right outputs reduces the apparent width of the reverberation.

Ambisonic Reverberation

Such an approach gets costly in RAM and CPU time when working in surround. Even after ignoring the LFE and front-centre channels, a 7.1 channel implementation would need six times more RAM buffers and proportionately more CPU time. But Ambisonics can help here, as soon as the number of channels is more than two. We've seen how horizontal first-order Ambisonics represents sounds from any direction *in a plane* with just three channels; processing reverb that way halves the cost of implementing pantophonic reverb on a conventional 5.1 or 7.1 home cinema setup. One extra height channel gives fully 3D reverb, for 3D7.1, 8.0 cube or larger speaker arrays, at only half or two-thirds the cost of treating each speaker separately.

This is the approach taken by Derby University's AmbiFreeVerb. The first version, rather like Freeverb's stereo implementation, just decorrelated the W mono component of the input spatially; AmbiFreeVerb 2 uses all four B-Format channels to support directional reverb with separate left/right, front/rear and top/bottom dimensions.[14]

There's some loss of spatial acuity unless higher-order processing, with more channels, is performed, but given the nature of reverb we'd not expect pinpoint positioning in the wet component of the signal, so first-order reverb sounds directional enough, especially if higher-order processing is used for the direct path. Ambisonic techniques can be used similarly to reduce the cost of impulse-response reverb.

Reverberation by Convolution

If we were to generate a single very short click and then record all the times and levels at which it was reflected in a room, we'd end up with a linear map similar to Schroeder's ideal model of reverb. A short period of silence would precede a high spike for the first reflection, followed by more spikes corresponding to later, more distant reflections. The density of spikes would build up with time, as reflections were themselves reflected, but individual spike levels would decrease due to reflection and air attenuation losses.

This map of intensity with time is an impulse response and can be used to map the reflections of a real room onto any sample by FIR filtering, in a process known as convolution. It's a generalised version of the early-reflections modelled by a multi-tap delay, except that every sample interval from the first reflection to the end of the reverb tail must be considered, even if many of those intervals are empty, especially near the start.

Bill Gardner worked for Kurtzweil and Lexicon and wrote a great paper about fast convolution for the *AES Journal* while researching "perceptual computing" at MIT.[15] He points out that the "direct form" of convolution—multiplying each sample by the entire impulse response—is actually the fastest way for very short impulses, fewer than 100 samples in length, but as the reverb tail grows it becomes much quicker to convert the data from the time domain to the frequency domain, using a fast Fourier transform. His approach uses the direct form for the start of the impulse and progressively larger transforms later, all chosen to minimise overall delays and maximise the reuse of data.

Bill Gardner's paper explains how to eliminate the delay normally associated with conversions to and from the frequency domain, how to schedule the work across multiple CPU cores, how to manage the communication between those efficiently, plus speed-up tips. His way, a 3-second impulse response can be applied to a mono 44.1-kHz stream using a mere 427 FMA instructions per output sample (18,830,700 a second, mono, 37.7 MFLOPS) with a group delay of just 1.5 ms. The "direct form" of the same convolution would require over 300 times more FMAs, though it would be easier to parallelise.

There's a trade-off between efficiency and processing delay. If you're willing to convolve very large blocks and wait for the result, the number of FMAs per sample can be trimmed to 68—barely 1/2000 of the brute-force cost—but the 6-second turnaround makes this wholly unsuitable for real-time audio.

Compared with a synthetic reverb, convolution is relatively inflexible. You can alter the reverb level, but that's about it. Adding a variable delay line to the input allows the pre-delay to be adjusted, and hence the apparent size of the space—strictly speaking, the distance between source, reflector and listener—without having to convolve hundreds of zero samples. Ambisonics allow suitable impulse responses to be rotated smoothly around the listener, but the price you pay for the accuracy of IR reverberation is that there are not so many knobs for creative sound designers to twiddle.

Multi-Channel Convolution

Synthetic reverb can easily be spacialised and decorrelated between several speakers, but conventional convolution is strictly one channel in, one channel out. Most downloadable impulse responses are mono, and a realistic stereo one requires twice as much data, with separate responses for each channel. A 3-second stereo impulse response

Figure 21.5: Impulse response capture at Troller's Gill, Yorkshire, for openairlib.net

sampled at 48 kHz occupies well over a megabyte if stored in float format, the most efficient for further processing; unpacking it from Mu-law or short integers (scaled to unit range, to avoid stability problems) halves the footprint but slows down convolution.

When we started to use convolution seriously in games it soon became apparent that there was a severe shortage of 3D impulse responses. This led to a collaboration with Damian Murphy of the University of York Department of Electronic Engineering AudioLab, funded by the UK Arts and Humanities Research Council, and the creation of an online resource collating B-Format Ambisonic impulse responses, some specially created to meet the needs of games.[16]

Each of these impulse responses has four components, for the mono aspect and each 3D axis. To make a mono sound appear to be playing in that space, convolve it separately against all four components then decode the resultant channels like any other B-Format. To make the reverb turn with the listener, rotate the XYZ components before convolution, as explained in Chapter 19. This is a very cheap operation.

Many of the impulse responses are deliberately strongly directional, such as those made at the mouth of a tunnel or two sets made on a path through a Finnish forest, with open track in front and behind and trees either side, either in summer leaf or snowy winter, recorded in the same place but six months apart by Simon Shelley and Andrew Chadwick.

These and many other 3D impulse responses can be tested and downloaded from OpenAirLib. You're encouraged to share new ones under a Creative Commons license.

Spatial Editing of 3D Impulse Reverb

The directional soundfield manipulation technique explained at the end of Chapter 19 can also help make Ambisonic impulse response reverberation vary realistically with environmental changes. Let's imagine you're using one of the B-Format reverbs from the OpenAirLib which was captured along a forest path.

Typically you'll be oriented along the path and would expect directional reverberation from either side. But if mark-up or ray tracing indicate that you're about to pass some junction or cross-roads, the echoes from one side or both should momentarily be suppressed. This can be done by treating the B-Format reverberant output as if it was a pre-rendered soundfield.

Another runtime manipulation that can be applied to 3D impulse response reverb data is to vary the pre-delay, which involves no extra processing—just mix the wet signals in earlier or later. This determines the time between direct and reflected sound, and hence the perceived size of the reverberant space. It's only an approximation, and you may need to alter the reverb level at the same time, but it's a potentially neat trick for simulating a helicopter or rocket liftoff, especially in conjunction with directional manipulations.

Moving a source within that space is hard to fake with an impulse response, which captures one position of the listener relative to reflectors all around. In such circumstances consider cross-fading between soundfields, manipulating parts of the decoded speaker mix with directional delays or mixing synthetic and captured reverberation. The ideal approach depends upon the speed of the listener and relative sizes of the sources and spaces.

References

[1] *I3DL2*: www.iasig.org/wg/closed/icwg/21feb97.pdf
[2] *Virtual Audio Through Ray Tracing, Tom Zuddock; Dr. Dobb's Journal, ISSN 1044-789X, December 1996*
[3] *Reflector Source*: https://sourceforge.net/projects/thereflector/
[4] *Natural Sounding Artificial Reverberation; Manfred R. Schroeder, JAES, ISSN 1549-4950, Volume 10, Number 3, 1962, pages 219–223*: www.aes.org/e-lib/browse.cfm?elib=849
[5] *About This Reverberation Business; James Moorer, Computer Music Journal, ISSN 0148-9267, Volume 3 Number 2, 1979, pages 13–18*
[6] *Sceptre Xbox Bug*: https://randomascii.wordpress.com/2018/01/07/finding-a-cpu-design-bug-in-the-xbox-360
[7] *Loop Reverb History*: www.spinsemi.com/forum/viewtopic.php?t=3
[8] *Digital Delay Networks for Designing Artificial Reverberators; Jean-Marc Jot and Antoine Chaigne, 90th AES Convention, 1991*: www.aes.org/e-lib/browse.cfm?elib=5663
[9] *All-Pass Loop Reverb*: www.spinsemi.com/knowledge_base/effects.html#Reverberation
[10] *Crytek Reverb*: http://docs.cryengine.com/display/SDKDOC2/Using+Reverb+Volumes
[11] *Impulse Response Estimation for the Auralisation of Vehicle Engine Sounds using Dual Channel FFT Analysis; Simon Shelley, Damian Murphy and Simon N Goodwin, Proceedings of the Sound and Music Computing Conference 2013,*

Logos Verlag Berlin, ISBN 9783832534721: https://pure.york.ac.uk/portal/services/downloadRegister/25538798/Impulse_Response_Estimation_for_the_Auralisation_of_Vehicle_Engine_Sounds_using_Dual_Channel_FFT_Analysis.pdf

[12] *Freeverb Original Source*: http://stuff.bjornroche.com/freeverb.zip

[13] *Physical Audio Signal Processing; Julius O. Smith III, W3K Publishing 2010, ISBN 9780974560724*: http://ccrma.stanford.edu/~jos/pasp/Delay_Lines.html

[14] *AmbiFreeVerb 2, Development of a 3D Ambisonic Reverb with Spatial Warping and Variable Scattering; Bruce Wiggins and Mark Dring; AES International Conference on Sound Field Control (July 2016)*: www.aes.org/e-lib/browse.cfm?elib=18307

[15] *Efficient Convolution without Input-Output Delay; William G Gardner, JAES, ISSN 1549-4950, Volume 43 Number 3, March 1985*: www.aes.org/e-lib/browse.cfm?elib=7957

[16] *OpenAirLib Impulse Response Library*: www.openairlib.net

CHAPTER 22

Geometrical Interactions, Occlusion and Reflections

One of the unresolved challenges of interactive audio is the representation of reflecting and occluding objects in the simulated world. For decades it's been normal for game engines to support two sorts of 3D geometry, for graphics and physics, but neither of those is suited for audio. This chapter explains how the geometric needs of audio differ from those others and describes techniques to fill the gap.

Graphics Geometry

There are two main types of geometric object in a 3D world: *mobile* objects, like animals and vehicles, move around, while *static* objects are buildings or fixed parts of the terrain. Both types are authored in 3D graphics packages like Maya or 3D Studio Max and imported into a game in two parts—as a skeletal mesh of connected triangles which describe the outline of the object and as "maps" which cover the facets, superimposing colour images and surface details which improve lighting and smooth out the polygonal edges.

Audio can ignore the maps, needing only a representation of the reflective properties of each triangle to model acoustic reflections and occlusion. But there are far more triangles in a graphical object than we need for these purposes. The typical budget for a single character is about 100,000 triangles, racing game cars use around 250,000 polygons, and the total number of facets in a scene, rendered 60 times a second, is 10 million or more, including informational overlays, shadows, particles and similar decorations.

Such high-resolution meshes deliver finely detailed close-up graphics, using massively parallel rendering hardware, but they're gross overkill for audio. The computational expense of working out the interaction of each sound source and every potentially reflecting graphical facet far exceeds the real-time capability of PCs and consoles.

Physics Geometry

The same applies to detecting collisions between mobiles or between mobile and static objects. Physics systems use simplified object outline representations, just enough to stop objects intercepting or overlapping one another, and to report back which pairs are touching and the direction and intensity of their collision. Many visible objects are ignored, as long as they are or can be nudged within the bounds of another, usually larger one.

Audio runtimes take advantage of this information to play contact sounds. A collision will report the "materials" and a couple of sounds will be triggered—for instance, when a bike hits a post, wood and metal contact samples are played from the point of interception, with assets and volumes tailored according to the collision force and angle of incidence. If the angle is obtuse, a looping "scrape" sound might be chosen and updated in pitch and volume in subsequent frames while the contact continues.

Physics geometry is often used to model audio occlusion between game mobiles. In racing games, other cars nearby will reflect the sound of your own in varying ways as they try to pass. *RaceDriver Grid*'s audio system sent out a bundle of eight rays—two vertical, up and down, and six in a plane around the player—to detect which objects, static or mobile, were closest, and their direction. The vertical ray was called the "unicorn ray," as it was tilted forward to sense oncoming objects, such as bridges and gantries, a bit sooner.

Ray Tracing

Even eight rays are expensive, though some games use many more. The length of the rays was limited to save processing time by allowing the broad phase of the collision handler to sift out distant objects, and they were submitted together as a bundle to facilitate optimisations—it's quicker to handle a batch of rays from a common source than to test them individually.

Another big performance boost comes by not waiting for the results. Ray tracing is performed outside the main update thread of the game by separate threads which run asynchronously, taking advantage of specialist co-processors or multiple processor cores.

Even if the contact results arrive a frame late, that's soon enough to select and play suitable sounds without the player noticing the lag, and such relatively low-priority queries help to balance the processing load and maintain a steady frame rate—if there's time to spare, the results may arrive promptly, but if the system is fully loaded the audio work may be bumped off to the next frame, smoothing out the spike in demand.

It's practical to use ray tracing for the main player object, but unless you have a very small world, a super-fast processor or little else going on, it'll be unacceptable to cast rays from every source to every listener, every frame, just to find out what might be in the way.

There are other problems with using collision geometry for sound, which vary depending upon the type of game. A surface might obstruct sound but not bullets, so it lacks collision data in a shooting game. Racetrack designers don't mark up every trackside object but only the ones which are likely to get in the way of cars or large bits like tyres and bumpers.

An Armco crash barrier only needs to be a metre high to stop a car but allows sounds to pass above and below. Objects further from the road might still reflect sounds but won't have collision physics because it's expensive and not needed for anything except audio.

Some collidable objects, like fenceposts or wire mesh, might not obstruct audio at all. We don't expect to hear reflections from those, even if they're physical obstructions. Conversely, sound waves propagate around corners, so even if a path is blocked to light some audio leaks through.

Audio needs to know the reflectivity of a surface—a grassy knoll reflects less than a glassy shopfront—while level designers are usually happy just knowing if there's an obstruction. Getting the correct surface properties marked up and checking them becomes a big burden for the audio team. Even if the physics, graphics or level design teams are given a way to tag obstructions with their audio properties, they may lack the skill or the will to take on this unfamiliar responsibility.

Only two values are needed—one expressing the reflectivity as a proportion and another controlling the timbre of reflections, perhaps moderated by the angle of incidence. This could be an index into a table of materials and appropriate filter pre-sets or a direct tone control setting attenuation at a certain frequency. The indirect table approach is preferred—it allows each surface type to have arbitrary properties, such as a frequency response curve, and makes it easy to tweak all surfaces of a given type without selecting them individually. Often there's an existing materials database which can be used to select suitable "wood" or "metal" samples and extended to include audio reflectivity.

It's one thing including support for audio properties but quite another to get the artists and level designers to use it. It's often left set to the default or whatever the previous object required. It's common for such mistakes to go unnoticed until the object is in a busy simulation, and then it's hard to work out exactly what to fix, assuming someone notices the problem. Compared with sliding through a tree or falling through the floor, audio reflections are a minor concern. Even though the correct handling of occlusion is as important to sound as correct shading is to graphics, it's less immediately obvious.

Audio Geometry

Ray tracing is essential for mobile interactions, but neither graphic or collision meshes give the information we need for realistic volumetric audio. The obvious solution is to add a third class of geometry, in the form of "sound area meshes," to fill the gaps—or leave them open if they're porous to audio! A mesh is a group of spatially distributed control points with properties which can be interpolated for any position therein.

The best implementation of this concept I've yet used and heard was in the Saracen 2 world editor and game engine, implemented by Rob Baker at Attention to Detail for games like *Lego Drome Racers*. Rob took a system originally intended for lighting and adapted it to model the occlusion of volumetric sounds like alarms and the distribution and tone of reverberations.

The same codebase supported two types of audio mesh, as well as lighting effects. Meshes could just as well be used for any volumetric parametric control, e.g. occlusion. **Sound Area Meshes** determine where particular sounds may be heard, while **Effect Area Meshes** control standard I3DL2 reverb parameters, smoothly morphing between control values as the listener moved within and between the meshes.

Both these were drawn into the environment using a special PC version of the game which incorporated a "world editor" that allowed 3D objects to be added and moved around. Meshes are visualised as semi-transparent volumetric overlays, so they can be seen in context. Saracen used a pre-set alpha-blend, but in principle the degree of transparency can be varied to indicate the intensity of sound or reverberation, tailing off at the edges.

The illustration shows a mesh placed in a courtyard, used both to trigger and control the volume of an alarm sound played therein. The menu overlay shows the mesh properties, such as its location, visualisation and which sound sample or stream to play. Each corner of the mesh had properties associated with it—for a playing sound it could be as simple as the required volume at that point, whereas for reverb it might set early and late reflection levels or any of the other I3DL2 properties—with the proviso that not all of those could be changed at runtime without audible glitches. Overlapping cross-fades were sometimes needed to hide these.

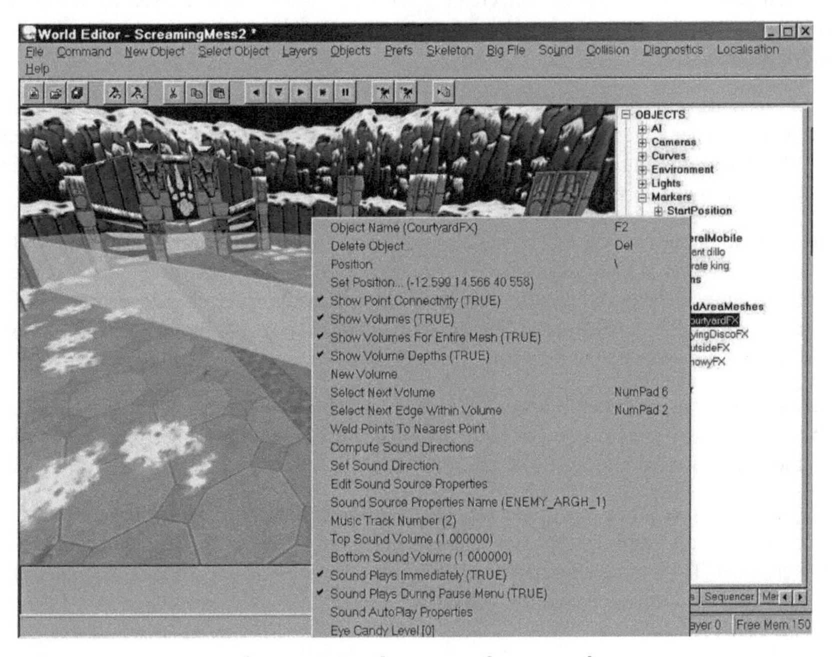

Figure 22.1: Alarm sound area mesh

At runtime the game calculates which triangle of the mesh the listener is inside and interpolates between those properties to get a representative set for that position. The simplest mesh has one full-on point in the middle and a ring of at least three points round the outside with controls set to mute the sound or effect. As the listener enters the ring and moves towards the middle, the sound varies accordingly.

Control Points

At first this does no more than create a sphere at MaxDistance around a source, albeit with a potentially irregular outline, and lacks the flexibility of custom rolloff curves. But as control points are added, it becomes far more capable. An intermediate ring of points allows the inner and outer volume curves to vary. Additional points approximate any curve, allowing different shapes depending upon location as well as distance—sound could fade off at the mouth of a cave, or a sheltered area could have a quiet point within it and varying levels around that corresponding to the occlusion caused by static geometry.

If that shelter is damaged, the mesh can be edited or replaced in the same way that a pristine building might be replaced with a shattered one by swapping physics and graphics geometry, usually under the cover of a cloud of smoke—or the audible equivalent, a big explosion. Editing is more powerful, just like deformable geometry, but a lot more expensive at runtime, which is why both techniques are commonly used.

Rob Baker explains:

> *Although the meshes were designed in 3D so the level designer could position them around the required objects for reference, the actual processing was all done in 2D with the vertical component ignored. Of course, this greatly simplified the processing as we were only dealing with 2D triangles rather than potentially complex 3D volumes which would have been a nightmare to author at a design level anyway. Because our race tracks were all essentially just 2D with no vertically overlapping areas this was never really a design limitation.*

Irregular meshes can be applied to reverberation, creating echoey pockets in an otherwise open world. Meshes can also be used to suppress sounds from outside—they're a general way to describe the properties of the space between objects, acoustic or otherwise.

Mesh control points can also embed directional information. To take a simple example in stereo, if the control points to the left of the listener have higher volume than those to the right, the sound can be panned left proportionately. This approach can be extended to surround by setting a suitable 2D or 3D panning vector. Since the sound is distributed around the mesh, Doppler should be disabled.

Hand-authoring of audio meshes is a big job, especially as designers are prone to move objects around at any point in development of a game, even at a late stage when they might be deliberately obstructing parts of the scene to keep a consistent frame rate. Testing and re-authoring audio metadata after such changes is costly and error prone. But just as "static lighting" offline processes can rebuild light and shadow maps automatically, similar systems can regenerate audio occlusion meshes before the game is tested.

It doesn't matter that they're using fine-grained graphics geometry, because the process doesn't need to run in real time. Indeed, it can be distributed across several build systems, which might be doing similar rework for the physics system. Most of the code you need may already exist.

It's wise for such processes to generate exactly the same data as manual markup, so the output can be tweaked by hand later if necessary. Write a simple system that catches the bulk of the common cases and can be tweaked later just where it matters most. That's more efficient than relying on an automatic one which aims to detect and handle all the possible edge-cases and leaves no escape mechanism.

Cheap Shapes

If you can't afford full-blown audio geometry or don't always need its flexibility, you may get by with an arrangement of simpler shapes, such as spheres, boxes and capsules, to control environmental audio. *DiRT* uses trigger boxes to time its speech, with adjustments for the player's speed. *Grand Theft Auto 5* uses almost a thousand spheres and arbitrarily shaped boxes to delimit its ambience zones. *GTA4* had only 87 such zones, marked up as axis-aligned boxes. These are quicker to process at runtime because there's no need to rotate them, but they're harder to author.

Spheres don't need rotating, which explains why pool and snooker games were amongst the first to go into 3D. Around 200 trigger spheres control the crowds, tunnels and local ambiences in *Colin McRae Rally* mobile. Freelance designer Tom McCaren placed these along the tracks for me, using a custom extension to the Unity editor.

I'll leave the last word, for now, to Rob Baker:

I think it's very true what you say about audio taking a back seat in game dev. While graphics march ever stridently forward I can see now that what we were doing fifteen years ago was considerably more progressive than what we're seeing even today. Remembering the sound tech that was developed at ATD and the Eagle SDK that was around then for geometric audio, it seems almost unbelievable that in 2017 in the world's premiere game SDK (Unreal Engine 4) which I work with now, there's no such notion of these things and we're back to spheres again, for the most part. What happened? It seems like a nonsense that we've actually regressed over the intervening decade or two.

CHAPTER 23

Audio Outputs and Endpoints

We've seen how the number of output channels expected from an audio system has increased from 1, mono, in the earliest days, via stereo, to 5.1 or 7.1 channel surround sound. Some of those surround channels are used for positional sounds, and some—like the front centre and LFE speakers—are effectively mono.

It follows that designers can get positional audio out of four or six speakers in a 5.1 or 7.1 rig, but some sounds are not panned at all, such as low-frequency effects and non-diegetic sounds, often including dialogue, customarily routed to the front centre speaker.

Just as we need separate, typically mono or stereo, outputs from each voice to the reverb units so we can give them varying amounts of reverberation consistent with their position, it's useful to have another mono output from each voice, directed to the optional LFE channel. This bypasses the panner, as there's only one LFE in consumer surround setups—when multiple LFEs are installed in large venues, that's done to reduce the effect of room resonances rather than to allow sounds to be panned between them. Each of these additional outputs or "sends" needs its own volume control, defaulting to muted.

Another mono channel is implemented for front-centre sounds, but this doesn't generally have its own volume control as the panner is bypassed and the front centre output selected rather than the surround array because of the lack of a Listener, which marks such sounds out as non-diegetic.

Synthetic Front Centre

This output is intended for the dialogue speaker in a surround system, but even if there's no such speaker it can usefully be made to stand out from a spatialised stereo mix by routing it through a very short delay. This trick, discovered in the 1950s, can make it sound separated and elevated from the other sounds in the mix; Bruce Wiggins of Derby University explains this perception by considering the two paths from a relatively highly placed sound to a less elevated listener.

The direct wave arrives first, followed a few milliseconds later by a wave reflected off the ground. The amount of delay depends upon the additional path length. For a source and listener 16 feet apart, both 6 feet off the ground, the shortest reflecting path will be 25% longer than the direct path, and it will arrive 4 ms later, at a slightly lower intensity.

It follows that adding a single delay, too short to be perceived as an echo, can make good use of the non-diegetic voice property to differentiate non-positional sounds from positional ones which happen to be in the "phantom centre" of a stereo soundfield. If you try this, remember to adjust the mixed output level to account for the added signal—comb-filtering at mid and high frequencies means the total gain will be less than the +6 dB boost introduced by direct superimposition, characteristic of still, dry air and a perfectly reflective floor.

A better way to maintain balance and minimise tonal artefacts is to add the delayed signal at a substantially lower level, perhaps 10 dB below the direct path—this is still enough to get the separation effect. You may also like to experiment with the delay time to suit the scale and acoustic properties of your world. A surface-dependent low-pass filter should not be necessary, as this sound is not meant to bed in to the 3D mix, so its timbre should be location-independent.

Controller Speakers

The challenge, these days, is more often to cope properly with extra speakers rather than synthesise missing ones.

Since Nintendo introduced a tiny speaker in each Wii remote controller, it's been possible to play short samples, or long streams on later systems, directly to individual players of a game. Sony followed by adding an individually addressable mono speaker in each PS4 controller and a stereo headphone socket. Headsets can be configured to relay just network voice chat or a copy of the game mix as well.

The original Xbox One controllers had no audio capability except via an add-on headphone adapter mainly intended for voice chat, but the controller was updated in June 2015 to add an audio socket.

The Wiimote speaker has rendered the sound of *Zelda*'s bow, collecting bricks in *Lego Star Wars* and player-directed speech like phone calls and pit radio in several games. It fits whenever a sound is specific to a player—related to something they've done, a message just for them, or something hitting their avatar, quite likely associated with a sub-sonic force-feedback jolt. Sometimes you'll also play it, at a lower level, in the main mix—it depends how private the sound is considered to be. Separate master-level and per-output send level adjustments give you the freedom to design this.

Even in single-player games it's fun to use the controller speaker for player-local sounds. These may include network chat and co-driver comments—it helps to know if speech is meant for you in multi-player games, and multiple endpoints are another solution to the challenges of delivering split-screen audio.

Multiple Endpoints

Sony's PlayStation 3 introduced a "secondary audio output" option. This uses analogue stereo or optical connectors to deliver a mix potentially distinct from the HDMI and Bluetooth headphone support. Nintendo's Wii-U introduced a larger controller, known as the GamePad, with its own screen and stereo speakers, so audio could be arbitrarily routed to that or to the console's HDMI output.

Additional mono or stereo endpoint sends allow all or any subset of a mix to be routed to a headset or controller independently of the main mix. The same code used to implement the non-positional front centre and LFE channels can generate a mono feed for each controller—games support up to four—or headset. These can be implemented as stereo sends if a pan control is added, as for stereo reverb, sidechain and DSP effects. It's up to you to decide how far to take this; platform vendors look kindly on cross-platform titles which support their unique extras, and this gives extra leverage when haggling over publishing deals and co-marketing.

In short, you can't assume that your game will generate a single mix. Even a VR demo might require two mixes: a first-person HRTF stereo headphone mix for the player and a separate speaker mix, in anything from mono to full surround, for observers and players waiting their turn. Such capability is common across platforms and should be supported by a competitive cross-platform audio engine.

Brick Wall Filters

Our preferred implementation performs all the mixing at 96 kHz so that we can pitch-shift 48-kHz samples by up to an octave without risk of aliasing distortion and play 96-kHz music unmediated. If our endpoint is HDMI we can probably output the final mix directly, since the HDMI specification directly supports both 96 and 192-kHz audio. But if the output is a 48-kHz DAC or optical TOSLINK, we'll need to downsample our mix to filter out frequencies above 24 kHz, the corresponding Nyquist limit, before we write it out.

If we just decimated the 96 kHz data to 48 kHz by skipping alternate samples (2:1 decimation) we'd introduce audible artefacts. So-called "brick wall" filtering smooths the output before decimation to suppress frequencies higher than a 48-kHz stream can represent.

There are the usual two ways to implement this feature in the time domain—IIR feedback filters and convolution-style FIRs. The IIR approach involves fewer mathematical operations but is harder to optimise, as each stage must wait on output from the previous one. The IIR approach is relatively fastest on a processor with a short pipeline—if you must wait more than half a dozen cycles for the result of an FMA operation, use FIRs instead.

The IIR filter involves less processing delay but affects the phase of signals passed through it in a frequency-dependent way, while the FIR filter is linear phase. Either way, the quality of the results—in terms of the amount of aliasing noise rejected—depends upon the length of the chosen filter. The longer the filter, the lower the noise.

Range Adjustments

As we recommend setting the "noise floor" below which samples are considered effectively silent somewhere in the -40..-60 dB range, 48 dB of aliasing noise rejection seems a sensible upper target. Shorter filters mean less work but more noise, so we'll consider simpler ones with a shallower slope or only 24 dB rejection for when processing time is scarce.

Note that filters like this may momentarily overshoot, generating output outside the original input range. You may wish to scale the input to avoid this or apply clamps as shown later, especially if your final output is an integer rather than floating-point DAC.

Anti-Aliasing with IIR Filters

The following function demonstrates a fourth-order IIR filter which processes BLOCK_SIZE input samples from **pIn** and outputs half as many at **pOut**. The six static floats with names starting with **w** hold history, so they should be separately maintained for each channel to be processed. The suffix **a** denotes the current sample; **b** and **c** are previous ones, in stages **1** and **2**, respectively. The loop has been unrolled to consume two samples and write one at each pass, eliminating four redundant operations which would condition the skipped output value. The asserts make sure that BLOCK_SIZE is even and non-zero. Substitute static_asserts if it's also constant.

```
void LowPassOrder4(float pIn[], float pOut[])
{
  assert(BLOCK_SIZE);
  assert((BLOCK_SIZE & 1) ==0);
  static float w1a = 0.f; static float w1b = 0.f;
  static float w1c = 0.f; static float w2a = 0.f;
  static float w2b = 0.f; static float w2c = 0.f;
  const float b = 0.33784654f;

  for (unsigned n=0; n<BLOCK_SIZE; n+=2)
  {
    w1c = w1b; w1b = w1a;
    w1a = (w1b * -0.32906416f) - (w1c * 0.15980034f) + *pIn++;
    float acc = (w1b * 0.63141218f) + ((w1a + w1c) * b);
    w2c = w2b; w2b = w2a;
    w2a = (w2b * 0.14352094f) - (w2c * 0.69742181f) + acc;
    w1c = w1b; w1b = w1a;
    w1a = (w1b * -0.32906416f) - (w1c * 0.15980034f) + *pIn++;
    acc = (w1b * 0.63141218f) + ((w1a + w1c) * b);
```

```
        w2c = w2b; w2b = w2a;
        w2a = (w2b * 0.14352094f) - (w2c * 0.69742181f) + acc;
        *pOut++ = ((w2a + w2c) * b) + (w2b * 0.46769287f);
    }
}
```

Data dependencies limit this function's speed; for instance, there's only one multiplication between the lines where **w1a** is calculated and the points where it's needed. As the graph shows, this is not a steep filter and lets quite a lot of signal through at frequencies up to 35 kHz. These will be aliased down to 13 kHz and are likely to be audible in isolation, though other sounds may mask them. The graph also shows an intermediate sixth-order version, which will certainly sound better than the fourth-order one.

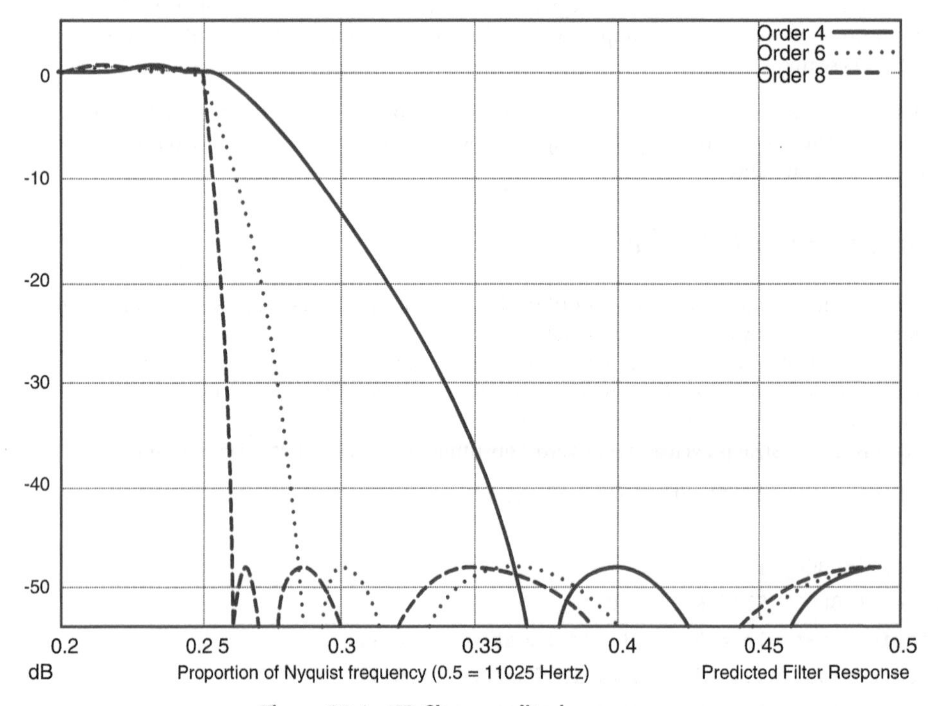

Figure 23.1: IIR filter amplitude responses

Pitch changes such as those caused by Doppler tend to expose such leaks, as the aliased signal falls in pitch as the source ascends. The brain is quick to spot such patterns. To make the IIR acceptably steep we need to add more stages and delays. The following code implements an eighth-order IIR filter which blocks frequencies above 25 kHz. An ideal arrangement would set the transition band a little lower, but in practice this works well enough; any aliasing is reflected down to the inaudible range 23..24 kHz and filters downstream will further suppress this. Here's the steep-slope eighth-order implementation, in full:

```
void LowPassOrder8(float pIn[], float pOut[])
{
  assert(BLOCK_SIZE);
  assert((BLOCK_SIZE & 1) ==0);
```

```
    static float w1a = 0.f; static float w1b = 0.f;
    static float w1c = 0.f; static float w2a = 0.f;
    static float w2b = 0.f; static float w2c = 0.f;
    static float w3a = 0.f; static float w3b = 0.f;
    static float w3c = 0.f; static float w4a = 0.f;
    static float w4b = 0.f; static float w4c = 0.f;
    const float b = 0.47005698f;

  for (unsigned n=0; n<BLOCK_SIZE; ++n)
  {
    w1c = w1b; w1b = w1a;
    w1a = *pIn++ - (w1b * -0.66792155f) - (w1c * 0.21867144f);
    float acc = ((w1a + w1c) * b) + (w1b * 0.80899057f);
    w2c = w2b; w2b = w2a;
    w2a = acc - (w2b * -0.30681346f) - (w2c * 0.60075078f);
    acc = ((w2a + w2c) * b) + (w2b * 0.34740806f);
    w3c = w3b; w3b = w3a;
    w3a = acc - (w3b * -0.069345591f) - (w3c * 0.85710198f);
    acc = ((w3a + w3c) * b) + (w3b * 0.13816219f);
    w4c = w4b; w4b = w4a;
    w4a = acc - (w4b * 0.015168767f) - (w4c * 0.96656985f);
    if (n & 1) // Output only alternate samples
    {
      *pOut++ = ((w4a + w4c) * b) + (w4b * 0.075951890f);
    }
  }
}
```

The rhythm-action game *Dance Factory* uses a similar algorithm in MIPS assembly language. To exploit the RISC architecture and avoid slow access to variables in RAM, my hand-optimised loop uses almost all the 32 available floating-point registers: 13 constants, 12 for delays and 5 for temporary results.

Anti-Aliasing with FIR Filters

Equivalent FIR filters can use direct convolution, as outlined in the section of Chapter 18 about HRTF panning. Unless you're mixing in the frequency domain anyway (don't forget the overlapped windowing!) it's not worth using FFT techniques by clearing high-frequency output bins before applying the iFFT to recover time-domain samples, as FIR low-pass filters that meet our specification are quite short. Here's the longest:

```
static float decimate48[45] =
{
  -0.0045475421f, -0.0022454010f,  0.0074317267f,
   0.0143589178f,  0.0065254288f, -0.0072311130f,
  -0.0056379248f,  0.0101162899f,  0.0125205898f,
```

```
    -0.0076429952f, -0.0185166744f,  0.0036046886f,
     0.0259228157f,  0.0048146027f, -0.0331057409f,
    -0.0188528516f,  0.0397183621f,  0.0429861476f,
    -0.0449985036f, -0.0929812900f,  0.0484192693f,
     0.3138084759f,  0.4503897057f,  0.3138084759f,
     0.0484192693f, -0.0929812900f, -0.0449985036f,
     0.0429861476f,  0.0397183621f, -0.0188528516f,
    -0.0331057409f,  0.0048146027f,  0.0259228157f,
     0.0036046886f, -0.0185166744f, -0.0076429952f,
     0.0125205898f,  0.0101162899f, -0.0056379248f,
    -0.0072311130f,  0.0065254288f,  0.0143589178f,
     0.0074317267f, -0.0022454010f, -0.0045475421f
};
```

That uses 45 elements to produce a frequency response flat within 0.7 dB from 20 to 20,000 Hertz, with at least 48.8 dB rejection for frequencies above 24 kHz.

The shorter one uses 29 elements and has at least 26.3 dB rejection, as follows:

```
static float decimate26[29] =
{
     0.0256327664f,  0.0007508599f, -0.0215196900f,
    -0.0000919109f,  0.0198065999f,  0.0149294605f,
    -0.0310143276f, -0.0226925755f,  0.0304790255f,
     0.0504837306f, -0.0390568930f, -0.0940461168f,
     0.0370495612f,  0.3170809594f,  0.4578110500f,
     0.3170809594f,  0.0370495612f, -0.0940461168f,
    -0.0390568930f,  0.0504837307f,  0.0304790255f,
    -0.0226925755f, -0.0310143276f,  0.0149294605f,
     0.0198065999f, -0.0000919109f, -0.0215196900f,
     0.0007508599f,  0.0256327664f
};
```

These were both designed using an online FIR filter calculator that uses the Parks–McClellan algorithm to optimise Chebyshev FIR filter coefficients.[1] You just enter the input sample rate, the range of pass-band and stop-band frequencies in Hertz and the maximum ripple you can tolerate in each band. The calculator works out the minimum number of coefficients needed to deliver that performance. It even generates implementation functions in C.

After applying either filter, 48-kHz output is derived by plucking out alternate samples of the result for each output channel. The FIR filter is easiest to optimise if the data is packed for each channel, while the IIR version can take advantage of SIMD vector operations if the input is interleaved in stereo pairs or, ideally, frames of four or eight contemporaneous samples, so it can process the channels in parallel.

48 kHz is the most common consumer output sample rate, but the same calculator can readily design filters for 32- or 44.1-kHz output if needed. It's not limited to simple integer ratios. A similar calculator for IIR filters is online.[2]

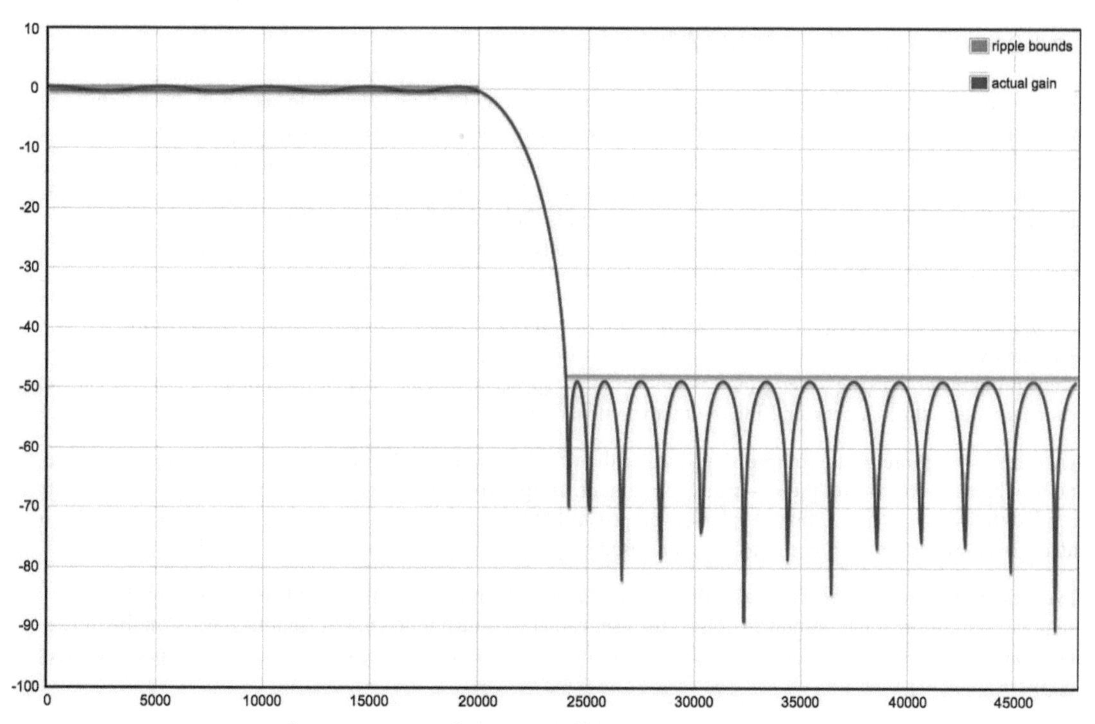

Figure 23.2: 45-element FIR filter amplitude response

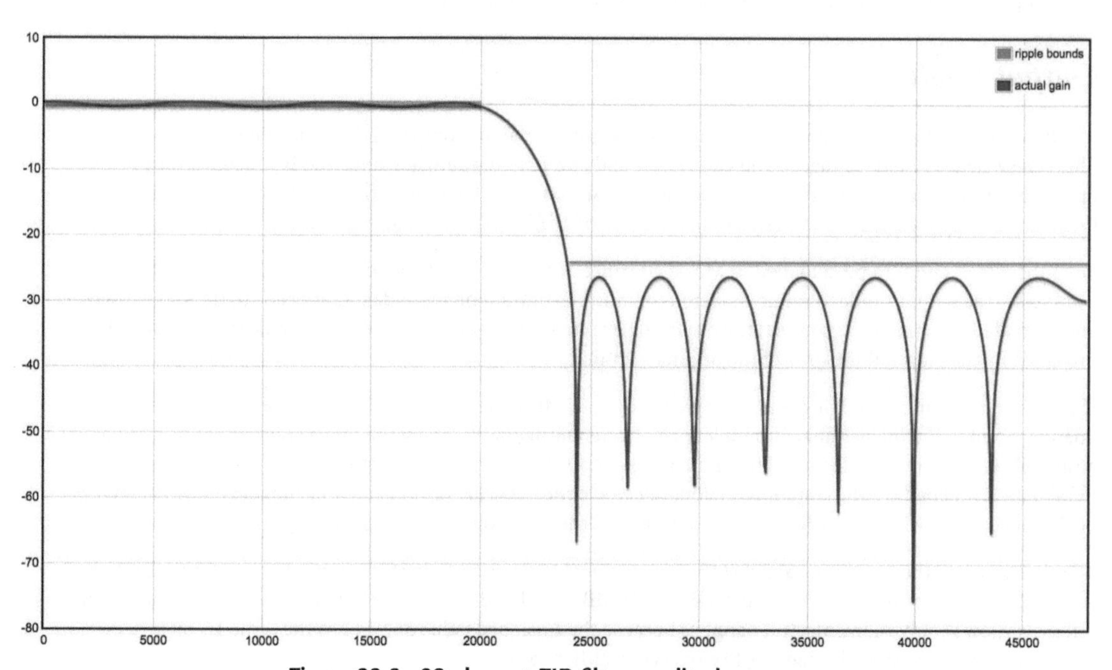

Figure 23.3: 29-element FIR filter amplitude response

Compression and Loudness Metering

Compressors adjust the mix level dynamically, metering output and setting volumes to keep it in check. Limiters prevent it exceeding a certain level. The difference is what happens close to the limit—a compressor continues to allow the level to rise, but at a reduced rate, above a set threshold. Low-pass filters on the controlling signal determine the Attack and Delay times, in a simpler version of the voice ADSR envelope. A fast attack traps onset distortion but may limit the impact of sudden loud sounds. Release times are typically a fraction of a second, fast enough not to hide later sounds but not so quick as to reveal audible "pumping"—unless that's an effect you deliberately want, like a 1960s pop mix.

Compression is a key part of the passive media sound, so you might want it in your game for that reason, but we've seen that dynamic control is often implemented more directly in interactive media by suppressing sound triggers at key moments or ducking lesser groups when you know speech is playing. Whole-mix compression is a relatively crude technique, though peak limiting is vital to control digital distortion.

Sub-mixes allow more subtle compression—rather than compress the whole mix, let key parts through unaltered and use their level to control the less-crucial remainder. Output from one group may be used as a sidechain control, rectified to RMS (root mean square) values, averaged and then used to control another group. Filters in the audio sidechain allow certain frequencies to be emphasised or ignored.

Given the limited headroom of digital audio, ideally the outputs from your title should be as loud as you can get them without distortion, to minimise noise—the user can always turn it down. But "loudness wars" in the music and broadcast industries have led to standards to minimise level jumps between films, music and especially adverts, and these apply to games too.

It's recommended, and sometimes required, that the loudness and dynamic range of games should match the ITU recommendation BS1770.[3] Some modern console firmware automatically computes logarithmic "loudness units" for relative comparisons and absolute LUFS measurements over time.[4] Otherwise, the references explain how to make appropriate measurements of your output and tweak it accordingly.

Output Capture

It's often useful to be able to capture the output of the mixer as an uncompressed multichannel WAV or AIFF file. This can be handy when testing, so you can review and compare output later, and vital if you allow the player to upload exciting replays to YouTube or similar sharing sites, where it will help to promote the title as well as give them extra bragging rights. This was a popular feature of *DiRT3* and *Showdown* games.

YouTube allows audio to be uploaded in Mu-law format, which can be encoded from PCM very quickly. You may want to set the output mode to mono or stereo to limit the amount of audio data that needs to be compressed and uploaded and maximise compatibility with the eventual replay system, which need not be the same as that preferred by the player at the time of the original capture.

If you're doing your own mixing this is simple enough. Capture the output of the mixer then convert it from floating-point to the integer upload sample format, attenuating or clamping the range appropriately. If you are absolutely sure the output floats will be in the range +1..-1 you can simply multiply them by 32,767 to get 16-bit PCM data for upload or Mu-law compression, but woe betide if values go outside this range, even momentarily—a peak value of +1.1 will map to 36,044, which is beyond the range of a signed 16-bit integer.

The overflow into the sign bit means that this excessively large positive offset is interpreted as a large negative value, -29,492. The result is a very loud click, over 90 dB, which will ring in the listeners' ears for seconds even if the overflow is only momentary. It happens because the unsigned value 36,044 has the same 16-bit binary representation as the signed value -29,492: 0x8ccc in either case.

A problem like this was embedded in the developer video capture option on one of the main consoles—normal output was mixed to 18-bit format for the hardware DAC, but only the low-order 16 bits were output in the capture. Quiet recordings were undistorted, but anything within 12 dB of maximum level wrapped around from top to bottom of the 16-bit range, causing gross distortion. The quick fix was to temporarily adjust all the Group volumes down by 12 dB to prevent the overflow. This could be done without code modifications, thanks to the "exposure" system primarily intended to allow designers to save and restore mix pre-sets.

In capture code, this fix corresponds to scaling the output by 8191 rather than 32767 when converting from floating-point to 16 bits. Later I incorporated this and a check for out-of-range values. As it's a mistake to equate the unlikely with the impossible, especially for something under customer control in your absence, it's still wise to clamp the possible range of results to eliminate the possibility of wrap-around—a momentary flat top on a wave is less aurally intrusive than a single sample spike between opposite ends of the maximum output range.

This is especially relevant on ARM32 platforms, where integer arithmetic tends to be the fastest choice. Use integers for mixing and metering on ARM32 but floating point for filters and reverb. On later ARM32 chips, with NEON, consider using the fp16 "half float" format, though that's not useful on some ARM64s. Most other platforms are quick enough at floating point that it's best to stay in that format throughout—indeed, floating-point DAC output is increasingly common, though the supply voltage still limits headroom.

This code maps the float sample value **f** between -4 and +4, allowing 12 dB of headroom beyond the usual +1..-1, to **j** in the 16-bit signed integer range and reports—once only—if the clamp limits are exceeded:

```
static bool moaned=false;
int j = (f * 8191.f) + 0.5f;
bool clamped=false;

if (j>32767)
{
  j = 32767;
  clamped = true;
}
else if (j<-32768)
{
  j = -32768;
  clamped = true;
}

if (clamped && !moaned)
{
  Moan("Sample value %f clamped to 16-bit range", f);
  moaned = true;
}
```

When embedding this in a class **moaned** should be declared as a private static member in the class header. This example also assumes that sizeof(int) is 3 or more, so there's room for the temporarily overflowed value to be accurately represented. A cast should be used to squeeze **j** into a 16-bit datatype, which may be called "short"—check your compiler options and documentation for appropriate datatypes and consider adding assertions to check this at compile time, e.g. this snippet for C++11 or later:

```
static_assert(sizeof(short)==2, "16-bit datatype needed");
static_assert(sizeof(int)>=2, "larger datatype needed");
```

This range clamp looks trivial, but it's important. One designer asked me to reimplement the "nice compression on the Xbox 360" on the PS3, which has a floating-point DAC output. I soon discovered the Xbox uses a simple 18-bit integer clamp! If reporting is not required, vector min and max or "saturating arithmetic" instructions,[5] or the branchless SIMD selection used in FastSinCos(), are effective ways to clamp integer or float output buffers in parallel.

Mobile Phone Latency

For years Android phones had exceptionally long audio output latency, generally over 100 ms and sometimes two or three times more, which degraded their interactivity.[6] Contrast this with typical Apple mobile devices' audio latency of just 10 ms, effectively an instant response.[6]

From Android 5 onwards Google added improvements and optional mechanisms to limit this, though there are still Android 7 devices with latency over 200 ms, introducing an obvious delay between screen and sound, precluding VR and making rhythm games almost unplayable. New Androids capable of sub-50 ms latency expose PackageManager.FEATURE_AUDIO_LOW_LATENCY, while the rarer FEATURE_AUDIO_PRO indicates a continuous round-trip latency of 20 ms or less.[7]

To minimise jitter and the need for intermediate buffering, your output buffer sample rate and size should match the AudioManager.PROPERTY_OUTPUT_SAMPLE_RATE and PROPERTY_OUTPUT_FRAMES_PER_BUFFER values.

Cinematic Lag

Before HDMI there were several attempts to squash 5.1 surround-sound channels down pipes made for stereo, including DTS and Dolby AC3 compressed encoding. Those old codecs were designed for passive video consumers rather than interactive applications and often introduce more delay than all other parts of the signal chain put together.

It's easy enough for a passive video player to preserve sync between pre-baked sound and pictures, from a film or TV programme, by delaying the display a few frames (not that they always get this right, as lip-sync slips attest), but it's unacceptable to do this for a real-time game or VR application! The speed and regularity of every step in the input-output loop—from reading the controller axes and buttons through interpreting them, triggering appropriate responses and getting them back out to the player—is vital to the sense of control and immersion.

The extra 85 milliseconds of encoder delay, plus a similar amount in the decoder at the receiver/amplifier, is likely to be the greatest source of latency in a 5.1 channel home cinema setup repurposed for PC or console gaming. If team members or customers complain that the sound is noticeably lagging the rev counter in a driving game or nearby gunshot sounds lag behind muzzle-flashes, the first thing to check is if they're playing in stereo or surround. If the problem goes away, along with rear-channel output, when they switch from 5.1 to 2.0 channels, the delay is due to the encoder used to squash six channels into hardware—typically a TOSLINK optical connection—only designed for two.

At the time of writing this potential snag affects all Microsoft Xboxes (2001–2018) and original Sony PS4 hardware. The Wii-U only supports uncompressed PCM 5.1 via its HDMI connector—no bitstream 5.1 or, sadly, uncompressed 7.1. Earlier Nintendo consoles are limited to stereo or analogue matrixed surround sound, similar to

Pro Logic or SRS 5.1 (an FMOD option) and compatible with Dolby's analogue surround or even simpler Hafler-style decoders but without the licensing overhead.

HDMI Tuning

The "slim" version of Sony's PS4 reduces the risk of this by eliminating the optical output connector, but even if the connection to the receiver/amplifier is over HDMI it may carry either a compressed bitstream in the slot reserved for 16-bit stereo or eight channels of uncompressed 24-bit linear audio—six times as much data, even before allowing for higher sample rates. Even if you can't hear the quality difference you'll almost certainly notice the reduced latency.

HDMI is potentially great news for interactive surround sound—its uncompressed 24-bit eight-channel replay capability, at up to 192 kHz, gives us 24 times the audio bandwidth of previous coax and TOSLINK consumer digital audio connections. But its legacy support for 1990s cinema compression technology means consumers may not get the benefit, even if their receiver supports uncompressed low-latency audio, like all since HDMI 1.3.

Backward compatibility with old DVDs requires that even modern HDMI receivers include optional compressed 5.1 channel so-called bitstream support, and if your console is configured that way it will be doing extra work, shedding quality and adding latency at both ends of the HDMI cable. To fix this, go to the Xbox console Settings menu, select All Settings, then Display & Sound, then Audio Output (sheesh!) and under DIGITAL AUDIO make sure HDMI audio is set to either 5.1 uncompressed or 7.1 uncompressed depending upon the extent of your speaker collection. Avoid the bitstream option unless you have a really ancient (HDMI 1.2 or earlier) receiver.

A similar dance is required on HDMI PlayStations. This won't help customers with ancient receivers; some sound designers still fall into that category, as a decent 7.1 setup is a substantial expense for most studios, often delivered to QA and producers long before the lower reaches of the audio department. But at least it explains the problem and gives them a choice of solution: revert to stereo or analogue surround, accept the lag, or get a modern HDMI receiver.

References

[1] *FIR Filter Calculator*: http://t-filter.engineerjs.com
[2] *IIR Filter Calculator*: https://www-users.cs.york.ac.uk/~fisher/mkfilter/
[3] *ITU-R-BS.1770: Algorithms to Measure Audio Programme Loudness and True-Peak Audio Level BS Series, Technical Report, International Telecommunications Union, 2006*: www.itu.int/dms_pubrec/itu-r/rec/bs/R-REC-BS.1770-4-201510-I!!PDF-E.pdf
[4] *EBU Loudness and Dynamic Range Documents*: http://tech.ebu.ch/loudness/
[5] *Saturating Arithmetic*: https://locklessinc.com/articles/sat_arithmetic/
[6] *Tablet and Phone Latencies*: http://superpowered.com/latency
[7] *Android Latency Fixes*: https://developer.android.com/ndk/guides/audio/audio-latency

CHAPTER 24

Glossary and Resources

This final chapter serves two purposes—it documents buzzwords and brands mentioned previously and introduces other useful resources for interactive audio system developers.

AAA (or triple-A) industry shorthand for a full-priced globally marketed product aimed at the top ten sales charts. Easier said than done. See also triple-B.

ADC Analogue to Digital Converter—a component that regularly measures an analogue input voltage and generates a stream of proportional digital values. See DAC for the reverse process, which is easier.

ADPCM Adaptive Digital Pulse-Code Modulation—a computationally inexpensive lossy audio compression scheme, typically capable of compression ratios around 3.5:1. There are different versions of ADPCM, variously used on Microsoft, Nintendo and Sony platforms. These are similar in principle and effect but incompatible in detail.

AES the Audio Engineering Society, a worldwide association of audio industry professionals. www.aes.org

AI Artificial Intelligences—label applied to characters or mobile objects in a game with similar freedoms to human players but controlled by the computer. AI might control fielders in a cricket match, non-player cars in a race or enemies and other squad members in a single-player shooter.

AIFF audio interchange file format, a file wrapper for uncompressed PCM audio samples. Microsoft's RIFF/WAV format is equivalent but incompatible as it uses different metadata and swaps the order of bytes in the sample data.

Ambisonics a mathematically rigorous and scalable technique used to capture and recreate sound from any direction, regardless of the number of eventual speakers or sound sources. Ambisonics was developed by Oxford academic Michael Gerzon in the 1970s, supported by the National Research Development Corporation, which also encouraged work on early Manchester computers, fuel cells and hovercraft. It generalises Alan Blumlein's 1930s concept of binaural sound to 3D.

AmbiX a common container for Ambisonic files, based on Apple's. CAF Core Audio Format, using SN3D channel order and normalisation factors.

AMD Advanced Micro Devices, a California semiconductor company, rival to NVIDIA in GPU and Intel in CPU manufacturing, responsible for the integrated chipset of the PS4 and Xbox One consoles and some PCs.

API acronym for "Application Programming Interface," a software specification intended to expose functionality to client programmers while encapsulating implementation details.

Apple][an 8-bit home computer with 1-bit sound capability, designed in California. The original 1978 "red book" manual includes handwritten pages. Apple has come a long way since.

AR Augmented Reality, a form of virtual reality which augments rather than replaces conventional perception of the world.

Area mesh a group of spatially distributed and editable control points, with properties such as reverberation, loudness and filtering, which can be interpolated to smoothly adjust the properties of sounds therein.

ARM a low-power processor architecture invented by Steve Furber and Sophie Wilson at Acorn Computers, now used in most modern smartphones and tablets and billions of embedded devices. The original 32-bit integer-only "Acorn RISC Machine" design has since been augmented with NEON SIMD floating-point co-processing, making it suitable for most of the techniques explained here. The ARM64 architecture (sometimes Aarch64) is a more conventional RISC design similar to MIPS, with ARM32 backward compatibility. Such chips run in ARM64 mode on Apple devices and ARM32 mode on Android. Arm is a trademark of Arm Limited. www.arm.com

Blue Ripple Sound London-based makers of Rapture3D interactive audio runtimes and O3A Ambisonic mixing plug-ins. www.blueripplesound.com

Clockstone Audio Toolkit an inexpensive collection of extensions for Unity which adds voice pooling, logging, grouping, scheduling and rationing. http://unity.clockstone.com/assetStore-audioToolkit.html

Co-driver the in-car companion who warns a rally driver about upcoming junctions and hazards. Accurate co-driver commentary is a key feature of rally games.

Console a highly customised domestic appliance originally dedicated to the playing of proprietary copy-protected games, sometimes extended to play other media such as CDs and DVDs. Consoles are made in large volumes with a view to shipping millions of identical hardware units worldwide. The hardware is increasingly bought in from subcontractors, which drives down prices and development delays but limits backwards capability. Compatibility is maintained for a period of five to seven years before a new "console generation" supplants the previous one, refreshing the processing, graphics, sound and copy-protection systems, often with help from

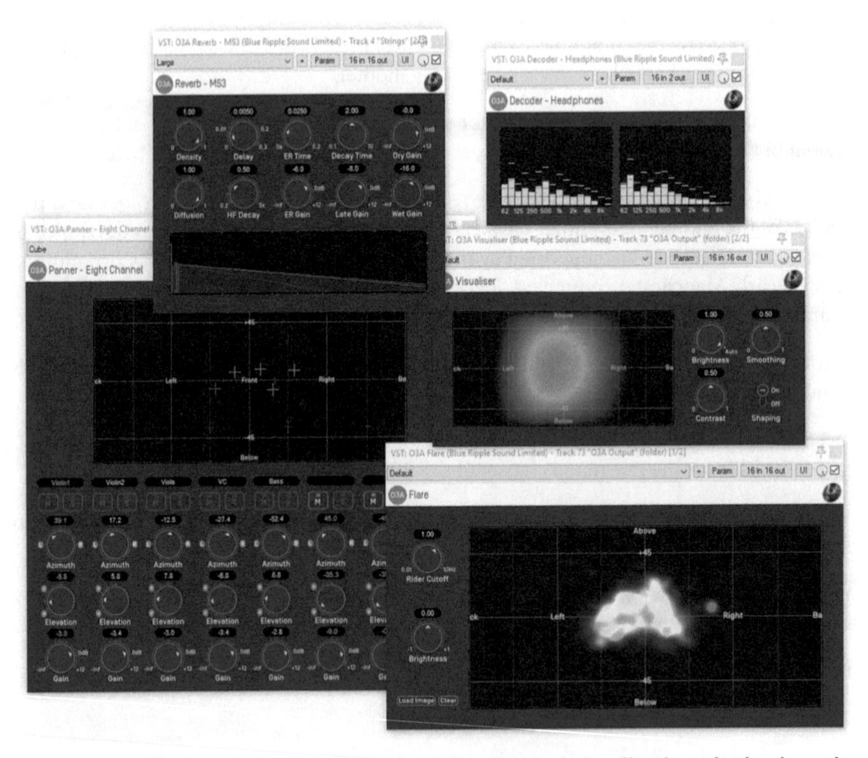

Figure 24.1: Blue Ripple Sound's O3A spatialisation and visualisation plugins in action

other firms. Hence Sony's original MIPS-powered PlayStation, PS2 (MIPS/Toshiba), PS3 (IBM/NVIDIA) and PS4 (AMD); Nintendo's ARM-based Game Boy Advance, DS and Switch handhelds; Microsoft Xbox (designed by NVIDIA, with an Intel processor and Motorola audio DSP programmed by Sensaura), Xbox 360 (made with IBM and ATI) and Xbox One (AMD); Sega's defunct Master System, Megadrive, Saturn and Dreamcast series; the Nintendo NES and Super NES, the Silicon-Graphics/MIPS-based N64, the ATI/IBM-made GameCube and Wii, and the Wii-U, which mixes IBM and AMD parts.

DAC Digital-to-Analogue Converter—a component that converts a number expressed digitally into a proportionate analogue voltage. See ADC for the reverse process.

Decibel or dB a unit of relative intensity—one tenth of a Bel. This unit was named after Alexander Graham Bell, who patented the first practical telephone in 1876. Since a difference of one Bel is a factor of ten in either voltage or power, and a centibel or millibel is too slight a difference for human ears to discern, one-tenth part of a Bel, or deci-Bel, is most convenient for comparing measurements of loudness. As each Bel, or fraction thereof, expresses a *ratio* rather than an absolute *difference* in level, the scale is logarithmic, with ever-increasing power needed to express an equivalent change in loudness while ascending the scale. Human hearing works the same non-linear way. So does the power-to-speed performance of motorised vehicles. It's all about curves.

Devkit a strictly licensed short-run version of a console adapted for game development by adding extra debugging capabilities and limiting the copy protection to allow prototype code and data to be loaded and tested.

DiRT a range of console, Apple Mac and PC rally simulations spawned by the earlier *Colin McRae Rally* series. Both are trademarks of Codemasters Software Company Ltd.

DMA acronym for "Direct Memory Access," referring to peripheral hardware which accesses RAM independently of the central processor. Game consoles derive much of their performance and parallelism through DMA.

DSP acronym for "Digital Signal Processing," a set of computer-based techniques used to analyse and manipulate digital representations of continuous analogue signals. Eponymous dedicated processors support these techniques.

Ducking dynamic adjustment of volume levels to allow a higher-priority sound to cut through. For instance, context-sensitive speech may cause momentary ducking of background ambience. This works best if the ducking starts a short fraction of a second before the priority sound, as the moment of relative silence draws the listener's attention.

EAX acronym for Environmental Audio eXtensions. A trademark of Creative Labs, which introduced these influential extensions to DirectSound3D to allow programmers to take advantage of the capabilities of PC sound cards incorporating E-MU DSPs. EAX went through five versions, steadily adding effects and support for more hardware-accelerated voices—from eight mono ones in the original release up to 128 stereo voices in EAX 5. Final hardware supported up to four simultaneous directional reverbs with I3DL2 standard controls; a software fallback offered only one simplified EAX 2.0 reverb but worked with any sound card, doing the work on the host processor. When Microsoft released Vista and disabled support for audio hardware acceleration, Creative re-enabled EAX through an add-on called Alchemy and by incorporating EAX reverb, filter and effect processing into OpenAL, renaming it EFX.

ECC Error-Correcting Code—a recording scheme which uses redundant data to detect and correct small errors in a block of data or to detect and signal larger errors even if they cannot be corrected.

Fabric Tasman Audio's Fabric extends Unity's basic audio to expose features of the underlying FMOD runtimes, adding a new event system and high-level components like random, blend and sequence containers that allow the creation of complex and rich audio behaviours. www.tazman-audio.co.uk/fabric

Facebook 360 Spatial Workstation a free software suite for designing spatial audio for 360 video and cinematic VR on macOS or Windows, with second-order Ambisonic import capability. https://facebookincubator.github.io/facebook-360-spatial-workstation/

FMOD Designer & Studio—commercial audio authoring tools and runtime systems from Australia. FMOD is a trademark of Firelight Technologies Pty, Ltd. It started out in 1995 as a simple player for MODule files. In releases up to version 3.75, former Electronic Arts and Criterion Games programmer Brett Paterson developed this into a cross-platform audio runtime system licensed to many developers. FMOD 4, also known as *FMOD Ex*, added the PC-hosted authoring tool *FMOD Designer*. FMOD 5 or *FMOD Studio* is a redesign of this. As of 2018, both are still in use. FMOD 5 has optional integration for Unity and Unreal Engine. In 2017 FMOD gained limited support for Ambisonic decoding, via the GoogleVR plugin, for non-console platforms. www.fmod.com

GPGPU abbreviation for General-Purpose Graphics Processing Unit, used in the context of leveraging the high parallel performance of a GPU for non-graphics purposes. Despite some impressive offline demonstrations, GPGPU techniques are rarely used in interactive media due to the extensive data marshalling needed to re-arrange audio input and output to match the expectations of the graphics chips and the difficulty of resolving contention between audio and traditional graphics applications. However, FMOD's convolution reverb can use the PS4 and Xbox One GPU.

Granular synthesis the creation of waveform data in near-real-time by the concatenation of "grains," short cyclical samples, into a continuous stream.

Group a collection of voices which can be mixed and prioritised together, akin to the submix groups of an analogue mixing desk. Group controls are used to coordinate sub-systems, balance levels and apportion DSP effects in interactive audio titles.

GTA *Grand Theft Auto*—the best-selling modern game franchise, a witty, provocative parody of urban gangster living, each release of which has grossed billions of dollars for Rockstar Games' studio in Scotland. At the time of writing, *Grand Theft Auto 5* has sold over 85 million copies, making it—along with Electronic Arts' FIFA soccer series—the one to beat if you're set on global domination of the interactive entertainment industry.

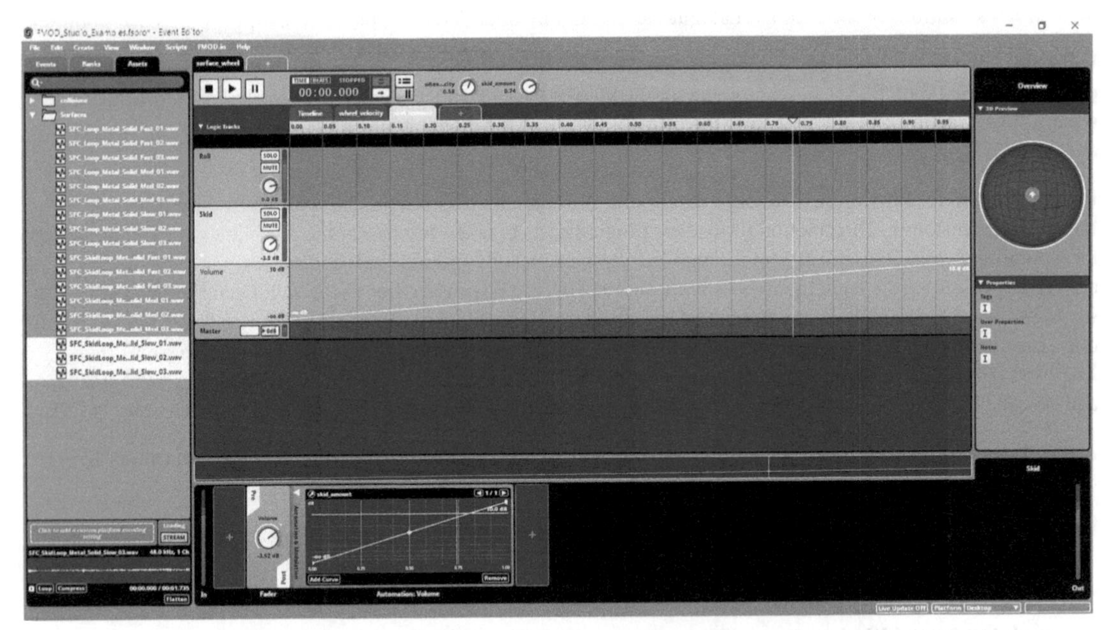

Figure 24.2: FMOD Studio configured by Jethro Dunn to cross-fade surface roll and skid sounds

Harpex a commercial spatial enhancer plug-in which works in the frequency domain. It can analyse first-order soundfields and track a limited number of sources therein to derive additional spatial information. HARPEX is a trademark of Harpex Ltd. Several 3D audio companies have licensed it as a way to enhance B-Format rendering. https://harpex.net

HDMI High-Definition Multimedia Interface, a trademark of HDMI Licensing, LLC. HDMI is a breakthrough for consumers, as it allows high-definition digital video and up to eight channels of uncompressed low-latency high-quality (24-bit, 192-kHz) audio to be connected via a single cable.

HRTF Head-Related Transfer Function. The unique shape of each person's head and ears determines how they determine the source of each sound around them. Personalised HRTFs reproduce this effect so sounds appear to be realistically positioned and moving around the listener even when they're being played directly into the ear canal via headphones. At the time of writing it's expensive to have such measurements made, and only a few game engines (notably those from Blue Ripple Sound) allow them to be integrated into a game or XR title. More common generic HRTFs approach this effect with filters chosen to approximate the response of "most people." These are less effective than personalised HRTFs and prone to misinterpretation, especially between front and rear. Generic HRTFs can be learned, but there's no standard implementation and none likely, as human physiology varies greatly. An extension of this technique, cross-talk cancelled HRTFs can be used to deliver surround sound via a single pair of speakers, rather than headphones, providing the listener's ears and the speakers remain stationary at a known distance. The cancellation technique introduces anti-phase components of each speaker's output to the opposite side so that a listener in the small sweet spot hears approximately what headphones would deliver directly to each ear, despite the fact they're in earshot of both speakers. The phase difference cancels (or at least reduces) the cross-talk. This is nice when it works but problematic if the listener is not tied down in relation to the speakers, and it still requires a custom HRTF for consistent results.

I3DL2 short for interactive 3D audio rendering guidelines, level 2.0, a set of platform-independent guidelines for 3D audio, created in 1999 by the Interactive Audio Special Interest Group, most widely cited for the reference set of parameters for control of synthetic digital reverberation it defines. www.iasig.org/index.php/component/edocman/?task=document.viewdoc&id=8, www.iasig.org/wg/closed/3dwg/3dl2v1a.pdf

IBM trademark of US computer manufacturers International Business Machines Corporation. IBM PowerPC microprocessors power 3DO M2, GameCube, PS3, Wii, Wii-U and Xbox 360 game consoles.

IOP Input/Output Processor. The part of a PlayStation 2 console that deals with data transfers to and from peripherals to other processors such as the SPU (sound processing unit), graphics and emotion engine memory. Essentially a PlayStation 1 MIPS R3000A.

Kickup sounds generated by stones and similar objects being thrown up against the underside of a vehicle moving at speed across rough terrain.

LFO Low-Frequency Oscillator—a subsonic wave used to vary the pitch, volume or position of a sound to introduce vibrato, tremolo or shimmer effects. Infinite feedback filters are a cheap source of sinusoidal LFO control waves.

Listener a 3D position and orientation relative to which 3D sound sources are measured and placed. The listener performs like a microphone placed in the simulated world.

Loop samples which play continuously without interruption are said to "loop" and are described as "loops" to distinguish them from one-shot sounds which play from start to finish without repetition. Careful programming and extra authoring steps are needed to make highly compressed samples loop seamlessly.

LPCM Linear Pulse Code Modulation—uncompressed audio data represented as linear (as opposed to logarithmic) pulses at a regular interval, often but not necessarily 48 kHz. AIFF and uncompressed WAV files use LPCM format and it is the preferred default for HDMI audio.

Matrix a multidimensional array of numbers. In 3D systems 2D matrices are commonly used to reposition or re-orientate coordinates. A nine-element square matrix of three rows and three columns can rotate and scale the X, Y and Z coordinates of a point in 3D space; 16 element 4 × 4 matrices are commonly used to encode translation, scaling and rotation (and sometimes perspective) transformations in ways which can quickly be combined and applied to 3D vectors. Rectangular matrices can be used to derive speaker outputs from B-Format Ambisonics and vice versa.

MCtools a suite of free command-line tools for manipulating multi-channel audio files, developed by Richard Dobson. MCtools are a fine complement for SoX. www.rwdobson.com/MCtools.html

MIPS Millions of Instructions Per Second (or sometimes, cynically, Meaningless Indicator of Processor Speed) —a measure of processor throughput; also, with conscious ambiguity, the name of a Stanford University spin-off company which provided processors for Nintendo and Sony game consoles, Unix workstations and some Android phones, under the acronym *microprocessor without interlocked pipeline stages*. MIPS is a trademark of Imagination Technologies Limited.

Mu-law a simple lossy codec, part of the 1972 International Telecommunications Union G-711 standard, which yields a consistent 2:1 compression ratio.

NaN an abbreviation for "not a number," used in the context of floating-point arithmetic standards to denote special values that are not computable. While part of the full IEEE-754 standard, these exceptions are best excluded from data that will be processed in real time, as it often takes a long time to process them and yields no audibly useful result.

NVIDIA trademark of NVIDIA Corporation, makers of graphics cards, responsible for the hardware design of the original Xbox console. Sensaura did the audio; Microsoft paid the bills.

O3A plugins a collection of high-quality free and commercial plug-ins for third-order Ambisonic manipulations, developed and published by Blue Ripple Sound. www.blueripplesound.com/products/o3a-core

Offline reference to a process performed as part of the preparation of data for an interactive system rather than in real time later. Offline processing is less sensitive to resource limitations than real-time work. Codecs typically require more resources to encode than to decode samples, since encoding is almost always done offline, and only decoding is time critical.

One-shot a sample that plays from start to end (unless stopped sooner) rather than as a continuous loop.

OpenAL an extensible positional audio application programming interface first developed by Loki Software in 2000, later adopted by hardware manufacturers like Creative Labs, Apple and NVIDIA, and software developers Blue Ripple Sound (Rapture3D) and OpenAL Soft, who offer an open-source cross-platform reference implementation. OpenAL is a trademark of Creative Labs, Inc. www.openal.org http://openal-soft.org

OpenGL a trademark of Silicon Graphics International, OpenGL is an application programming interface which sets out to do for graphics what OpenAL does for sound.

OpenSL ES a license-free "sound layer" programming interface for mobile devices, influenced by OpenAL. OpenSL ES is a trademark of the Khronos Group Inc. www.khronos.org/registry/OpenSL-ES/specs/OpenSL_ES_Specification_1.1.pdf

Order a measure of the number of processing layers, e.g. the slope of a second-order filter is twice that of a first-order one. Second- and third-order Ambisonics uses five and twelve additional spherical harmonics, respectively, compared with B-Format's four. Order is also the term for "instruction" in early computers.

Panner a panner distributes a sound between stereo or surround speakers to give it an apparent location in space.

PlayStation trademark of Sony Interactive Entertainment Inc.'s game console range, usually abbreviated to PS. PS1 is the original 1994 PlayStation, PSP the first portable and PS4 the 2013 update.

Profiling the measurement of resource usage in a running system. Usually a prelude to identification of bottlenecks which can be optimised to make the system faster, less memory hungry, more capable or more predictable.

Pro Logic a commercial analogue surround-sound matrixing technique, licensed by Dolby Labs, which approximately extracts additional channels from a stereo mix.

Psychoacoustics the perceptual interpretation of auditory properties according to the characteristic, typically non-linear and multifactorial, human responses to sound.

Rally Rally is a motor sport which pits cars individually against the clock in demanding terrain, crewed by a driver and map-reading "co-driver." 1995 World Rally champion Colin McRae went on to endorse Codemasters' Rally-racing games for PCs, consoles and mobile devices.

Random not apparently predictable.

Rapture3D advanced interactive audio runtimes developed by Blue Ripple Sound for iOS, Android, PC and macOS platforms, with Unity 5 integration. www.blueripplesound.com/gaming

Reaper a low-cost digital audio workstation especially well suited to the development of multi-channel and interactive audio projects and the recommended host for Wigware, O3A and similar plug-ins. http://reaper.fm

Rig shorthand term for an arrangement of loudspeakers. ITU 5.1 is a rig designed for cinema, 3D7.1 is a rig designed for 3D games; mono, stereo and quad (four speakers arranged in a floor-aligned square) are simpler rigs respectively capable of 0D, 1D and 2D sound delivery.

SIMD Single Instruction, Multiple Data instructions (sometimes known as vector extensions) that perform arithmetic operations on several values in parallel.

SN3D a standard order and weighting for Ambisonic channels, most commonly used in VR.

Soundfield a representation of all the sounds around a listener independent of a specific delivery system—headphones, speakers, bone-conduction headsets, etc.

Source a potentially audible rendered sound sample with pitch, volume and spatial properties.

SoX short for Sound eXchange, a powerful open-source command-line utility aptly described as "the Swiss army knife of sound processing programs." SoX can analyse, convert, mix, filter, play and record audio in dozens of formats. Along with MCtools for Ambisonics and multi-channel manipulation, SoX provides most of the offline DSP tools needed to prepare audio samples for interactive use in an easily scriptable wrapper ideal for experiments or batch manipulation. https://sourceforge.net/projects/sox/

S/PDIF Sony/Philips Digital Interchange Format, a serial bitstream carried on a coaxial phono connector, capable of conveying uncompressed stereo or compressed 5.1 channel digital audio.

SPE Synchronous Processing Element—these eight co-processors provide most of the DSP power of the PlayStation 3. Sometimes confusingly called SPUs—synchronous processing units—overloading the acronym of earlier PlayStation audio chips.

SPU Sound Processing Unit. Sony's PlayStation had one of these to mix 24 audio channels at 44.1 kHz, PS2 daisy-chains two at 48. Each unit offers ADPCM decompression, four-way mixing, sample rate conversion from 0 to 4x and stereo reverb.

SSE SuperScalar Extensions, a set of SIMD instructions for parallel vector processing.

Streaming continuous playing of audio samples which are only partially held in main memory and read piecemeal from a slower storage device to maintain uninterrupted output.

Sursound a well-established mailing list and resource for surround-sound researchers and content-creators. https://mail.music.vt.edu/mailman/listinfo/sursound

Sweep a sound sample running through a wide range of pitches. Full-bandwidth sinusoidal sweeps are a good test of linearity and alias rejection in an audio system. Engine sound sweeps, from tickover to maximum revs and vice versa, are the raw material from which internal combustion engine sounds are simulated.

TOSLINK Toshiba's optically connected implementation of the S/PDIF serial digital audio protocol, supported on many game consoles and AV receivers. Toshiba and TOSLINK are trademarks of Toshiba Electronics Devices & Storage Corporation. The generic name for the standard is *EIAJ optical.*

Triple-B or BBB—Alex Tyrer's term for a game that ostensibly aims for AAA ratings but fails at many hurdles. Triple-B games sometimes could have been AAA if they'd had more time for development. More often they're better released at a budget price point or scrapped altogether.

TRS-80 one of the earliest home computers, introduced by the Tandy/Radio Shack retail chain in the late 1970s. TRS stands for Tandy/Radio Shack (or trash if you're feeling cruel), while -80 references the 8-bit Z80 processor.

Unity a trademark of Unity Technologies, Unity is a commonly used low-cost cross-platform 3D game development platform with bare-bones audio built upon FMOD and strong support for indie developers and plug-ins. Clockstone Audio Toolkit, Fabric and Rapture3D (q.v.) greatly extend Unity's sound capabilities. https://unity3d.com

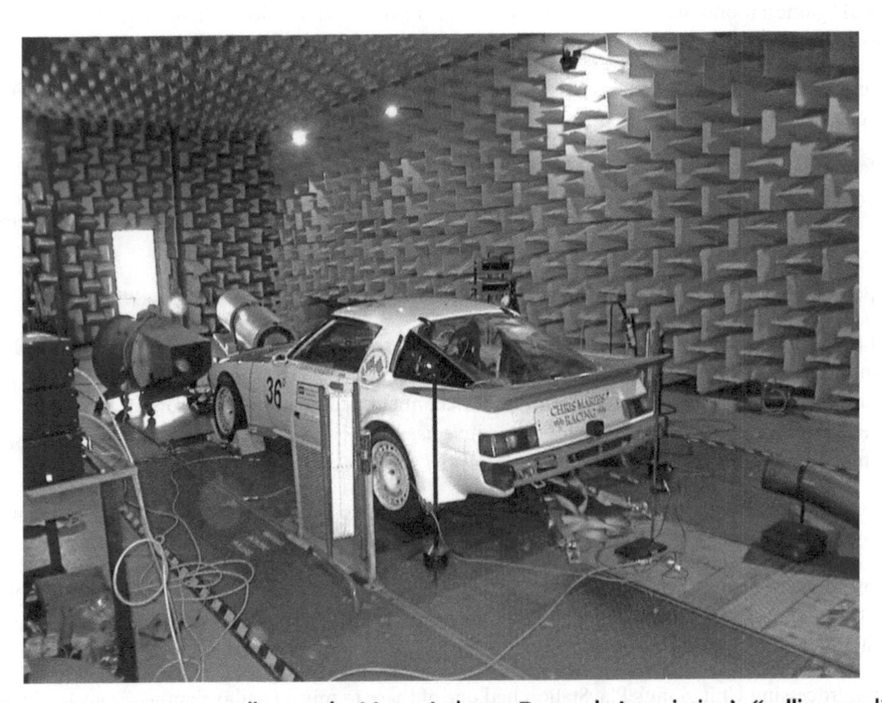

Figure 24.3: Sweep recording on the Motor Industry Research Association's "rolling road" chassis dynamometer

Unreal Engine a well-known cross-platform game engine with a powerful Windows-hosted editor, trademark of Epic Games. Unreal originally layered over OpenAL or XAudio runtimes but now includes its own software mixer and optional Wwise integration. Unreal4 features visual scripting through "blueprints," which allow audio assets and events to be connected graphically. Like its main rival Unity, Unreal is well-suited for rapid prototyping on PC but can require extensive tuning to ensure decent performance on embedded systems, especially in terms of its memory handling. www.unrealengine.com

Vector a 3-dimensional numeric representation of a position, velocity or direction, composed of three orthogonal components, X, Y and Z. Vectors often occupy four words, with a dummy component W, for alignment reasons.

VGM the Video Games Musician's mailing list—a world-wide collaborative community of hundreds of working game audio professionals. A peerless source of advice if you qualify to join. http://server.bobandbarn.com/mailman/listinfo/vgm

Voice a candidate sound for potential inclusion in a mix, with pitch, volume and spatial properties. A Virtual Voice may or may not be heard depending upon its relative priority.

WAV short for WAVe, the customary filename extension used for uncompressed audio samples on Windows/Intel platforms. Metadata such as sample rate and channel mapping is—somewhat inconsistently—held in "chunks" alongside the sample data, mimicking the Electronic Arts IFF (interchange file format) specifications popularised by Commodore Amiga in the 1980s.

Web Audio API The Web Audio Application Programming Interface allows audio mixing and rendering systems to be composed using ECMAScript in many web browsers, particularly those based on the Webkit framework. It is based on the idea of linking audioNode instances into a directed graph. For compatibility it relies upon impulse response reverb, with support for mono, stereo and four-channel impulses, the latter compatible with B-Format Ambisonics. www.w3.org/TR/webaudio/ https://developer.mozilla.org/en-US/docs/Web/API/Web_Audio_API

Wigware free Ambisonic audio plug-ins for Windows and macOS developed at Derby University for Ambisonic panning, metering, decoding, reverberation and format conversions. Wigware was used to author soundfields for *F1* and *DiRT* games, among others. www.brucewiggins.co.uk/?page_id=78

Wwise a trademark suite of interactive audio authoring tools and cross-platform runtimes, developed in Canada by AudioKinetic Inc. Optional integration with Unity and Unreal4 game engines and very strong toolchain support

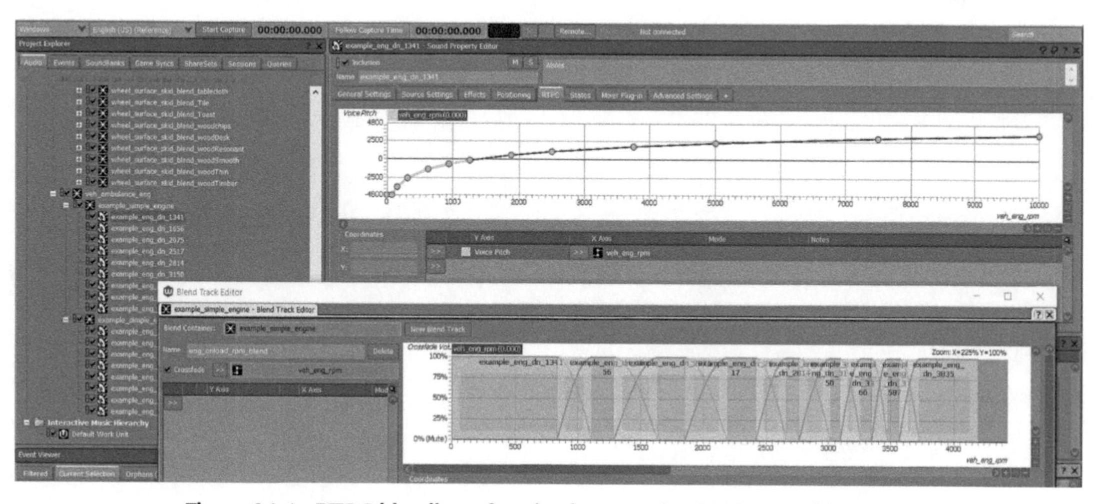

Figure 24.4: RTPC blending of engine loops under RPM control in Wwise

for sound designers running either Windows or macOS make Wwise an increasingly common choice for big commercial products. AudioKinetic's console licensing model demands substantial monthly payments to keep the tools running, but full-time professional sound designers will find lots to play with, even before there's a game to host their work. The generic runtimes are competent but less exciting, especially compared with platform-specific libraries, but perform reasonably even in large projects once late-binding RTPC (real-time parameter control) bottlenecks are replaced with custom code. A major revamp of internal routing in 2017 added good Ambisonics support to Wwise. As game audio programmer Jon Holmes puts it, "*Wwise has enough pinch-points to allow customisation of anything you don't like.*" www.audiokinetic.com

x64 shorthand for the 64-bit Intel/AMD microprocessor architecture built upon x86.

x86 shorthand for the 8/16/32-bit Intel architecture built upon their 1972-vintage 8008 microprocessor design.

XAudio Microsoft's replacement for DirectSound3D, introduced for the Xbox 360 console. The successor XAudio2 runs on Windows PCs as well as Xbox consoles. https://msdn.microsoft.com/en-us/library/windows/desktop/ee415762(v=vs.85).aspx

Xbox trademark name of Microsoft's series of home gaming consoles.

XR short for "eXtended Reality," a category encompassing VR, AR and their ilk.

ZX Spectrum an 8-bit Sinclair home computer, popular in Europe in the 1980s.

Additional Trademarks

Trademarks appear herein just for descriptive purposes, implying no proprietorial claim. Apple, iPad, iPhone and macOS are trademarks of Apple Inc. F1 and Formula 1 are trademarks of Formula One Licensing B.V. Android, Google and YouTube are trademarks of Google Inc. GameCube, N64, NES, Nintendo 64, Super Nintendo, Wii, Wii-U and Nintendo are trademarks of Nintendo Corporation. DirectSound, Microsoft and Windows are trademarks of Microsoft Corporation. ATRAC, DualShock, PlayStation, PS2, PS3, SIXAXIS and Sony are trademarks of Sony Corporation. Codemasters, Dolby, DTS, Eventide, Intel, IBM, Motorola and Sega are trademarks of the respective companies.

Acknowledgements

This book is dedicated to the memory of audio visionaries Alan Blumlein (1902–1942) and Michael Gerzon (1945–1996), to my father Nick Goodwin (1939–1976) and to my son Ingo Lyle-Goodwin (2000–).

I've worked with and learned from scores of designers, programmers, researchers and suits in my 40-year career making interactive audio. I can't list them all, so with apologies to the others, here are the names of quite a few who helped bring this book into existence:

Mike Kelly, Heather Lane, Damian Murphy and Mark Yonge at the **Audio Engineering Society**.

Fleecy Moss, researcher Don Cox and audio systems programmers Davy Wentzler and Simon Jenkins at **Amiga Inc.**

Fred Gill, Andrew J A Wright, Rob Baker, John Davies and Lyndon Sharp at **Attention To Detail Ltd.**

Dan Peacock, Carlo Vogelsang and Jean-Marc Jot at **Creative Labs.**

Codemasters: Audio Manager Tim Bartlett; my Central Tech Audio "dream-team" Aristotel Digenis and Pete (no relation) Goodwin; long-time CTO Bryan Marshall, Audio Director Stephen Root, audiophile producer Pete Harrison; game audio programmers Justin Andrews, Jon Holmes, Hugh Lowry, Jon Mitchell, Rob Pattenden and Adam Sawkins; sound designers Stafford Bawler, Jethro Dunn, Dan Gardner, Andy Grier, Oliver Johnson, Stuart Ross, Pete Ward and other colleagues in game studios, QA, and the audio department.

Former **DTS** "games team" colleagues: Pedro Corvo, Leslie Jensen-Link, Niall Douglas, Ethan Fenn, Pietro Marcello.

Ambisonics experts: Dave Malham of York University, Bruce Wiggins at Derby, Andrew J Horsburgh at Southampton Solent, Richard Furse of Blue Ripple Sound.

Console audio experts: Scott Selfon, Bryan Schmidt, Jason Page, Paul Scargill.

Kindred gurus of interactive audio: Robert Bantin, Anastasios Brakis, Chuck Knowledge, Alistair MacDonald, Andy Mucho, Tim Southorn, Col Walder and Sandy White.

Benevolent professors: Colin Blakemore, Mark Eyles, Adrian Hilton, David Howard, Chris Kyriakakis and Barry C Smith.

Bookmaking helpers: Tom Lean, Dave Reeves, Dean Bilotti, Dave "Newt" Newell, Tony Horton, Stephan Schütze, Imdad Shah; Bryony Hall and Ellen Robertson at the Society of Authors and my editors Claire Margerison and Lara Zoble.

Special thanks are due to these others who gave us permission to reuse their photos, quotes or diagrams in this book: Chris Burton, Matthew Logue, Don Norman, Francis Rumsey, Dan Smith, Franz Zotter and Tom White of the MIDI Manufacturers' Association. Graphs were smoothly redrawn by Chris Oxenbury of https://theartofokse.com.

Index

Note: Figures page numbers are in *italic*; tables page numbers are in **bold.**

3D7.1 96, 198, 203, 205, *213*, 214–220, 244
5.1 surround 144, 147, *154*, 154, 183–184, 192–194, 197–203, 264–265
7.1 surround 184, 193, 197–198, 205, 207, *208*, 211–213, 219, 243

AAC 48, 178
A:B comparisons 121, 146, 204
Adaptive Digital Pulse-Code Modulation (ADPCM) 30–33, 48, 63–69, 80, 98–99, 176–178, 180–181
ADSR ("Attack, Decay, Sustain and Release") 18, *18*, 23, 108, 262
AES69 personal spatial data 192
A-FORMAT 203–204
AIFF files 262
AI players 14, 81–83, 95–96
alignment in memory 58, 82, 127, 148, 165, 191
all-pass filters 231–232, 240, 244
all-pass loop reverb 240–241, *240*
AllRAD 217
alSourcePlayv 33, 70
AltiVec 159, 165–166, 191
AMB 184, 205–220
ambience 4, 40, **48**, **49**, 57, 61, 65, 70–72, 80, 113, 139, 221, 253; B-Format 198, 219–222; continuous looped 57; crowd 67, 199; environment(al) *95*, 59; local 35, 67, 80, 122, **123**, 253; paddock 219; weather 29, 67–69, 115
ambience zones 67, 253
AmbiFreeVerb 245
Ambisonics 6, 192, 194–195, 197–222, 244–245; B-Format 154–155, 173, 198, 201–204; classical 215; first-order *154*, 205, 211–216, 221–222, 244–245; higher-order 154, 204; horizontal 204–213, 215; hybrid 201, 206, 217–219; second-order 154, 215–217; third-order 154, 201, 206, 215–219
Amiga *23*, *27*, 27–30
analogue surround 31, 67, 72, 80, 183, 265
Android 4–5, 41, 45, 47–48, **48**, 87, 147, 234, 241, 264
angles 87–90, 128, 133–136, 167, 186, 189, 193–194, 198, 203–204, 206–208, 210–211, *210*, 214, 217–219, 229, 238, 242, 249–250; speaker 193–195, 200, 204
animation 42, 168, 235
anti-aliasing 29, 154, *154*, 157–158, 167–168, 256–261
anti-phase 67, 189, 210, 243
Apple 4–5, 8, 11, 23, 76, 147, 169, 176, 194, 204, 264
ApplyMatrix 130–132, 220
arcade games/arcade cabinets 3, 16, 20, 59, 140, 143–144, 189, 198
ArcTangent 89–90, 133, 167, 209, 212
arithmetic 85–94, 149, 153, 165, 170, 263–264; 3D 126–137; floating-point 85, 91, 162; saturating 264
ARM/ARM32/ARM64/ARM7e 8, 44, 47, 76, 87, 89, 91, 159, 165–166, 169, 174, 178, 191, 263
arrays of structures 106, 151–152
asynchronous processing 46, 82, 96, 102, 148–149, 164, 250
Atari 14, 16–17, 19–20, 22, 25, 27; POKEY 20
atmosphere 111
ATRAC 30, 33, 40, 49, 61, 72, 74–75, 98, 178–179, **178**, 183
attack time 168
attenuation 16, 21, 82, 103–105, 108, 112, 114–116, 121–122, 133, **151**, 231, 236, **241**, 245, 250; dB 21, **22**, 86, 104, 113, *113*, 115, 133, **146**, **151**, 186, 200; distance 81, 85, **107**, 111, 113–116, 121, 128, 142–143; factor 241; linear **22**; listener 103, **107**, 116, 133–135, 137, 186; logarithmic *21*, 21–22, *22*; rearward 133, 140
audio, digital 176, 197, 262, 265; hardware 11–25, 177
audio, interactive 1–9, 13, 16–17, 20, 27, 43, 61, 77, 81, 91, 95, 117, 163, 183, 198, 249; codecs 173–181; development roles 35–42
AudioConfig 114, 136, 145–146, 148–149
AudioDeviceInfo (Android) 147
AudioManager (Android) 264

audio runtime *7*, 41–42, *77*, 100, 111–112, 126, 140, 249; 3D 85; architecture of 77–83; game 29, 141; system 1, 3, 6, *7*, 8–9, 68, *77*, 126, 145, 153
audio systems programmers 30, 35–37, 40, 44–45, 57, 59, 63, 78
Aureal3D 185, 190
AVAudioSession (Apple) 76, 147
AVX/AVX2 159, 165–166, 170, 191
AX/AXFX/AXPBLPF/AXVPB structures 152
AY-8910/AY-8912 chips 16, *17*, 19, 24
azimuth 89, 189, **190**, 208–209, 211–212, 215, 217–219

B2X 216
backfiring 37
background (ambience/music) 13, 29, 35, 52, 57, 61, 67, 76, 99, 101, **123**, 219, 231
BAFTA 44, 233
band-pass 23, 105, 227–228
bandwidth, audio 28, 65–66, 74–75, 105, 139, 154, 157–158, 183, 203, 205, 228; HDMI 265; memory 43, 59, 68, 79–80, 107, 174–175, 181, 239
bass 15, 20, 29–30, 72, 105, 111, 139, 192, 207, 216, 225; controls 227–228; drums 29–30; management 5; sub- 154, *154*
batch processing 82, 177, 250
BBC 11, 16, **168**, 174, 198, 200
B-Format *154*, 155, 173, 192, 203–204, **203**, 206, 214–215, 217, 246–247, 271; ambience 198; channels 173, 214, 245; components 221; horizontal 198, 215; mixes 198, 202; Mu-law 173; reverbs 247; sculpting 221–222; soundfield 146, 159; soundfield 219; stereo voice 146; tracks 155
binaural 8, 80, 117, 183–192, 198, 210; headphones 194, 198; mixing 155; spatialisation 204; stereo speakers 198
bitstream 14, 58, 264–265
blending 106, 116, 251
Blumlein, Alan 183
Blu-ray 6, 40, 48, 51–55, 57–58, 69, 72
bottlenecks 30, 44–45, 51, 54, 85, 96, 107, 230
brick wall filters 157, 256–257
buffering 57, 64; deep 75–76; double 59–61, 76, 151; intermediate 264
BuffersQueued 151
bugs 45–46, 52, 122, 148
bullets 117, 250

C++ 8, 77–79, 87, 90, 166, 176, 243
C++11 8, 90, 174, 263–264
C64 23, 175
C99 109
CAF (Apple Core Audio Format) 215
CalculateLFE 211
callbacks 47, 82, 147–149, 151
cardioid 133–134, *134*, 214–216
Cartesian 89, 218; -to-polar conversion 89–90
CAV (constant angular velocity) 54
CBR (constant-bit rate) 75
CD 38, 40, 51–55, 58, 61, 158–159, 170, 173, 177–178; audio 29, 31, 158, 168; beats 29; format 29–30; players 14; rate 155; sample rate 28, 66; sections 29; standard rate 31, 179; tracks 29, 43
CD-ROM 54, 61
chain, sample 99–101, 108, 122, 151–152
chaining/ChainSample/ChainStream 100–101, 149–152
Chamberlin, Hal 157–159, 162, 226–227
chromatic intervals 168
cinema 1–5, 70, 81, 140, 183–184, 192, *193*, 195, 197–204, 207, 219, 243, 264–265; and gaming *208*
cinema 5.1 *193*, 202, 214, 217, 244
cinema 7.1 147, 202–203, 207, *208*, 244
cinematic lag 264–265
cinematic surround 183–184, 194, 207
clamp/clipping 32, 88, 109, 137, 140, 142, **146**, 161–162, 211, 257, 262–264
Classical Ambisonics 215–218
clicks 11, 13, 153, 177, 213; and interpolation 162–165
CLV (constant linear velocity) 54–55
codecs 6, 30, 33, 35, 40, 46, 48–49, 52, 72, 75, 78, 80, 145, 150, 183–184, 202, 205, 264; ADPCM 176; ATRAC 61; console 180–181; frequency domain 178–179; interactive audio 173–181; VBR 74–75
Codemasters 1, 14, 56, 72, **96**, 126, 193, 198, 233
co-driver 4, 37, 39–40, 65, 70, 80, 256
coefficients 157–159, 184, 189, 201, 205–208, 215, 218–220, 222, 228, 260
collisions 6, **48**, **49**, **69**, 72, 81, 83, 96, 103, 119, **123**, 135, 139, 144, 235, 249–251
comb-filtering 240, 255
combinations 20, 29–31, 58, 67, 122, 178, 192, 216
Commodore 19, 28, 54; Amiga 1000 27; SID 22–23
comparisons 68, 146, 204, 262
compatibility 63, 79, 193, 198, 202–203, 205, 231, 234, 262, 265
compensating 56, 177, 207, 215
compilers 78, 82, 90–91, 164, 166, 174
compression, dynamic 41
compressors 146, 178, 181, 262
consistency 3, 38, 105, 143, 206, 230
console(s) 1–3, *4*, 5–6, 16, 40, 43, 45–48, 52–57, 61, 63, 69, 71, 76, 81, 91, 95–96, 107, 147–148, 173, 184, 194, 197, 201–202, 226, 249, 263–265; Ambisonic standards in 200; codecs 180–181; drives 63; *F1*

2011 96; firmware 262; flash-based 59; game(s) 3, 5, 14, 16, 43, 76, 139, 183, 264; *GRID* 73; handheld 16; hardware 159, 180; HDMI output of 256; Intellivision 14, 16; Japanese 133; loading times 56; menus 144; Microsoft 5, 53–54, 73, 181, 183; middleware 201; N64 144; Neo Geo Pocket 16; Nintendo 5, 53, 78, 264; optical-disc 52; output rate 61; PlayStation 1, 78; PS4 48, 55; Sony 5, 40, 178, 181, 183; update cycle 140; VCS 2600 14; Vectrex 16; Wii 54; Wii-U 67; Xbox 1, 78, 80, 178, 265; Xbox One 48, 54–55, 159
constant-Q transforms 168–169
constants 38, 87, 89, 91, 142, 187, 226, 259; compile-time 123; FLAG 102; floating point 91; NO_SOURCE 108; VOICE_STATE 106
controller(s) *7*, *77*, *95*, 119, 139, 144, 147, 256; add-on 144; axes 264; Dreamcast 144; DualShock 144; envelope 31; game 63; handheld 144; input 139; Microsoft 139; player 139; PS4 139, 256; -rumble 143; SIXAXIS 144; speaker 104, 139, 149, 155, 256; state changes 147; Wii remote 256; Xbox 144; Xbox One 256
conversion 25, 175; analogue to digital 157; asset 74; byte-order 174; Cartesian-to-polar 89–90; code 82; complementary 86; decibels 86–87, 94; frequency domain 245; pitch 7; sample rate (SRC) 29, 32, 154–155, *154*; time domain 167, 169
convolution 155, 166, 234, 236, 257; direct 259; and fake speakers 191–192; filter 189; HRIR 191; multi-channel 245–247; reverberation by 245; simple 190–191; vector 191
crossfading 66, 76, *188*
crowds 36, 67, 72, 96, 253

damage 36, 59, **69**, *95*, **123**, 219, 242–243, 252
Dance Factory 29, 43, 55, 158–159, 170, 259
DC 137, 167
debugging 38, 45–46, 106, 114, 121, 148–149
decay time 143, **241**
decibel 16, 21, *22*, 86–87, 94, 113, 168, 173, 231; attenuation 86, 104, 114–115, **151**
decoder 75, 154, 168, 173–174, 179–180, 201, 204, 214, 218, 220, 264; AAC 61; Ambisonic *154*, 202, 213; analogue surround 67; B-Format 155; Cardioid (in-phase) 214; Hafler 67, 265; hardware 62, 63, 68; Layer III 179; matrix 155; music 47; Opus 61; Pro Logic 67; spatial 202, 210, 213–214, 218–221 ; WMA 71; XMA 180
DecodeSoundfield 220
decompression 8, 56–57, 69–70, 149, 152, 180–181
decorrelation/decorrelated 32, 44, 229, 231–232, 240–241, 243, 245
default voice 101, 105
delay line 233, 236–237, 239–241, *240*, 244; acoustic 53; binary 20; mono 236; multi-tap 239; variable 245
density 111, **241**; echo **241**; frequency **241**; reverberation **241**; spikes 245
Derby University 193, 219, 245, 255
designers: audio 198; game 6, 19, 42, 53, 231; level 250; microphone 203; sound 2–3, 19, 29, *31*, 35–36, 42, 48, 66–68, 70, 74, 78, 80, 92, 112, 121, 148, 155, 173, 179, 192, 197, 200–201, 219, 230–231, 233–234, 239, 242, 245, 265; specialist 37
determinacy 93
DFT 167, 169
diametric decoder theorem 204
diegetic/non-diegetic 69, 83, 96, 101, 108, 116, 119, 122–123, **123**, 137, 139, 154, 185, 204, 211, 255
digital-to-analogue converter (DAC) 25, 27–28, 30, 32, 256–257, 263–264
dipole, stereo 189
Direct3D 133, 142, 187, 221
directional reverberation 247
directory lookup 58
DirectSound3D 47, 71, 112–113, 145, 150, 185
DiRT/DiRT2/DiRT3/DiRT Showdown games 1–2, 30, 43–44, 72, **96**, 96, 176, 214, 217, 219, 233–234, **234**, 243, 253, 262
dirty discs 53, 58–59, 73
dirty flag 102, 107–108, 113–114
disc layout 52–54, 56–59
discrete Fourier transforms (DFT) 167, 169
distance attenuation 81, 85, 111, 113, 115, 121, 128, 142–143
distance compensation, loudspeaker 207
distance filtering 230
distortion 30, 32, 173–174, 204, 228, 230, 256, 262–263; aliasing 168, 256; codec 177; deliberate 177; digital 262; engine 71; high-frequency 170; low-frequency 174; small-signal 174
distribution: of data 73; frequency 74, 170; of game data 176; of long sequences of values 91; normal 93; of reflections *235*; of reverberation 251; of sounds 197; spatial 154, 242; of spectators 67
dividing 36, 164, 235
division: floating-point 92, 114, 137, 164; by zero 90

Doppler effect 8, 75, 105
DotProduct 128, 136
double-precision 88, 91
downmixing 200, *200*, 241
downsampling 47, 168, 173
DSP effects 78, 119, 145, 153, 256
ducking 37, 262
dummy voice 97, 105–106
duty cycle 11, 23
DVD 38–39, 51–57, 63, *64*, 73, 194, 265; capacity of 54–55, 58, 68, 72, 74; drive 40, 52; dual-layer 54–55; ECC blocks 65–66; HD- 54; media 54; PlayStation 80; -R 73; -ROM 173; single-layer 54; standard 31; streams 71
dynamic adjustment 5, 75
dynamic range 22, 24, 28, 30, 85, 146, 174, 187, 262

early reflections 235–238, 241, **241**, 243
ear-related transfer function (ERTF) 192
EAX 185, 231, 241–242
ECC 58, 64–66
echoes 1, 104, 115, 119, 233, 235, 239, 247
effects *see* DSP effects; sound effects; spatial/directional effects; split-screen, effects; volume effects
efficiency 36, 44, 48, 51, 56–59, 63, 70, 82, 99, 104, 106, 144, 152, 163–164, 166, 175–177, 180, 191, 218, 243, 245–246, 252; Ambisonic 202; cache 239; data copying 177; exhaust 36; implementation 99; of loading 61; streaming 178
elevation 89, 133, 135, 189–190, **190**, 192, 211, 217–219; Z- 207
embedded system 63, 76, 91, 107, 150, 173, 240
encoder(s) 67, 74–75, 154, 179, 201, 210, 211–213, 264; AAC 178; Ambisonic 154–155, *154*, 211; delay 264; IMA 176; third-order 217
encoding 74, 176–177, 200–201, 206, 208–209, 218; 2D 211–213; 3D7.1 211–213; ADPCM 61; Ambisonic 87, 173, 201, 208; Ambisonic UHJ 155; analogue surround 31; Cardioid in-phase 215; Dolby AC3 compressed 264; Dolby-compatible 67; flag 177; front/rear 188; HRTF headphone 243; hybrid third-order Ambisonic 217–218; linear 173; looped sample 71; positional 215; speculative 74; standardised 202; vertical 188
endian byte-order 174, 176
endpoints *95*, **104**, 119, *120*, **120**, 125, *125*, 139, 145, **146**, 147, 149–150, **151**, 153–155, *154*, 163, 201, 255–265; HDMI 256; multiple 4, 8, 119, 149, 201, 255; properties 147; stereo 155
envelopes 18–19, *18*, 22, 24; ADSR 23, 262; controller 31; effects 19; generator 24–25; hardware 19; sawtooth 24; volume 18, 108
environment, listening 1, 3–4, 201–203
environmental modelling 5, 53, 59, 119, 168, 247
equalisation: parametric 8, 153
ERRNOs 79, 97, 148–150
error-correcting code 58, 64–66
ERTF 192
event sounds 81
exhaust, vehicle 36–37, 67, *95*, 123, **123**, 129, 135, 233, 242

F1/Formula 1 1–4, 96, **96**, 214, 233, **234**, 239
Facebook 189, 192, 215
fading 21, 28, 76, **107**, 119, 121, 169; amplitude 186; automatic 108–110; cross- 66, 76, 104, 115, 129, 168, 187–188, *187*, *188*, 247; distance-related 109; smooth 199
FastAtan2 90
FastCosine 89
fast Fourier transforms (FFTs) 167, 169–170, 245
FastLog2 86
FastLog10 86
FastPower 86–87
FastSinCos 87, 218, 264
FastSine 87, 89
Fast trigonometry 87–89
feedback, force/haptic 5, 143
feedback comb filtering (FCF) 111, 233, 236, 240, 244, *244*, 255
Feedback Delay Networks 240
FFT 167, 169–170, 245
filtering, comb 240, 255
filters 5–6, *7*, 20, *77*, 95, 104–105, 115, 119, 122, 133, 141, **146**, 146, 154, 215, 257, 262–263; all-pass 231–232, 236, 240, 244; anti-aliasing 29; AXPBLPF 152; brick wall 256–257; compensating 215; digital 225–232; digital low-pass 20; feedback comb (FCF) 244; FIR 166, 170, 232, 259–261, *261*; group 78, 119, **146**, 154; hardware analogue 28; high-pass 241; I3DL2 231; IIR 244, 257–260; low-pass 119, 222, 225–227, 241, 262; passive analogue 12; per-voice 8, 149, 154; resonant 228; shelf 153; shelving 213–214; short 188; slope 105; state-variable *227*, 227–229; voice 78
FindNearestSound 81
finite impulse response (FIR) 166, 170, 225, 232, 245, 259–260, *261*
floating-point division 90, 92, 114, 137, 162, 164, 237–239
fluids 115
FMOD 29–30, 112, 115, *116*, 179–180, 241–242, 265
FM radio 203
FM synthesis 27
fold-down 197, 199–202
footprint, memory 38, 44, 48–49, 59, 180
footstep sounds 37
force feedback 5, 143–144

Fourier transforms 167, 169–170, 245
frame rate 42, 230, 250, 252
Freeverb 243–245, *244*
frequency domain 157, 166, 169, 178, 191, 245, 259
frequency response 5, 188–189, 225, 231, 241, 250, 260
Front Centre 255
FuMa 206, 215
fused-multiply-accumulate (FMA) 159, 191, 245, 257

game audio programmers 36–37, 40–41, 80, 144
GCC 8, 78, 91, 99, 127, 164–166
gear wobble 158
geometry *7*, 39, 56, *77*, 204, 221, 229, 235–238; 3D 54–55, 249, 252; audio 236, 251–253; collision 250; graphics 249, 252; physics 249, 252; static 252
Gerzon, Michael 155, 183, 204, 213–214, 240
GetTime 109–110
G-Format 203–204, 214
glitchfinder/glitch/glitches 3, 52, 71, 135, 160, 164, 177–178, 208, 231, 233, 242, 251
Grand Theft Auto (GTA) 2, 36, 47, 52, 56, 69, 206, 233, 235, 253
granular synthesis 29, 43, 61, 69, 99, 104, 235
Graphics Processing Unit (GPU) 180, 239
GRID game 41, 44, 72–75, 144, 239
groups of voices 1, 29–30, 32, 78, 96, 104, 119, 190
gunshots 129, 264

Haas precedence effect 206, 238
Hafler 31, 67, 265
handedness of coordinates 133, 142, 187, 205–206, 221
haptic feedback 5, 143–144
harmonics 23, 25; cylindrical 219; spherical 183, 204, 210, 215, 217, 219
Harpex 202, 214
HDMI 3–4, 43, 147, 154–155, 184, 197, 256, 264; reverb 243; tuning 265
headphones 1, 4–5, *4*, *7*, *77*, 80, 95, 139, 147, 149, 155, 183–195, 197–198, 201, 220, 243; binaural 198; constant power stereo panning 184–187; cross-talk cancellation 189; fake speakers 191–192; HRIR 189–190; HRTF 188–189, 243; simple convolution 190–191; symmetry and personalisation 192–193; vector convolution 191
head-related impulse response (HRIR) 189–191, **190**
head-related transfer function (HRTF) *7*, *77*, 80, 133, 155, 184, 188, 192–193, 201–203, 230, 256; and cross-talk cancellation 189; headphones 188–189, 243; and HRIR 189–190, **190**; panning 259
headroom 22, 32, 37, 88, 146, 168, 211, 242, 262–263
headroomScale 146, **146**, 154
headsets 65, *95*, 139, 192, 256
height 111, 133, 139, 141, 155, 168, 188, 198, 203, 205, 211, 214–215, 219, 244; encoding 183; resolution 217
Hermite interpolation 156–157
high-definition multimedia interface *see* HDMI
higher-order Ambisonics 154–155, 204, 215, 218
high-pass filters 139, 227
horizontal surround 183, 203
humidity 111
hybrid Ambisonics 154, 198, 206, 215, 217–219

I3DL2 reverb 149, 231, 236, 240–242, **241**, 251
IEEE-754 floating-point 85, 91–92, 109, 153, 164–165
IFF files 56
IIR filters 244, 257, 260; anti-aliasing with 257–259, *258*
important priority 41, 70, 81, 86, 98, 100–101, 103, 108, 119–121, 143, 148
impulse response: head-related 189–191, **190**
infinity/infinitely 21, 85–86, 90, 105, 112, 153, 157–158, 164–165, 170, 204, 211, 225, 228–229
initialisation **60**, 237
integers 85, 162, 246, 262–264
interactions 15, 35–36, 45, 93, 114–115, 119, 244; geometrical 249–253
Interactive Multimedia Association (IMA/IMA4) 176–178, **178**
interface 6–8, 35, **48**, 64, 71, 77–79, 82, 100, 103, 106, 148, 151, 174; audio 231; cross-platform 241; Physical Layer 145; processor 16; programming 8; public 105; software 77; user 48, **49**, **123**; virtual 106
interior panning 206–210, 213, 217
interleaved/interleaving 40, 61, 63, 65, 68, 154, 159, 170, 260; audio 33; channels 74, 80, 100, 178, 208; coarse 68; data 66; de- 64, *65*; mono streams 67; samples 66, 68, 177; spatially 67; stream buffers **69**; streaming 61; streams 61–62, 68, 100–101, 150
interlock 69, 216, *216*
interpolation 28, 110, 155–156, 159–164, 166, 189, 233, 241; clicks and 162–165; cubic spline 156; digital 28; Hermite 156–157; linear 155–156; mesh 251–252; Sinc 157–159; triangular 190
interrupt, timing 18–19, 28
iOS 4–6, 41, 47–48, **48**, 76, 78, 99
iPhone 5–6, 242
IsSourcePlaying 108, 150–151
IsValid 97
ITU 5.1 layout *193*, 194, 198, *199*, 204, 206, 217
IXAudio2SourceVoice 152

Japan (Japanese) 16, 19, 38, 70, 133

kickup 93, 234–235

latency 59, 66, 73, 75, 89, 163, 264–265; mobile phone 264
layer switching 55
leaderboard 37, 67, 80–81
linear feedback shift register 17–18, 20, 92
linear interpolation 155–156
Linear Pulse Code Modulation (LPCM) 98, 173, 178, **178**
lip sync 42
listener attenuation 103
loading sample assets 98
localisation of speech assets 37–39, 41–42, 183, 198
logarithmic curves 13, 15, 20–22, 25, 85, 164, 167, 173, 262
looped: ambiences 57; music 73; sample 71, 162; streams 65; voices 97
loops 11, 36, 43, 51, 66–67, 69, 72, 80, 99–101, 103, 121, 179, *188*, 240; all-pass 242; engine *275*; music 66
loudness 41, 81–82, 188, 219, 262; metering 262
loudspeaker distance 207
low-frequency effects (LFE) 5, 104, **104**, **107**, 108, **151**, 154, 193, 197, 203, 205, 207, 211, 214, 218–219, 244, 255–256
low-frequency oscillator (LFO) 115, 229–230, 241
low-pass filters 17, 20, 105, 119, 132, 157–158, 222, 225–227, 241, 255, 259, 262; implementation 225–227, 257–258

macOS 6, 8, 49, 176, 178
matrix 81, 126–127, 129–131, 201, 208, 214, 219–220; 2D 208; decoder 155, 214–215; identity **126**, 219; inverse 129, 131; listener 81, 129, 131, 136, 143; orientation 126; output 208; panning 210; parallel 126; rotation 131
MatrixInverse 129, 131
MaxDistance 79, **104**, 105, 112–114, 236, 252
MaxImportantVoices 119, **120**
meshes 249, 251–252
metadata 69, 96, 98–99, 252
metering 153, 263; loudness 262; peak 168
Micro Machines 47, **48**, 49, **49**
MIDI 11, 16, *17*, 19, 29, 94, 231, 242
MinDistance 79, **104**, 105, 112–114
mixing 6–8, *7*, 21, 27, 29–32, 44, 69, *77*, 78, 96, 143, 146, 149, 151, 153–170, *154*, 180, 189, 192, 197, 209, 225, 256, 259, 262–263; block size 153, 161–165, 191; buffers 47; clicks and interpolation 162–166; customised 4; desks 105, 122; down- *200*, 200, 241; pairwise 199, 201, 222; re- 199; reverberation 247; un- 199; up- 201; vectorised 165
mobile platforms 47
modulation 81; LFO 229; over- 22; phase 24; pitch 19; pseudorandom 14; pulse-width 28; regular 14; ring 23
momentary effects 32, 55, 57, 73, 117, 136, 179, 262–263
MP3 6, 12, 33, 75, 129–130, 178–180
Mu-law 61, 72, **178**, 246, 262; codec 173–176
multichannel assets 78, 108
multichannel phasing 32–33
multi-player 4, 35, 44, 68, 81, 95, 113, 117, 119, 140, 233, 256
music 2, 6, 11–16, 19, 29, 35–36, 43, 48–49, 52, 54–56, 59, 61, 63, 67, 69–76, *95*, 99, 113, 119, 122–123, 139, 148, 150, 158–159, 170, 177, 179, 208, 235, 256, 262; background **123**; chip 25; files 48; foreground 143; front-end 69; game 19, 29; interactive 37, 72–73; layers/layered 74, 235; loops/looped 66, 73; mobile 76; MP3 179; non-looping 180; player 28, 76; resident 48, **48**; seamless 57; stereo 32, 47, 72, 76, 154–155, *154*, 178, 180; streams/streamed/streaming 57, 72–73, 105, 181, 234
muting sound 16, 58, 136
NaNs 153, 164
nested groups 122
network play 51, 99, 141, 256
non-directional bass 192
noise floor 82, 112–114, 257
Nyquist limit 157, 167–168, 231, 256

objects 69, 95–110, 112, 201, 249–252; collidable 250; dummy 106; mobile 249; occluding 249; playing source 79; reflecting 249; static 249; visible 249
occlusion 112, 119, 199, 249–252
octahedron 205, *213*, 214, 218
offset, listener 126, 141–142
Ogg Vorbis 33, 48–49, 61, 75, 178
omnidirectional 203–206, **203**, 215, 220, 242; pattern 133–134
OnVoiceDeath 98, 102, 109
OnVoiceHandleDeath 98, 102, 110
OpenAL 7, 70, 78, 104–105, 112, 116, 127, 133, 143, 145–146, 150–152, 156, 192, 199, 241
open-world 36, 52, 56
optical disc performance 54–56
optimisation 7–8, 54–55, 57, 59, 70, 75, 78, 126–127, 136, 154, 164, 169, 178, 201, 211, 217, 230, 239, 242, 250, 259; Ambisonic coefficients 206; ARM NEON 88, 91, 165, 178, 191; cache 82, 106, 127, 239; codecs 80, 181; code-generation 166; compiler 90; decoder implementations 173; development build 164; DSP effects 78; index 58; interactive media 197; joint stereo 205; loop 259; platform-specific 78, 80, 169, 242; premature 107; reverb 153, 242–243; single-player 4; stream 63–75; transcendental 85–94; vector 88, 91, 165, 178, 191; velocity inferring 131
optional output 5, 93, 102–103, 139, 144
optional priority 41, 102–104, **104**, 121, 148

Opus codec 33, 181
orientation 81, 115, 121, 126–127, **126**, 130, 198, 219–221, 247; AI character 81; camera 141, 198; landscape 140; listener 81, 126, 131–135, 140, 149, 201, 219; matrix 126, 220; microphone 135; soundfield 214, 219–220
oscillators 105, 115, 229–230, 241
overlapping windows 170
overmodulation 22
oversampling/oversample/oversampler/oversampled 158–160

PackageManager (Android) 264
panning, Ambisonic 8, 189, 197–221
panning, constant-power 186–187, *187*
panning, headphones 188
panning, interior 206–210, 213, 217
parametric equalisation 8, 23, *24*, 153, *227*, 227–229
passband *24*, 227–228, *227*
passby reflections 239
Paula chip 27, *227*
pausing/pause/paused 25, 46, 56, 101, 104, 106, **120**, 121, 137, 140, **151**, 151, 153, 164, 204, 243; menu 69, 184; Pause Audio Strangles Console 153
peak metering 153, 168, 262–263
personalised HRTF 192–193, 271; *see also* head-related transfer function (HRTF)
phase 31, 43, 67, 111, 166–167, 213, 229, 231, 240, 250, 257; -accumulator technique 12; anti- 67, 189, 210, 243; cancelation 189; configurable 229; envelope generator modulation 24; flipped 177; frequency-dependent 232; in- 214–215; linear 257; relative 115, 210, 213; reversals 243; -shift filter 155
phrase/phrases/phrasing 30, 38–42, 51
pipeline, processor 89, 159, 257
pitch changes 136, 258
platform capabilities 5–6, 20, 23, 35, 40, 61, 80, 102
PlayOneShot 100–101
PlaySample 100, 149
PlayStation (range) 1, 31, 39, 43, 47, 51, **53**, 53–54, 67, 67, 78, 99, 144, 177; *see also* PS1 (original PlayStation); PS2 (PlayStation 2); PS3 (PlayStation 3); PS4 (PlayStation 4); PSP (PlayStation Portable)
PlayStream 101, 150
POKEY chip 20–22, 25
polar pattern 81, 132–134, *134*, 217
PosInListenerSpace 114, 131–133
pre-roll stubs 71–72, 150
preserving punch 179
priorities 36, 41, 44, 98, 108, **120**; PlayingVoices **120**; static 41; WaitingVoices **120**; *see also* important priority; optional priority; Virtual Voice; vital priority
private members 46, 79, 107–109, **107**, 113, 119–121, 135–136, 152, 263
propagation delays 116–117
proximity 36–37, 121, 154, 190, 211, 230, 236; ProximityBoost **120**, 121, 211
PS1 (original PlayStation) 31, 39, 43, 47, 51, **53**, 54, 67, 137, 241
PS2 (PlayStation 2) 29–32, 39–40, 43, 47, 52, **53**, 54–55, 63–64, 67, 69–70, **69**, 72–73, 80, 85, 137, 144, 158–159, 170, 185, 239, 241
PS3 (PlayStation 3) 30–31, 40, 43, 52, **53**, 54, 59, 72–73, 76, 89, **96**, 96, 144, 147, 159, 180, 214–215, 217, **234**, 234, 241, 243, 256, 264–265
PS4 (PlayStation 4) 40, 48–49, **49**, **53**, 55, 57, 76, 139, 144, 159, 166, 180, 239, 241, 256, 264–265
pseudorandom 14, 16–18, 20, 91–93
PSP (PlayStation Portable) 40, 48–49, 62, 80
psychoacoustics 2–3, 14, 21, 25, 29, 73, 85, 93–94, 167, 178–179, 230, 238
pulse-width modulation 28

QA testing 76
quadraphonics 183, 197
Quaternions 219
queuing buffers 60, 64, 100

RaceDriver GRID 41, 44, 72–73, 144, 208, 249
radius, source 217
Rally 5, 30, 113; *Colin McRae Rally* 4–5, 37, 39–41, 63–64, 68, **69**, 70, 72–73, 80, 234–236, 242, 253; *DiRT Rally* 243
random-access seeking 239
random curves 93–94
randomisation 94, 239
random noise 17, 23, 27; pseudo- 16, 20
Rapture3D 75, 173, 192, 215–216, 233, 241
ratio, compression 74–75, 173–174, 177–180, **178**
ray-tracing reflections 43, 250–251
reflections/reflectivity, audio 250
RelativePosition 102, **107**, 131–132, 136
RelativeVelocity **107**, 131–132, 136
resampling 66, 75, 153, 155–156, 162, 176; offset 156; runtime 74; systems 153–170
reverb/reverberation 5–8, *7*, 31, 36, 43, 47, 70, *77*, 78, 95, 104, 115–116, 119, 141, 143, 146–147, 150, **151**, 152–154, 170, 181, 207, 221, 231–232, 251–252, 255, 263; 3D 233, 243–244; 3D impulse 247; active units **146**; all-pass loop 240–241, *240*; ambient 117, 155; Ambisonic 244–245; available units **146**; B-Format 247; buffer considerations 238–239; by convolution 245; costs and mitigations

242–243; decorrelation 229; -delay 31; delay lines 236–237; directional 243, 245, 247; *DiRT2* 44; early and late 235–236, *235*; early reflections 236–237; environmental 191, 243; Eventide 44, 153, 242; filtering 78; filters 232; Freeverb 243–244; hardware **69**, 70; HDMI 243; I3DL2 149, 231, 236, 240–242, **241**, 251; immersive 44; impulse-response 152, 166, 234, 240, 242–243, 245, 247; interactive 233–248; local 192; *Midiverb* and *Microverb* 44; multi-channel convolution of 245–247; natural 239–240; numReverbs **146**; on-the-cheap 234–235; pantophonic 244; passby reflections 239; REVERBHI structures 152; ReverbsMask 119, **120**, 146, **146**, 154; soundfields 222; spatial reflections 238; spatialisation/spatialised 154, 233–234, 239, 241, 243; stereo 43–44, 242–243, 256; synthetic 231, 243, 245; systems 232; tuning the loop 241; usage 233–234, **234**; Xbox 360 44, 153
Richard Burns Rally 3
Rolloff 79, **104**, 105, 108, 111–115, **146**; curve 113–115, *113*, **146**, 252
RolloffDirty 79, 102, **107**, 108, 114
RolloffFactor 79, **241**, 242
RolloffHF 79, **107**, 108, 113–114
rumble 5, 143–144, 154

SAA1099 chip 24, *24*
sample rate conversion (SRC) 29, 32, *154*, 154–155
sample replay 3, 14–15, 20, 25, 27–33; and headroom 32; and mixing samples 30–31; and multichannel phasing 32–33; and pulse-width modulation 28; and sample rate tuning 29–30; and tracker modules 28–29; and voice management 29
SamplesQueued 149
saturating arithmetic 264
scalability, platform 2, 5–6, 216
scale, pitch 13
seamless loops and chains 57, 75, 108, 152, 180
secondary output 256
secondary storage 51, 53
seek/seeking time 40, 52, 56, 58–59, 71, 148
SetAmbiGains 220–221
SetEffect 149
SetEndPoint 147, 149
shelving filters 213–214
ShutDown 145, 149
shutdown 148
SID chip 22–23, *23*, *24*, 25
sidechain processing 256, 262
silence 57, 71, 75, 105, 111–112, 153, 179, 245; percussive 113
SIMD (Single-Instruction, Multiple Data) parallel processing **88**, 127, 165, 169, 260, 264
Sinc interpolation 157–162
single-player 4, 43, 121–122, 139–140, 142–143, 256
sliding window 39, *64*, 162
smart pointers 96–98
smooth panning 186, 199, 210
SN3D 206, 215
SN76477 chip 16
SN76489 chip 16
soundbank 48–49, 70, 79–80; sizes **48**, **49**
sound designer 1–3, 19, 29, 35–36, 42, 48, 66–68, 70, 92, 112, 121, 148, 155, 173, 179, 192, 197, 200–201, 219, 230–231, 233–234, 239, 242, 245, 265; roles 19, 35, 80
sound effects 16, 19, 23, 29, 37, 56, 96
soundfields 3–4, 6, 80, 108, 139–140, 154–155, 173, 198, 201, 219, 221, 243, 247; ambient 222; AmbiX 215; B-Format 146, 159, 219; first-order 215; FuMa 215; pre-authored 219, 202; reverberant 222
sound processing unit (SPU) *31*, 31–32, *64*
SourceID **107**, 108, 145, 149–150
spatial/directional effects 183, 229
spatial editing 247, 251–253
spatialisation 4, 188, 191, 193, 197–199, 204, 210, 217, 233–234, 239, 241, 243, *268*
speaker positions 4, 155, 198–199, 202–204, 210
speech 25, 32, 35–42, 49, 52, 56–57, 64–65, 69–70, 74, 80, 83, *95*, 96, 119, 139, 154–155, *154*, 158, 173, 177, 179–180, 204, 211, 234–235, 253, 256, 262; background 101; band 21; buffer 40, **69**; channel 42; co-driver 70, 80; compression 176; custom 38; diegetic 83; editing 179; front centre (non-diegetic) 154; input processing 180; lip sync 42; localisation 37–38, 48, 54, 59, 206; memory-resident 65; mono 32; off-screen 119; player-directed 256; resident 80; samples 30, 40–41, 99; sequential 30; splicing 39; streaming of 39, *64*, 64, **69**, 80; synthesis of 25; volume 119, 177
speed of sound/speedOfSound 136–137, **146**, 236
spherical harmonics 183, 204, 210, 215, 217, 219
split-screen 4, 44, 122, 125, 139–144, 233, 256; effects *141*, 143; voice management 143
SSE 88, 159, 165–166, 178, 191
StartCapture 149
state-variable filter 227–229, *227*
StealSample 100, 108, 149–150
stereo panning 24, 141, 184–187, 207
stopping sounds 39, 109
streaming 5–6, 8, 48, 70–71, 78–79, 149–150, 178–179; buffers 31; case studies 63–76; and channel synchronisation 69–70; and coarse interleaving 68; concepts 51–62; and

custom soundbanks 70; and deeper buffering 75–76; and disc layout 52–53, 57–58; and double buffering 59–61, **60**; and error correction 58–59; interactive music 72–73; and interleaving 61–62; and memory map 68–69; of mobile music 76; need for 51–52; of optical media 59; and optimal disc performance 54–56; and pathological seeking 56–57; and pre-roll stubs 71–72; reasons for 53–54; samples 151–152; and sharing the spindle 73; of speech 39, 64, *64*, **69**, 80; strategies 80; and sucking 61; threads 64–65; and variable bit rate 74–75; *see also* streams
streams 39–40, 48, 51, 61, 63, 65, *65*, 67–74, 79–80, 99–101, 106, 108, **146**, 149–150, 155, 226, 256; ADPCM 68; ambient/ambience 70, 80, 234; audible 67; interactive music 72–73; interleaving/interleaved 61–62, 67, 101, 150; mono 67; music 234; stereo 67–68, 80; sub- 80; VAG 68
streamSpec 99–101, 108, 150–151
structure of arrays 106, 151–152
surround sound 1, 8, 72, 80, 183–184, 192, 202–203, 255, 264–265; Ambisonic 155, 197–222
swapping resources 43–44, 47, 59, 145, 152, 243
symmetry/asymmetry 12, 66, 137, 144, 157–160, 173, 186, 190–194, 192–193, 204, 206, 214, 229, 243
synthesis, granular 29, 43, 61, 69, 99, 104, 235, 239
synthetic front centre 255

Tabu search 206
tap delay/delaytap 230, 236–237, 239, 245
temperature 111, 136
threads 40, 43–45, 63–65, **77**, 82, 96, 102, 239, 250
thunder 117
TIA (television interface adapter) chip 14, 20, 25
tracker modules 28–29
transcendental functions 85–91
transforms, Fourier 166–170, 179, 245
TransformToListenerSpace 114, 131, 136
translations, speech 38–39, 174
translucent file-system 59
treble control 105, 188, 225, 227–228
triggers, event 1, 19, 31, 36–38, 41–42, 70, 80–81, 262
TrueAudio 180
tuning, performance 20, 119, 240–241; HDMI 265; platform 71, 78
tuning, sample rate 13–14, 29–30

underflow 153
Unity 6, 37, 49, 116, 127, 241, 253
unloading sounds 46
Unreal Engine/Unreal4 6, 37, 133, 253
UpdateDelayLine 237
UpdateFade 108–110
Update methods 19, 82, 96, 108, 126
UpdateMix 145, 149
UpdateSource 151

VAG 30, 32–33, 49, 64–65, 68, 99, 177, **178**
VBR (variable bit rate) 74–75, 178–179
vectorising critical code 170
VectorLength 114, 128, 136
vectors 81–83, 87–89, 108, 127–128, 130, 154, 180, 211, 218, 220–221, 234, 264; 2D 167, 252; 3D 126, 130, 252; butterfly 169; convolution 191; co-processor 30; de-clicking 166; direction 211, 242; directional 115; Doppler 135–137; energy 204; fast Fourier transforms 170; fields 211; graphics 73; length 114; listener adaptation 131–132; listener transformation 126–127; mixing in 165; offset **126**; old position **126**; panning 252; posInListenerSpace 114; position **104**, 135–136, **151**; relative position **107**; relative velocity **107**; scale **126**; scales and offsets 141–142; SIMD 260; Sinc 159; velocity **104**, 106, **107**, **126**, 131, 135–136
VectorSub 127–128, 131–132
vehicle sound 35, *95*, 135, 239
velocity 102, **104**, **126**; air 115; constant angular (CAV) 52; constant linear (CLV) 54; Doppler implementations 135–137; listener adaptations 131–132; listener static methods 126; relative **107**, 131, 136; vector **104**, 106, **107**, **126**, 131, 135–136
Virtual Voice 6–8, 46, 95–98, 100, 102, 105, 143, 145, **146**, 149, 235; and audio runtime *77*, 79, 81–82; and automatic fading 108–109; and modelling distance 111–112; and noise floors 113–114; priorities of 102–103; private members of **107**, 107–108; processing 126; public properties of 103–104, **104**
vital priority 32, 36, 67, 96, 102–103, 108, 148
voice groups 79, 119–123, **123**, 136
VoiceHandle 97–98, 101–102, **107**, 108–110
voice handles 96–98
voice leaks 46–47
volume effects 132, 142
volumetric sound source 104, 112, 117, 207, 213, 216, 251
Vorbis codec 33, 48–49, 61, 75, 98, 152, 178
Vu meter **168**

WASAPI 147
WAV 56, 68, 176, 215, 262

wavefield synthesis 202, 215
weapons 35, **49**, 56, 83, 96, 117
weather 29, 39–40, 53, 61, 66–68, **69**, 70, 76, 80, 96, 113, 115, *188*, 234
WetSend 104, **104**, **107**, 115, **151**, 154
windowing 169–170, 259
WMA 70–71, 74–75, 80, 178
wrapping 64, 148–149, 162, 237
writeOut 160–161
Wwise 49, 112, 148, 200, 206, 241

X3DAudioCalculate 116, 204
XADPCM 33, 66, 70, 80, 98–99, 176–177
XAudio/XAudio2 7, 43, 96, 145, 150, 153; and codecs 180; and config callbacks 148; and first-order horizontal Ambisonics 204; and interior panning 207; and linear interpolation 156; and maxed distances 112; and reverb bottlenecks 44; and reverb costs 242–243; and reverb parameters 241–242; and soundfield decoding 220; and source updates 151; and source voices 152; and spatial flags 116; and surround sound 205
Xbox 63, 99, 145, 197; and analogue surround 67; and audio memory allocations 47; and audio RAM 39; and audio resource management 43; and audio runtimes 78, 80; and cinematic lag 264; and cinematic surround 183; and force-feedback 144; and HDMI tuning 265; and mono 187; and output capture 263; and pathological seeking 56; and physical voice counts **96**; and reverb parameters 153, 241; and reverb usage **234**; and streaming 74–75; and variable bit rate 74; and XADPCM 176; and XMP 76
Xbox (original) 39–40, 56, 63, 70–71, 80, 144, 176, 183, 187, 189
Xbox 360 40, 43–44, 52, 54–55, 67, 72, 74, 76, 89, 96, 99, 153, 180, 187, **234**, 239, 242, 264
Xbox One 40, 48–49, **49**, 52, 54–55, 57, 74, 76, 99, 159, 166, 180, 239, 256, 265
X-Fi 44–45, 96, 104–105, 197
XM 29
XMA 40, 43, 49, 61, 72, 74–75, 98, 178, 180, 205
XMP 76, *95*, 148, 150
Xorshift 92
XWMA 179–180

Yamaha YM2149 chip 16
YouTube 51, 176, 192, 201, 262

ZX Spectrum 11, 15–16, 30, 57